A Dictionary of

Agriculture and Land Management

T0177694

Will Manley is a Principal Lecturer at the Royal Agricultural University. He is a Chartered Ecologist and a Fellow of the Institute of Ecology and Environmental Management. In association with lecturing responsibilities he has over 50 publications covering a range of agricultural, environmental, and rural policy evaluative research.

Katharine Foot is a Senior Lecturer at the RAU. She is a Rural Practice Chartered Surveyor, a Fellow of the Central Association of Agricultural Valuers, and a qualified teacher. She worked in private practice before joining the academic staff at the RAU, and has a particular interest in planning and sustainability.

Andrew Davis farmed a large mixed farm for over 15 years before becoming a regional director for the then Country Landowners Association. Subsequent to this he joined the academic staff at RAU where his lecturing responsibilities include farm and estate management.

 SEE WEB LINKS

For recommended web links for this title, visit www.oxfordreference.com/page/agric when you see this sign.

VISIT ONLINE

OXFORD QUICK REFERENCE

The most authoritative and up-to-date reference books for both students and the general reader.

Accounting
Agriculture and Land Management
Animal Behaviour
Archaeology
Architecture
Art and Artists
Art Terms
Arthurian Literature and Legend
Astronomy
Bible
Biology
Biomedicine
British History
British Place-Names
Business and Management
Chemical Engineering
Chemistry
Christian Art and Architecture
Christian Church
Classical Literature
Computer Science
Construction, Surveying, and Civil Engineering
Cosmology
Countries of the World
Critical Theory
Dance
Dentistry
Ecology
Economics
Education
Electronics and Electrical Engineering
English Etymology
English Grammar
English Idioms
English Language
English Literature
English Surnames
Environment and Conservation
Everyday Grammar
Film Studies
Finance and Banking
Foreign Words and Phrases
Forensic Science
Fowler's Concise Modern English Usage
Geography
Geology and Earth Sciences

Hinduism
Human Geography
Humorous Quotations
Irish History
Islam
Journalism
Kings and Queens of Britain
Law
Law Enforcement
Linguistics
Literary Terms
London Place-Names
Marketing
Mathematics
Mechanical Engineering
Media and Communication
Medical
Modern Poetry
Modern Slang
Music
Musical Terms
Nursing
Philosophy
Physics
Plant Sciences
Plays
Political Quotations
Politics and International Relations
Popes
Proverbs
Psychology
Quotations
Quotations by Subject
Rhyming
Rhyming Slang
Saints
Science
Scottish History
Shakespeare
Social Work and Social Care
Sociology
Statistics
Synonyms and Antonyms
Weather
Weights, Measures, and Units
Word Origins
World Mythology
Zoology

Many of these titles are also available online at
www.oxfordreference.com

A Dictionary of

Agriculture and Land Management

WILL MANLEY, KATHARINE FOOT,
AND ANDREW DAVIS

Great Clarendon Street, Oxford, OX2 6DP,
United Kingdom

Oxford University Press is a department of the University of Oxford.
It furthers the University's objective of excellence in research, scholarship,
and education by publishing worldwide. Oxford is a registered trade mark of
Oxford University Press in the UK and in certain other countries

© Oxford University Press 2019

The moral rights of the authors have been asserted

First edition published in 2019
Impression: 1

Published in the United States of America by Oxford University Press
198 Madison Avenue, New York, NY 10016, United States of America

British Library Cataloguing in Publication Data
Data available

Library of Congress Control Number: 2018968296

ISBN 978-0-19-965440-6

Printed and bound in Great Britain by
Clays Ltd, Elcograf S.p.A.

Contents

Preface

Rural areas are complex and ever-changing. As well as providing food and other physical resources such as timber and minerals, they hold enormous wells of biodiversity. The way they are managed impacts on our air and water quality, our landscapes, and our heritage. They also provide places for people to live as well as opportunities for recreation. They are areas where businesses operate and employment is provided. As technology evolves and society changes, rural areas also adapt and change; for example, as rural land is potentially used for renewable energy generation, or for the provision of more housing, or infrastructure.

Consequently, the scope of this *Dictionary of Agriculture and Land Management* is wide. Agriculture, including both husbandry and management aspects, necessarily makes up a significant proportion of the entries; however, we have also sought to cover wider aspects of the management of the natural and built environment in rural areas. This includes environmental issues and their management, woodland and forestry, landlord and tenant and land law, heritage, rural planning and development, finance and business, marketing, and equine and field sports. The dictionary also addresses general terms, policy areas, and organizations which are relevant to rural areas. Where changes in the law or policy relating to an entry have yet to take effect, we have tried to indicate the probable direction of travel in future.

This dictionary has therefore been written with a wide range of readers in mind. Key amongst these are students, at all levels, on agriculture, agricultural science, animal and crop management, land, property, and estate management, farm and agri-business management, countryside conservation and management, equine management, rural tourism, and rural economy courses. However, the dictionary should also be of value to students on a wider range of programmes where agriculture or rural land and property management form a component part; for example, town and country planning, accountancy, law, or environmental management. In addition, those people undertaking professional examinations in directly related sectors may also find the dictionary of assistance, as well as other professionals with a rural client base, elected representatives, and policy and practice advisers from the private and public sectors. It may also be of more general appeal to others who perhaps have a wider interest in what goes on in rural areas.

Each entry provides a short definition of the word at the beginning and then goes on to expand upon it within the context of agriculture and land management. Where the entry relates to a concept, the entry will often be longer, providing an overview of the concept's principles and application. In the print edition, an asterisk (*) before a word within an entry indicates that this word has its own entry elsewhere in the dictionary which will provide further information. These cross-references are links in the online and ebook editions.

Acknowledgements

The authors would like to thank all those present and past colleagues at the Royal Agricultural University who have provided support and assistance during the process of writing the dictionary; notably Nicola Cannon, Rachael Foy, Sue Smith, Sally Story, and Matthew Smith. We would also like to thank Rebecca Lane, who originally commissioned the dictionary for Oxford University Press, and Joanna Harris, Senior Editor at Oxford University Press, whose invaluable, and very patient, support throughout the whole process of writing the dictionary has been greatly appreciated. Finally, we would like to thank our families and friends for their understanding and forbearance during what has turned out to be a somewhat longer-term project than we had originally anticipated.

Will Manley, Katharine Foot, and Andrew Davis
Royal Agricultural University, Cirencester, UK

abatement 1. In land and property law, the reducing, or cancelling, of a requirement to pay a sum, or sums, of money; for example, under a *lease a *tenant may not have to pay *rent for a time period in which they were physically unable to occupy a property damaged by fire. **2.** Removing, limiting, or ending, a nuisance. A *statutory nuisance is likely to be addressed initially through the serving of an **abatement notice** by the *local authority. In the case of a *private nuisance, action through the courts under the *tort of nuisance is likely.

abatement notice *See* STATUTORY NUISANCE.

Aberdeen Angus Breed of cattle predominantly used for beef production. Usually black in colour but can be red, the breed originates from Aberdeenshire in Scotland but is now found across the globe, and is prized for its docile nature, premium carcass quality, and strong brand reputation.

(⊕) SEE WEB LINKS
• UK breed society website for the Aberdeen Angus.

abiotic Non-living component of the environment that exerts an influence over a life-form. Usually a reference to factors affecting the ecology of an organism, examples include temperature, water content, light intensity, and gas concentration. *See also* ABIOTIC STRESS; BIOTIC.

abiotic stress The detrimental impact *abiotic factors have on a living organism. In order to have a demonstrable effect on the organism, the abiotic factor or stressor must influence the environment beyond the normal range for that factor and environment such that the living organism exhibits symptoms of stress. Abiotic factors are unavoidable and at their most damaging when they occur in combinations together.

abomasum Third of four chambers in the *ruminant stomach, most similar in terms of conditions and function to that of a single-stomached animal, being acidic and involved in *protein digestion. *See also* RUMINANT.

abort To lose a foetus before the completion of a full pregnancy. May be brought about deliberately through veterinary intervention, but more commonly it is as a result of disease, stress, or malnutrition, and represents a significant area of financial loss to the livestock sector.

absolute freehold *See* FREEHOLD; TITLE.

absolute leasehold *See* LEASEHOLD; TITLE.

abstraction *See* ABSTRACTION LICENCE.

abstraction licence Form of permission required in the *United Kingdom to **abstract** (take) water either temporarily or permanently from rivers, lakes, canals, reservoirs, or underground sources. Licences are issued by the relevant government environmental agency (the *Environment Agency in England, *Natural Resources Wales, the *Northern Ireland Environment Agency, and the *Scottish Environment Protection Agency), and are commonly held by farmers for purposes such as *irrigation. The requirement to apply for an abstraction licence before taking water means that the relevant environmental agency can control how much water is being removed from each water *catchment, to prevent issues such as water shortages and environmental damage.

The thresholds (in terms of volumes of water that can be abstracted without triggering a requirement to either apply for permission from, or to notify, the relevant government environmental agency) vary slightly between the different parts of the United Kingdom. England and Wales operate a single-tier system only using abstraction licences to control water abstraction. However, Northern Ireland and Scotland both operate multi-tier systems.

Very small volumes of water abstraction in Scotland are managed through a system of **general binding rules** with which all abstractors must comply. At higher volumes an abstractor must register with the Scottish Environment Protection Agency, whilst only those abstractors taking the largest volumes of water have to apply for an abstraction licence.

A similar tiered system also operates in Northern Ireland, with all abstractors being expected to comply with specific rules on abstraction. All but those abstracting the smallest volumes of water must notify the Northern Ireland Environment Agency, and abstractors taking the largest volumes of water must apply for abstraction licences.

See also IMPOUNDMENT LICENCE.

acarid Describes members of the taxon acari which is a subclass of the class arachnida. They include ticks and mites, many of which are parasitic organisms. *See also* MITE; PARASITE; TICK.

acceptable daily intake The amount of a substance that may be ingested for either nutritional or veterinary/medical reasons through either food or water on a daily basis over a lifetime without an appreciable health risk. Values are commonly provided in a measurement per kilogram of bodyweight. Often abbreviated to ADI. Substances awarded an ADI have since been extended to include *pesticide residues.

accommodation land Term used to describe land (usually next to existing settlements, and historically often used for grazing animals prior to slaughter) that may have some potential for *development at a future date, but usually not current *planning permission. Accommodation land in rural areas often commands a higher value than purely agricultural land, because whilst purchasers may buy the land for use for agricultural purposes in the short to medium term, they may also do so in the hope that there may be development on it at a future date, and it will then be worth significantly more than they have paid for it. In accommodation land sales and valuations the additional element of value over agricultural value which is attributable to the hope of future development is referred to as *hope value, and it will usually be greater the greater the likelihood of development happening in the foreseeable future.

accommodation works *See* COMPULSORY PURCHASE.

accounts A record of transactions; the term is also commonly used to describe a set of **financial statements** detailing the financial position and performance of an organization at a given date—in essence, how much *profit the organization has made; how much it owes to other people or organizations; how much it owns; and how much other people owe to it. This is set out in a series of documents, also known as financial statements, including, in the United Kingdom, a *balance sheet (also known as a statement of financial position), *cash flow, *profit and loss account (also known as an income statement), and a **statement of changes in equity** (showing how much has been paid to shareholders as a dividend, and any other contributions made by shareholders during the period covered by the accounts). The requirements of, and information shown in, financial statements are controlled by accounting standards and principles, and hence there is some variation between countries depending upon which accounting standards and principles are in use in that country.

accrual *See* BALANCE SHEET.

acidosis An increased level of acidity in the blood and tissue fluids. Can be sub-acute (symptoms not detectable) or acute (detectable symptoms), and can itself be a symptom of more complicated digestive disorders. It is of particular concern in *ruminants where it is caused by feeding large amounts of readily fermented *carbohydrates. The condition puts extreme pressure on the kidneys and other organs involved in maintaining blood and tissue water balance. Symptoms can include all or some of: muscle weakness, renal failure, reduced appetite, increased heart and respiration rate, diarrhoea, and sudden death. *See also* DISEASE.

acid rain Rainwater that has a pH below 5.6 can be considered as acid rain. Natural rainwater normally has a pH of about 5.6, i.e. very slightly acidic under normal circumstances due to carbon dioxide in the air forming carbonic acid, which is a weak acid. However, increasing levels of atmospheric carbon dioxide are mainly responsible for the increased levels of acidity in rainwater, thus lowering the pH. Acid rain has many damaging environmental effects, including acidification of oceans and lakes, which can impact on biodiversity, degradation of soils, and erosion of building materials such as limestone and marble. *See also* CLIMATE CHANGE; GREENHOUSE GAS.

ACOS *See* ADVISORY COMMITTEE ON ORGANIC STANDARDS.

acre An imperial and US unit of measurement of land. Defined by the area of 1 **chain** (66 feet) by 1 **furlong** (660 feet). An increasingly more common unit of measurement on UK *farms and in *forestry is the metric *hectare, which equates to approximately 2.47 acres. *See also* APPENDIX 1 (METRIC–IMPERIAL CONVERSION).

Actinomycetes Describes members of the taxon Actinobacteria, a group of largely anaerobic, gram-positive bacteria that typically live in large colonies, or mycelia, with long branching filaments. The majority of species are soil-living, and have limited or no impact on the health and welfare of plants, animals, or livestock; however some are pathogenic, some have significant beneficial impact on *organic matter decomposition and nutrient recycling, and some produce beneficial bioactive compounds that can be harvested and utilized, including *antibiotics, antifungals, *insecticides, *herbicides, and immune-moderators. It is generally recognized that there are eight groups of actinomycetes; however, studies continue to discover further differences between the members of this taxon, and hence changes in their classification occur from time to time.

active ingredient The component of a drug, *pesticide, *insecticide, or other biochemical agent that is biologically active and causes the desired effect in the target. Sometimes referred to as the active substance in pesticide development, or active pharmaceutical ingredient in drug development. There may be more than one active ingredient in any given product.

Act of Assembly Law (equivalent to an *Act of Parliament) approved, through vote, by the elected representatives of either the *National Assembly for Wales, or *Northern Ireland Assembly. A law established through approval by a legislature (parliament or assembly) is also known as a *statute.

Act of Parliament Law (known as an *Act of Assembly in Wales and Northern Ireland) approved, through vote, by the elected representatives of the *Houses of Parliament, *National Assembly for Wales, *Scottish Parliament, or *Northern Ireland Assembly. Acts of Parliament normally set out the general principles of a new law, and give legal powers to ministers of the respective governments to put in place the detail of the new law through issuing pieces of *delegated legislation which do not have to be subject to a further vote. Once agreed through the respective Parliament or Assembly, the law must be formally agreed by the monarch in a process called *Royal Assent. A law established through approval by a **legislature** (parliament or assembly) is also known as a *statute.

(⊕) SEE WEB LINKS

• Website of the UK Parliament: includes information on how laws are made, and also the progress of legislation currently before Parliament..

adaptation Term used in *climate change policy in the *United Kingdom to describe policy approaches which aim to change future mechanisms for the delivery of goods and services so that they are best adapted to suit the changing

*climate as a result of climate change. For example, climate change models
indicate that rainfall patterns in the United Kingdom are likely to change in the
future with wetter winters and drier summers predicted, which will mean that the
way water is managed will need to be adapted to deal with greater numbers of
incidents of flooding in the winter and possible water shortages in the summer.
Climate change policy in the UK is structured around adaptation approaches,
alongside *mitigation policies.

ADAS *See* AGRICULTURAL DEVELOPMENT AND ADVISORY SERVICE.

adjudication *See* ALTERNATIVE DISPUTE RESOLUTION.

ad lib feeding A system of providing food for animals that allows them to
eat to appetite. They are free to choose how much of the feed they eat and are
not restricted to a predetermined quantity. The term comes from the Latin
ad libitum, which translates as 'at one's pleasure' or 'at liberty'. Advantages of ad
lib feeding are that individual differences in animal requirements are catered
for through the self-regulation by the animal, and where groups of animals are
at differing performance levels this method can ensure that nutritional deficiencies
are not the cause of any lack in performance. Disadvantages of the system can be
higher feed costs through wastage and through difficulties in accurately predicting
quantities of feed required. Ensuring that each animal has an equal opportunity
to access the feed is also important, as social dominance may result in inferior
animals being prevented access to food if not carefully managed.

***ad medium filum* rule** Legal presumption that states that where the boundary
between two properties lies either side of a highway or non-tidal river or stream,
the boundary runs down the middle line of the highway, river, or stream.
See also HEDGE AND DITCH RULE.

admixtures The result of breeding between two populations of previously
separate individuals. The resulting offspring is therefore a *hybrid of the two
parent populations.

adverse possession The act of occupying land without the permission of the
person who holds the *title to it (the **paper owner**), with a view to depriving
them of their legal rights of ownership. Someone who is occupying land without
the owner's permission is said to be *squatting. In England and Wales, under the
principle of **limitation**, if the owner of land does not assert their rights to their
land, and another party (the **squatter**) occupies that land without the owner's
permission for a period of time prescribed in law, then the squatter potentially
has the ability to take ownership of that land through adverse possession.
(For Northern Ireland and Scotland, see below.)
   The operation of adverse possession in England and Wales is slightly different
depending upon whether the land is registered at the *Land Registry or not.
For unregistered land, after twelve years of adverse possession the *title of the
paper owner is effectively extinguished. With registered land, a successful
application to the Land Registry (which can initially be made after ten years)

would result in the squatter being able to register as the proprietor of a registered estate (in effect, as the owner).

To successfully claim adverse possession with the Land Registry, a squatter must be able to show that they have:

- taken factual possession of the land. *Case law suggests that this, in effect, means that the squatter has a degree of physical control over the land and has treated the land in exactly the same way as an owner would have done, perhaps by excluding others, including the paper owner, for example, by erecting fencing (although this would not necessarily be sufficient on its own to demonstrate adverse possession). The factual possession also must be adverse to the paper title owner (in essence, without their permission).
- that the squatter has demonstrated an intention to deprive the paper owner of their rights (by taking actions that demonstrate that the squatter treated the property in the same way as they would have done if it was their own, including excluding the paper owner).
- and that this has been done over a sufficient period of time.

Once a squatter has applied to the Land Registry, the owner of the paper title and others with an interest in the land (if known) are notified. If they do not object, then the squatter will be registered as the owner. If, however, they object, then the squatter will have their application rejected, unless one of three grounds apply:

- Following the principles of *estoppel the squatter argues that they were acting on a representation made to them by the owner and that denying adverse possession to them would be to their detriment.
- That the applicant had some other reason to be registered as owner (say, for example, they have been left the property under a will).
- That the land is part of some boundary dispute, and the squatter believed it to be their own land and has acted accordingly.

In the case of registered land, if the squatter's application is rejected but they are still not evicted within two years, then at the end of that period they may reapply to the Land Registry to be registered as proprietor, and on this occasion the paper owner does not have any way of defending against the application.

It is also possible for someone who was a former tenant to claim adverse possession against their landlord if they remain in a property after their lease has ended and the tenant does not then pay rent or in some other way acknowledge that the landlord is the owner.

At the time of writing, the *Law Commission was reviewing the law in England and Wales around adverse possession, alongside other land registration matters.

It is also possible to gain ownership of land by adverse possession in Northern Ireland, although the distinction between registered and unregistered land does not apply in the same way as in England and Wales, and a twelve-year period applies to them both. In Scotland, the nearest equivalent to adverse possession is **prescription**.

See also TRESPASS.

advertising controls The planning control system which regulates how and where advertisements can be displayed in the United Kingdom. Advertisements, in this context, include a wide array of signs and notices, ranging from town and village name-signs, through to large billboards. They are divided into three different categories:

- Those which are permitted without requiring consent from the *local planning authority, providing certain criteria and conditions, as set out in the advertising *regulations, are met (for example, in relation to size, positioning or illumination). Advertisements which may fall into this category include those displayed inside a building.
- Those with **deemed consent**. These types of advertisements do not need specific consent from the local planning authority, providing they comply with the specific criteria and conditions set out in the advertising control *regulations for that particular class of advertisement (again, for example, in relation to size, positioning or illumination). Advertisements potentially falling into this category include a 'for sale' board placed on land that is being marketed, or a temporary notice announcing details of a local charity event.
- Those that require **express consent** from the local planning authority through the submission of an individual application. This category applies if the advertisement does not fall into either of the other two categories, or if it does, but does not comply with the relevant criteria and conditions (for example, it is too big).

Regardless of which category in which the advertisement sits, it must comply with the **standard conditions**. These include requirements that they must have the permission of the owner of the site, and that they must not cause safety issues for either the public or for transport (for example, by obstructing a road sign).

The advertising control system controls advertisements in relation only to their effects on amenity (primarily visual and sound-related amenity) and public safety (including road, rail, and aviation safety), and it is therefore less complex than the general planning system.

Local planning authorities have the power to request permission from central government to restrict deemed consent in a particular area, thereby increasing the number of classes of advertisements that are subject to the requirement to apply to the local planning authority for consent. Additionally, an **Area of Special Control** designation can be put in place in England, Wales, and Scotland, which puts additional restrictions on advertisements in a particular locality; for example, reducing size limits for some deemed consent classes. This designation is often used in rural areas.

Local planning authorities also have a number of *enforcement powers in relation to unauthorized advertisements, which include the power to issue a notice ordering that such an advertisement be removed (or, in some circumstances, they can remove it immediately themselves). If the person or persons responsible for the advertisement do not remove it after having been given notice to do so by a local planning authority, then the authority may remove it themselves and seek to recover the reasonable costs of doing this from the person or persons responsible for the advertisement (which may include the person erecting or maintaining it, the person who benefits from it, or the occupier of the land or building on which it is placed).

Whilst England, Wales, Scotland, and Northern Ireland each operate their advertising control system under their own regulations, the practical operation of each system is very similar.

Advice Notes (Scotland) *See* PLANNING POLICIES.

Advisory Committee on Organic Standards (ACOS) Previously the UK Register of Organic Food Standards (UKROFS), and set up because of the need to unify standards in *organic food production. These are now national standards which are regularly reviewed and updated. ACOS is an organization independent of government and is managed by individuals associated with the organic food sector. It approves a number of independent bodies which carry out inspections and certification procedures of *organic *holdings. *See also* SOIL ASSOCIATION.

aeration The process by which air is mixed with a liquid or other material. The term is often applied to soils where tillage by means of spikes or a *cultivation increases the air flow within the soil. This reduces compaction, and aids drainage and root growth and thus the take-up of nutrients by plants. It can also apply to the oxygenation of water such as a lake where the stocking rate of fish and high temperatures may result in natural oxygen levels becoming depleted.

aerobic Any process or organism requiring the presence and involvement of oxygen. Often used to describe an organism as being aerobic, meaning it requires oxygen to survive, or a biochemical pathway that requires the involvement of oxygen. *See also* ANAEROBIC.

affordable housing Category of housing available for sale or rent which is considered to be more affordable than housing available on the open market. In the decades following World War II, affordable housing in the United Kingdom was generally built by *local authorities, and was commonly known as **council housing**. However, the construction of affordable housing is now often provided for through the planning system in connection with the grant of *planning permission to private developers for open market housing. Planning permissions for open market housing are now frequently granted subject to the provision of some element of affordable housing.

As a result of the role now played by the planning system in the provision of new affordable housing, what constitutes affordable housing is defined within national/regional *planning policies for England, Wales, Scotland, and Northern Ireland. Whilst there is some variation in terms of what is considered to be affordable housing in planning terms between the different parts of the UK, in general terms, affordable housing can potentially include:

• Homes owned by local authorities, or registered providers (or in some specified circumstances, other owners), and made available at lower levels of rent than market rent.

- Housing purchased through affordable home ownership schemes; for example, **shared ownership** (where the purchaser buys a percentage share of the property, rather than all of it, and pays rent on the remainder).
- Market sales housing which is offered for sale at a discount against local market value.

'Affordability' is usually determined in relation to local market values or market rents, and planning policy generally specifies that provisions should be included to ensure that the property continues to remain affordable in future, or that any subsidies should be recyclable to provide alternative affordable housing.

affordable housing exception site (Wales) *See* RURAL EXCEPTION SITE.

afforestation The *planting of usually commercial tree species on land that is not already forested. Reforestation describes the same but on land that is forested. Normally associated with larger-scale programmes of *conifer trees on more marginal and often upland areas.

afterbirth The placenta and placental membranes that are passed after the expulsion of the foetus. Also sometimes referred to as the 'cleansings'. Following successful delivery of the foetus, the placenta and associated membranes should follow soon after, although a delay of a few hours is not uncommon. However, a retained placenta is a significant cause of metritis, particularly in cattle. This is an infection of the uterus following the initiation of decomposition of the retained membranes. If untreated, the infection could result in temporary or permanent infertility, or death of the animal.

age-class A means of measurement by categorization of trees or groupings of trees by specific age ranges. Normally included with or in relation to *size class and/or *standing timber volume.

aggregate A substance made up of a number of different component elements.

agrarian Relating to farmland and its economy, the cultivation of land or land tenure. For example, the agrarian revolution in the UK took place in the eighteenth and early nineteenth centuries, which saw significant improvements in productivity and agricultural reform. *See also* AGRICULTURE.

agribusiness Term referring to businesses directly or indirectly related to agriculture and rural activities. Examples of agribusinesses include food production, food processing, marketing and retail of food, agri-chemical manufacture and distribution, land management agencies, and tourism related to agriculture.

Agricultural Advisory Panel for Wales *See* AGRICULTURAL WAGES BOARD; MINIMUM WAGE.

Agricultural Census This census is conducted annually in June by the *Department for Environment, Food and Rural Affairs (DEFRA) in England, and by the respective agricultural departments for each of the devolved governments of the UK. Farmers are surveyed to collate information on farm business activities related to holdings and geographical regions, and the results are published in a range of publications. *See also* HOLDING.

Agricultural Development and Advisory Service Largest UK agricultural and environmental consultancy providing research and policy advice. It was formerly attached to the *Ministry of Agriculture, Fisheries and Food before eventually becoming privatized.

Agricultural Holdings Act 1986 *Act of Parliament governing the commercial letting of agricultural land in England and Wales if the *tenancy was granted before 1 September 1995 (and in some limited circumstances after that date, as discussed below). New tenancies granted on, or after, 1 September 1995 are, instead, governed by the *Agricultural Tenancies Act 1995, and are commonly referred to as **Farm Business Tenancies**. Tenancies governed by the Agricultural Holdings Act 1986 are commonly called **AHA tenancies**.

AHA tenancies are essentially lettings of agricultural land for the purposes of running a trade or business (although they can also include residential property, such as the farmhouse, providing that the main purpose of the letting is for agriculture). The Agricultural Holdings Act 1986 (AHA 1986) sets out the exact circumstances which must exist for an AHA tenancy to exist, and the definition of 'agriculture' is also detailed in the *legislation.

AHA tenancies do not have to be set out in writing; however, under the Act either the *landlord or *tenant may request that the other enters into a written agreement.

Where an AHA tenancy was granted for an initial *fixed term then (with a few exceptions) at the end of that term the tenancy automatically converts into a tenancy running from year to year under the same terms as the original tenancy. There are, therefore, now very few contractual fixed term AHA tenancies in existence, with most now being tenancies from year to year.

The responsibilities of the landlord and the tenant under an AHA tenancy for the maintenance, repair, and insurance of *fixed equipment are set out in law in the **model clauses** *statutory instrument. (England and Wales currently have different versions of this, and, at the time of writing, the *Welsh Government is proposing changes to the model clauses applying in Wales.) Fixed equipment is also defined in law and generally includes buildings and structures which are affixed to the land as well as works in, on, over, or under it (for example, roads and ditches). It also includes things which are grown on the land but not normally removed for use or consumption (for example, hedges). The model clauses are automatically implied into every AHA tenancy (unless any written agreement states otherwise). The model clauses set out the obligations of both the landlord and the tenant. For example, the landlord generally has responsibility for the repair and maintenance of many of the structural elements of the buildings, with the tenant being responsible for some specified types of maintenance (for example, cleaning eaves and guttering). The costs of repair and maintenance of some things are effectively split between the landlord and tenant (for example, doors and windows).

An AHA tenant has the right to remove *fixtures and buildings which they have erected or installed, as they remain the tenant's property (this is slightly different from the normal *common law principle that fixtures generally become part of the land and therefore would be owned by the landlord). If a tenant wishes to remove their fixtures or buildings then they must serve notice on the landlord. The landlord may, in turn, serve a *counter-notice electing to buy the fixture or building instead of the tenant removing it. Under the AHA, the tenant may also apply, in England to the *First-Tier Tribunal Property Chamber (Agricultural Land and Drainage) or the *Agricultural Land Tribunal for Wales, for a direction that the landlord must provide, alter, or repair fixed equipment which the tenant needs in order to comply with *statutory requirements (i.e. something set out in the law).

In the fixed term of an AHA tenancy the provisions for *rent reviews are as set out in the tenancy agreement. However, for tenancies that have continued beyond the fixed term then the AHA 1986 sets out provisions for reviewing the *rent. The landlord and tenant can reach agreement, or either can serve a **section 12 notice** on the other party requiring that the matter of the rent is referred to **arbitration** or other third-party determination (*see* ALTERNATIVE DISPUTE RESOLUTION). The provisions of the Act are such that, in essence, between twelve and twenty-four months' notice must be given of any change in rent, and, generally, the rent can only be reviewed once every three years. The Act also sets out complex provisions which the arbitrator or third party must use in determining the new rent. In essence, the guiding principle is that the rent must be the amount which a prudent and willing tenant might expect to pay to a prudent and willing landlord for the holding, taking into account all relevant factors, which include:

- The *terms of the tenancy.
- The character, situation, and locality of the holding.
- Current rent levels for *comparable (similar) holdings (although the Act sets out a number of aspects which much be disregarded when this is assessed, including elements of rents which are attributable to *scarcity value or *marriage value).
- The **productive capacity** and the **related earning capacity** of the holding.

The Act defines productive capacity, and it is, in essence, a measure of the potential opportunities presented by the holding for a reasonably competent tenant practising a farming system suitable for the holding (in terms of, for example, the nature of the land and the facilities and fixed equipment that are available). The related earning capacity is the extent to which a competent tenant could reasonably be expected to make a profit from undertaking such a system of farming, in light of the productive capacity. The Act also requires that a number of things are disregarded when the rent is determined, including the *tenant's improvements or their fixed equipment. The landlord is, though, in some circumstances, entitled to increase the rent to reflect *landlord's improvements.

a

There are *security of tenure provisions under the AHA 1986 which restrict the situations when a landlord's *notice to quit can operate. Generally, a landlord must give at least twelve months' notice to quit. There are two types of notice to quit that a landlord may serve: one which specifies one or more of the statutory *grounds for possession included in Schedule 3 of the Act (known as **cases**), and one that does not rely on any of the specific cases.

There are eight cases, including **Case B** (the land is required for a use other than agriculture—for example, for some form of *development); **Case D** (non-payment of rent and **remediable breaches**, i.e. ones that could be corrected); **Case E** (irremediable breaches, i.e. those which it is not possible to correct); **Case F** (insolvency, i.e. the tenant is bankrupt), and **Case G** (death of the tenant). The Act sets out specific conditions which must apply for each of these conditions to have effect. The cases also include **Case C** (certificate of *bad husbandry). Bad husbandry is defined in legislation and essentially covers a situation where the tenant is not considered to be farming the holding in such a way as to maintain a reasonable standard of efficient production and is not keeping the holding in a condition to allow for this standard to be maintained in future. In order to rely on Case C the landlord must have, first, applied to the relevant Tribunal (as noted above) for such a certificate, and they are rarely granted. Notice served under some of the cases can be challenged by the tenant demanding arbitration.

The second type of notice to quit that the landlord may serve is one that does not specify one of the statutory grounds for possession (cases). If this type of notice to quit is served then the tenant has the right to serve a counter-notice. This stops the notice to quit taking effect unless the landlord has obtained consent for it to operate from the relevant Tribunal. The Tribunal will only grant consent if it is content that one of the grounds set out in section 27 of the Act applies, and it is of the view that a fair and reasonable landlord would insist on possession. The grounds in section 27 include *good husbandry; that the landlord wishes to change the land to a use other than agriculture (but not covered by Case B); and that greater hardship would be caused by not allowing the notice to operate than will be by consenting to its operation.

Under the AHA 1986, tenants also benefit from potential *succession rights. Generally these only apply to tenancies granted before 12 July 1984 (unless the tenancy is, itself, a tenancy obtained through succession after that date, or in some other limited circumstances specified in the Act). The Act potentially offers the opportunity for two successions and, therefore, although new agricultural tenancies since 1995 have been Farm Business Tenancies, as a consequence of succession rights potentially covering two generations, AHA 1986 tenancies will continue to exist for some time in the future. Succession can happen either on the death or retirement of the existing tenant. In order to succeed to a tenancy on the death of the previous tenant, the person who wishes to succeed must apply to the respective Tribunal, and must meet three tests set out in the Act. These are that they must be a close relative of the deceased, that they meet the **principal livelihood** test, and that they do not already occupy a commercial unit of land. The Act specifies who can be considered to be a close relative and also

what is meant by the principal livelihood test. To meet the principal livelihood
test then the close relative must have, in essence, obtained most of their income
from work on, or from, the holding for at least five of the last seven years
(although there are some allowances in connection with this; for example, for
someone who has been a student during this period). The Act also specifies how
'occupation of a commercial unit' is assessed for the purposes of succession.
Once the three tests have been met, the Tribunal must also consider whether the
applicant is a suitable person; this will account for the degree to which the
applicant has experience of, or training in, agriculture, their age, health, and
financial standing, and the views (if any) of the landlord on the applicant's
suitability. If there are two potentially eligible applicants the Tribunal will usually
choose between them; however, if the deceased had designated one of them as a
successor in their will then the Tribunal will consider that applicant first (unless
they are not considered to be a suitable person). Generally the terms of the new
tenancy will be the same as those of the deceased's tenancy (though they will
automatically contain a *covenant preventing *assignment, *sub-letting or
otherwise parting with possession without the landlord's consent, even if the
original tenancy did not). As noted above, succession is also available, in some
circumstances on the retirement of the previous tenant. This can be done either
by agreement with the landlord or by following a procedure set out under the Act.
In order to rely on these provisions, the tenant (who must usually be at least 65
years old) must serve a retirement notice on the landlord indicating that they
wish for an eligible person to succeed to the tenancy, and an application must be
made to the relevant Tribunal. As with succession on death, the eligible person
must meet the close relative, principal livelihood, and non-occupation of a
commercial unit tests, and also be considered to be a suitable person.

   At the end of the tenancy the tenant is able to obtain *compensation from their
landlord for some tenant's improvements which they have carried out, as well as
*tenant-right matters. Those things which the tenant is able to claim
compensation for are set out in the Act (as is whether the consent of the landlord
to the improvement is required in order for the tenant to be able to claim
compensation). The tenant can also be entitled to compensation for **disturbance**
if they have left the holding as a consequence of a landlord's notice to quit. The
eligibility for this, and the way it is determined, is set out in the Act; it includes
compensation for losses incurred by the tenant as a consequence of having to
quit the holding (for example, from moving, or selling stock) and/or, in some
circumstances, a payment based on a multiplier of rent. If, however, the holding
has *dilapidations, damage, or has deteriorated because of the tenant's actions,
or inactions, then the landlord can, in turn, claim compensation from the tenant.

Agricultural Land Classification System for classifying agricultural land
in England and Wales on the basis of its potential for agricultural production.
Land is classified on a scale of 1–5, with the best-quality land which has the
greatest potential for growing a wide range of crops, including horticultural
products, at high yields being graded 1, and the poorest quality land, being
graded 5. The majority of land in England and Wales lies within the middle

of the grading bands. Scotland has a parallel system known as the **Land Capability Classification for Agriculture**.

Agricultural Land Tribunal for Wales Independent, publicly funded *tribunal which resolves disputes in Wales between landlords and tenants under the *Agricultural Holdings Act 1986, as well as disputes concerning land drainage under the **Land Drainage Act 1991**. It has a similar role to the *First-Tier Tribunal Property Chamber: Agricultural Land and Drainage in England.

agricultural occupancy condition *See* AGRICULTURAL TIE.

agricultural property relief (APR) *See* INHERITANCE TAX.

agricultural tenancies Tenancies of commercial agricultural land (including, often, farmhouses). In the *United Kingdom there are several different types of agricultural *tenancy, and they are each governed by different *statute. *See also* AGRICULTURAL HOLDINGS ACT 1986; AGRICULTURAL TENANCIES ACT 1995; SCOTTISH AGRICULTURAL TENANCIES.

Agricultural Tenancies Act 1995 *Act of Parliament governing the commercial letting of agricultural land in England and Wales if the *tenancy was granted on or after 1 September 1995. Tenancies under the Agricultural Tenancies Act 1995 (ATA 1995) are commonly referred to as **Farm Business Tenancies (FBT)**. Tenancies granted before 1 September 1995 (and a few, in some specific circumstances, after that date) are governed by the *Agricultural Holdings Act 1986 (AHA 1986) and are often called **AHA tenancies**.

Unlike the AHA 1986, the ATA 1995 creates much more flexibility for the parties to contract with each other on the *terms that they wish. FBTs have minimal *security of tenure, and, unlike AHA tenancies, under FBTs there are no rights of succession (*see* AGRICULTURAL HOLDINGS ACT 1986 for a discussion of agricultural tenancy succession). FBTs also provide tenants with greater opportunities to undertake *diversification without losing the protection of the Act in comparison with those that are available under the AHA 1986.

In order for a tenancy to be considered to be an FBT it must meet conditions set out in the ATA 1995; namely, the **business conditions**, plus either the **agriculture condition** or the **notice conditions**. The requirements of each of these conditions are set out in the Act. In summary, the business conditions are such that all or part of the land in the tenancy must have been farmed for the purposes of a trade or business from the beginning and throughout the tenancy (the Act sets out how this is assessed, and also how 'farming' and 'agriculture' are defined for the purposes of an FBT). In addition to meeting the business conditions, the tenancy must meet either the agriculture or notice conditions. The agriculture condition is such that, at the time of any proceedings which require a definitive assessment of whether the tenancy is an FBT, the tenancy is primarily or wholly agricultural. In order to avoid leaving it up to the *courts to assess this, if there are, for example, subsequent changes in the use of the land, the notice conditions can instead be used. To utilize this provision, before or at the beginning of the tenancy the *landlord and *tenant must have given written notice to each other that they intend that the tenancy, or proposed tenancy, is to be an FBT and remain so throughout. Providing the tenancy was also wholly or primarily agricultural at the beginning, and that it continues to meet the business condition and any *covenants within the tenancy agreement, it will then remain an FBT.

Unlike AHA tenancies, there are minimal *security of tenure provisions for an FBT. These, in essence, are limited to providing *statutory notice periods for a *notice to quit (in most cases at least twelve months). Providing the notice to quit is valid, and the right period of notice has been given, then it is not necessary to make out any *grounds for possession for the notice to quit to have effect.

An FBT will automatically continue as a tenancy from year to year at the end of its *fixed term if that fixed term is for two years or more (unless either the landlord or tenant has served at least twelve months advanced notice of their intention to terminate the lease at the end of the fixed term). However, it is possible to use *break clauses within the fixed term of an FBT, subject to the notice provisions set out in the Act. As with *business tenancies, *forfeiture can also be used to end the tenancy for *breach of covenant by the tenant if there is a right of re-entry in the tenancy agreement.

Under the ATA, the landlord and tenant are largely free to agree the level of rent and whether and how it can be varied (with some limited controls under the Act). However, the ATA does include statutory *rent review provisions which can operate in the absence of any written agreement to the contrary.

If relied upon, the statutory rent review provisions allow either the landlord or tenant to demand a rent review generally every three years. This is done by serving a **statutory review notice** between twelve and twenty-four months in advance demanding that the rent payable is referred to **arbitration** (*see* ALTERNATIVE DISPUTE RESOLUTION). Once such a notice has been served, the landlord and tenant can voluntarily agree a new rent between themselves, or agree to appoint a third party to determine the rent on a basis agreed between them, or agree to appoint an arbitrator (or apply to the President of the *Royal Institution of Chartered Surveyors to appoint an arbitrator). The basis on which an arbitrator will determine the rent is set out in the Act and is based upon the rent that might reasonably be expected if the holding were let on the open market by a willing landlord to a willing tenant, taking into account all relevant factors including the terms of the tenancy. This may result in an increase or decrease in the rent. Generally (with some exceptions set out in the Act) the Act also requires that a number of factors are disregarded when the new rent is determined, including the *tenant's improvements.

Under an FBT, the tenant has (subject to some conditions) the right to remove *fixtures and buildings which they have erected or installed (or purchased from the preceding tenant). They are also entitled to *compensation from the *landlord for tenant's improvements on quitting the holding on the termination of a tenancy. Improvements are defined in the Act and can include both physical improvements and intangible advantages which become attached to the holding (for example, *planning permission in certain circumstances). The landlord (or an arbitrator if the landlord has not given consent on acceptable terms and the tenant has demanded the matter be considered by arbitration) must have given consent for an improvement in order for it to be eligible for compensation. The Act sets out the basis by which compensation is to be determined; however, it is possible for a landlord and tenant to have agreed an upper limit in advance on the amount of compensation payable for an improvement.

agricultural tie Term used to describe a situation where, as a *condition of a *planning permission, a residential dwelling in a rural area must only be occupied by someone directly connected to agriculture or forestry (also known as an agricultural occupancy condition), or where the property must be jointly occupied with a specific area of land. The tie is often set out in a legal agreement, and they are usually found in situations where permission would not have been granted for the house for any reason other than as a farmhouse or as a house for a farm or forestry worker. It is possible to apply for planning permission to remove an agricultural tie from a property; however, to do so it is usually necessary to demonstrate that there is no need for such a property in the locality (regardless of the individual circumstances of the owner).

Agricultural Wages Boards Bodies setting minimum wages, and employment conditions, for agricultural and *horticultural workers in the United Kingdom. The **Scottish Agricultural Wages Board** and **Agricultural Wages Board for Northern Ireland** fulfil these functions in Scotland and Northern Ireland respectively, whilst until 2013 the **Agricultural Wages Board (England and Wales)** covered agricultural and horticultural workers in England and Wales. However, agricultural and horticultural workers in England employed after October 2013 ceased to be covered by these provisions, and instead must now be paid in accordance with the appropriate hourly national *minimum wage rate. Wales continues to operate a system of a minimum agricultural wages, and has an **Agricultural Advisory Panel for Wales** to carry out the functions of the former Agricultural Wages Board (England and Wales).

Agricultural Wages Order *See* AGRICULTURAL WAGES BOARD; MINIMUM WAGE.

agricultural waste regulations *Regulations which set out how *waste produced as part of farming activities in the *United Kingdom must be managed. The scope of the regulations includes the full spectrum of waste management from how it is managed at the site where it is originally produced (including whether it is to be re-used or recycled by the original producer), through to its storage, onto any transport and transfer to a licensed waste management organization, and through to its final disposal.

The regulations set out several obligations on farmers. These are based around a legal **duty of care** that applies to everyone who deals with waste from the point of its production to its final recovery, recycling, or disposal. The requirements falling under the duty of care are designed to ensure that waste does not cause harm or problems to the environment (such as the pollution of water). For farmers, these obligations include the following:

- Waste must be stored safely so that it cannot escape, and waste producers are required to apply the *waste management hierarchy to their waste.
- If waste is passed on to any third parties then there is a requirement that the person(s) passing on the waste ensures that whoever they are passing it to is correctly authorized by the relevant government body to transport or deal with the type of waste in question (the *Environment Agency in England, *Natural Resources Wales, *Scottish Environment Protection Agency, or *Northern Ireland Environment Agency).
- If waste is passed on to any third parties then a *waste transfer note must be completed confirming what waste has been passed on, where, and when. A copy must then be kept by both the party passing on the waste and whoever receives it (and they must then, in turn, do the same thing if passing the waste on again). This establishes a traceable chain which allows waste to be tracked back to its original point of production (with a view to reducing the risk of *fly-tipping).

Waste on farms generally cannot be buried or burnt (apart from in the case of some low-risk, mainly plant-based wastes, subject to holding the relevant *waste exemption, as discussed below).

Farmers are required to register waste exemptions with their respective government body. These exemptions cover a range of common activities that farmers may undertake in relation to managing their waste (including, for example, burning waste plant materials in the open from activities such as tree cutting). These exemptions are subject to conditions on, for example, the volume of waste being dealt with. If a farmer wishes to carry out potentially exempted activity, but is not able to comply with the respective waste exemption conditions (perhaps they have too large a volume of that particular type of waste) then the farmer must apply for an *environmental permit (in England and Wales), or a *waste management licence and/or *pollution prevention and control permit in Northern Ireland and Scotland.

The agricultural waste regulations form part of the wider *waste regulations system (and operate on the same principles), and it is possible that a farm that has engaged in *diversification activities may produce both agricultural waste as well as other forms of *controlled waste (such as household or commercial waste, for example, from a farm shop). In practice, whilst household waste from a farm is likely to be dealt with through the normal domestic waste chains, any other commercial waste will need to be managed following the same principles as agricultural waste (albeit under the general waste regulations).

Most farms are also likely to produce some forms of *hazardous waste (known as special waste in Scotland), and will therefore also be required to comply with the *hazardous waste regulations, which, although operating on similar principles, have more stringent requirements.

See also WASTE CARRIER REGISTRATION.

agricultural worker's dwelling *See* RURAL WORKER'S DWELLING.

a

agriculture A very broad term covering the production of food, fibre, energy, medicines, and other products, primarily from plants and animals. Farmland may be used for the growing of crops, including fruit and vegetables, which constitutes *horticulture. These crops may be for food, such as *cereals, for fibre, such as cotton or flax, or for energy, such as *oilseeds or *short rotation coppice. Liquids to drink also come from agriculture, whether natural, such as milk or fruit juices, or processed, such as wine, beer, or spirits. The crops grown also include *grass and other *fodder crops for consumption by animals kept for meat, milk, or fibre. Pigs and poultry may be reared intensively in buildings so that farmland is not used other than to grow the food for the animals. Plants may also be grown indoors, such as mushrooms, and may not require soil as in *hydroponic production. Many plants are grown for compounds they contain that are highly prized for their medicinal properties, such as opium poppies for morphine, or for use in pharmaceutical products such as lavender. Animals, too, may be reared for medicinal purposes, such as pigs for insulin.

Humans have been involved in agriculture for thousands of years, originally eating wild grains and animals, and gradually domesticating the wild species, resulting in plants and animals very different from today. Once nomadic, the *tilling of land to grow *crops enabled farmers to settle on a tract of land.

Agriculture and Horticulture Development Board *See* LEVY BODIES.

agri-environment General description of the schemes developed to support environmentally friendly farming practices including organic and low-intensity systems, using voluntary management agreements that reward participating farmers with grant aid. Later schemes have also included a focus on environmental enhancement. The financial incentives paid to participating farmers and landowners are mainly based on *income foregone, and capital items reflect a proportion of actual costs. The schemes or **measures** were developed following the introduction of environmental issues as part of the *Common Agricultural Policy (CAP) in several European countries in the 1980s. In the UK, *Environmentally Sensitive Areas (ESA) were introduced in 1986. The integration of environmental concerns into agricultural policy was further developed in CAP reform with the Agri-Environment Regulation, which acknowledged the link between intensive farming and environmental damage, and with the *Rural Development Regulation. All EU member states implement some or several forms of agri-environment measure. For example, ESAs occurred across the UK, but in addition the devolved UK countries developed their own agri-environment schemes:

  Countryside Stewardship Scheme and *Environmental Stewardship Scheme in England.
  Rural Stewardship Scheme and Countryside Premium Scheme in Scotland.
  Tir Cymen and Tir Gofal in Wales.
  Countryside Management Scheme in Northern Ireland.

These are all now closed or finished and have developed into new or retitled schemes operating as:

*Countryside Stewardship Scheme in England.
*Glastir in Wales.
*Northern Ireland Countryside Management Scheme.
Scottish Rural Development Programme.

agri-food Agricultural practice that specifically focuses on producing food products. *See also* AGRICULTURE.

agritourism *Farm-based visitor-attracting enterprises. These might include farm holidays that involve helping with farming activities, visitor-friendly farms open usually seasonally and often with an educational focus, or payments for farm trails or other forms of access. Like **ecotourism**, the enterprises are often low key, but can be relatively labour intensive and thus an important source of employment.

agrochemicals A term covering the range of chemicals that are used in agricultural practice. This includes chemicals used as *pesticides, *herbicides, *insecticides, and synthetic *fertilizers, as well as chemical growth promoters, hormones, and *growth regulators. Strict legislation is often in place that regulates their use, as they are often toxic, or have major ramifications for the local ecology and human population if used incorrectly.

agro-ecosystem In contrast to *agri-environment, agro-ecosystem specifically describes a unit or grouping of farming activity that includes the *biotic and *abiotic factors as well as the socio-economic. The farm itself can essentially function as a defined *ecosystem. **Agroecological zones** are geographical areas sharing similar climatic conditions and linked to their respective abilities to support agricultural enterprises. *See also* ECOLOGICAL ZONES.

agroforestry The integration and combined management of farming and forestry systems on the same piece of land. An example of **agri-** *silviculture is the intercropping of wheat and *trees, and an example of of **silvi-pastoralism** is the grazing of livestock under trees. *See also* WOOD PASTURE.

agronomist An expert in the area of producing food and other beneficial products from the cultivation of plant-based species. Often acting as a consultant, an agronomist will offer specialist advice to a farmer in a number of areas such as soil management, crop rotation, irrigation, plant breeding and selection, and management of pests and diseases. *See also* AGRONOMY.

agronomy The science and practice of producing food and other beneficial products such as fibre and biofuel from plant-based resources. Agronomy encompasses chemistry, biology, genetics, ecology, biotechnology, and economics in order to optimize yield.

AHO *See* ANIMAL HEALTH OFFICER.

AHVLA *See* ANIMAL HEALTH AND VETERINARY LABORATORY AGENCY.

AI *See* ARTIFICIAL INSEMINATION.

air pollution The presence of substances in the air that have the potential to cause damage to human health, property, or the natural environment. Air pollution can be either localized, having its effects very close to the pollution source, or can have effects over a much larger area, possibly internationally, often as a consequence of combining with pollutants from other sources. Examples of air pollutants include *ammonia, nitrogen oxides, and sulphur dioxide, as well as particulates. Localized air pollution can also cause smells, which may be the subject of *statutory nuisance actions. *See also* GREENHOUSE GASES.

air source heat pump *See* HEAT PUMPS.

alder *See* TREE.

alfalfa (lucerne) A *perennial *forage *legume and member of the pea family, extensively grown for its superior feed quality and soil improvement properties. As a forage crop, it is usually used to produce *hay or *silage and has one of the highest protein and digestible fibre contents of all of the available forages. Because of this it is usually used for feeding *dairy cows or *horses, but *beef production and increasingly dairy *goat production is incorporating alfalfa into their diets. Alfalfa is rarely grazed, for a number of reasons. Because of the high protein and digestibility, it can be a significant cause of *bloat in *ruminants. It can also be a difficult crop to establish, particularly in temperate climates, and therefore the risk of *poaching from animal hooves is undesirable. Finally, the degree of wastage from grazing livestock means that grazing is an inefficient method of using this valuable crop. In some countries alfalfa can be grown as a salad for human consumption as alfalfa sprouts or dried alfalfa leaf supplements. Alfalfa has soil improvement properties due to the *nitrogen-fixing nodules present within its root system. This reduces the need for nitrogen-based *fertilizers and greatly increases efficiency. Alfalfa has drought-resistant properties, which is becoming of increasing importance within the context of climate change.

algae A large and diverse group of water-dwelling eukaryotes. Organisms can be single-celled, such as Euglena, or multi-celled, such as kelp. Most species of algae are autotrophic, meaning they produce their own *carbohydrates through *photosynthesis and are therefore pigmented, with the majority having the green pigment chlorophyll, but they can utilize brown and red pigments. Algal species are found in both fresh and salt water, and some are highly specialized to very specific environmental conditions, occupying niches that would be toxic to other life-forms. There are three examples of where algae have adapted to form symbiotic relationships with other life-forms: lichen (a relationship between algae and fungi), coral reefs (a relationship between algae and marine invertebrates), and sea sponges (a relationship between algae and a sea sponge). Algae are often used as an indicator species for aquatic ecosystems, as they respond quickly to changing environments, as seen by their population density and their composition. Algae are also farmed, as they produce products that are of great current and potential economic benefit. Examples of algal-based products include agar (a medium used in the cultivation of bacterial colonies), gelling or stabilizing agents in food and pharmaceuticals, fertilizers and soil conditioners, and nutritional supplements for humans and livestock. Areas for potential and increasing importance include bioremediation (the use of biological agents to correct spillages and leaks of toxic chemicals), pollution risk control through the treatment of sewage and other waste products, and biofuel production.

alienation The transfer of an interest in land to another. For example, in the case of land which is subject to a *lease, the *lessee could potentially **assign** the lease (in effect, transfer the rights they have for possession of the property to another party, thereby meaning they do not have an ongoing interest in the land; *see* ASSIGNMENT, or its Scottish equivalent *assignation). A lease may have an **alienation clause** in it either expressly allowing assignment or *sub-letting (with or without specific permission from the *landlord) or forbidding it.

alienation clause *See* ALIENATION.

alien species *See* NON-NATIVE SPECIES.

alkaloid A chemical term for a group of naturally occurring compounds that contain nitrogen and are basic in property. In addition to nitrogen, they also contain carbon and hydrogen and commonly oxygen and sulphur. More rare forms of alkaloids may also contain phosphorus, bromine, and chlorine. They often have toxic properties; however, in controlled amounts they can have significant medical benefits, and are often found in drugs for controlling blood pressure, treating cancer, antibacterial compounds, vasoactive compounds, anaesthesia, and stimulants.

allele The form that a gene may take, usually dominant or recessive, although there may be many more than just two forms. The majority of higher life-forms are *diploid and therefore have two copies of each chromosome present. The form of the genes found at a specific location on these chromosomes (i.e. its genotype) dictates the phenotype of the organism, i.e. the physical interpretation of that gene. For example, the gene determining coat colour may have two alleles—that giving rise to black and that giving rise to red. The black form of the gene is dominant over the red form. Where the animal carries two black alleles, the animal is said to be homozygous dominant for that characteristic and will have a black coat. If it carries two red alleles, it is said to be homozygous recessive for that characteristic and will have a red coat. If the animal carries one of each allele, the animal is said to be heterozygous and will exhibit the dominant characteristic, i.e. its coat will be black. It will, however, be able to pass on the potential to have a red coat to any offspring it may have.

allelopathy The ability of one organism to produce biochemicals that have a negative effect on the growth, reproduction, or survival of another organism.

Usually a trait found in plants, it is the subject of much research interest due to the potential for manipulating and controlling weeds and pest species.

alley cropping A cropping technique that uses permanent planting in rows of specific *nitrogen-fixing tree species to provide nitrogen for staple crops that can then be harvested annually off the strips of land in between the rows of trees. Usually used in areas where soil is poor, or subject to significant *erosion pressures, the principle relies on the use of tree species with large tap roots that are able to draw nitrogen to the surface of the soil, thus supporting their own but also neighbouring plant species' growth. In addition to nitrogen, the trees provide further benefits through preventing the physical removal of soil; the deposition of leaves, which provides mulch that can help with water retention and *nutrient recycling; and the provision of shade. If shade is not required or is undesirable, coppicing the trees can remove unwanted canopy and provide wood as a by-product. *See also* SUSTAINABLE FARMING.

allium Genus of plants that includes onions, garlic, shallots, chives, leeks, and spring onions, as well as numerous species used as ornamental flowers. They have been cultivated for thousands of years, and are characterized by the distinctive onion flavour caused by chemicals largely derived from cysteine sulphoxides. Alliums are perennial plants, having a tuber that reforms at the base of the previous year's bulb.

all risk yield (ARY) *See* YIELD.

all-terrain vehicle (ATV) A vehicle specifically designed for use off road and in rough terrain, and may also be adapted for aquatic use. Also known as a *quad bike due to its design, the ATV is generally four-wheeled, with the rider straddling the seat and the use of handlebars to steer and operate the vehicle. The tyres are at low pressure to facilitate maximum grip on uneven terrain. They are often used in agriculture due to the low footprint which reduces soil compaction, their speed and agility, and their relevant low cost compared to other agricultural vehicles. They are able to transport lightweight loads, or can be fitted with a tow bar in order to tow a small trailer. ATVs operators are obliged by law to undergo training, and there is advice and guidance in the safe use of ATV operation on the Health and Safety Executive website.

(( )) SEE WEB LINKS

• Document issued by the Health and Safety Executive on the safe use of all-terrain vehicles.

alluvial Soil type typified by loose sediments and rock deposited through the action of water. Most alluvial deposits are geologically young. It usually contains a wide diversity of particle size, from fine clay and sand through to gravel, and is usually high in organic matter, making them highly fertile. Commonly alluvial deposits are found as flood plains or where rivers often burst their banks.

alpaca A species of domestic *camelid originating from South America. It is closely related and similar in appearance to a llama, but is a distinct species in its own right, and has two different types: the huacaya and the suri. The huacaya has a tight fleece with crimped fibres growing vertically from the body. The suri has uncrimped, spiral fibres growing out from the body which hang down in locks. Both types are bred predominantly for their fibres which are highly prized; however, they are also used as pack animals, particularly in their native South America. Alpaca farming is relatively new to the UK, with the first imports from Peru occurring in 1996, and it is largely viewed as 'hobby farming' due to the niche element of the fibre. However, demand for the best-quality fleece has stabilized the industry and there are now several successful alpaca studs producing quality stock for export. Despite this, however, the majority of alpacas in the UK are kept as pets or companion animals or as part of animal collections.

(⊕) SEE WEB LINKS
• Website of the British Alpaca Society.

alpha amylase (α-amylase) Predominant form of the hydrolytic enzyme found in humans and mammals, responsible for the chemical breakdown of starch and glycogen into maltose and glucose. It is also found in seeds that are rich in starch, and some fungi secrete amylase to aid their nutrition. In humans and other mammals there are two sites of secretion: the salivary glands and the pancreas. As a component of saliva it is the first part of chemical digestion and is active in the mouth and oesophagus, but is denatured when it reaches the stomach. Pancreatic amylase is secreted into the small intestine to continue chemical digestion of starch once the digestate leaves the stomach.

alternative dispute resolution (ADR) Process of resolving disagreements between parties without having to go to the law *courts. There are a number of different forms of ADR, and in the *United Kingdom (as well as in increasingly large numbers of other countries) courts encourage parties to use ADR approaches, where appropriate, rather than going through the courts, as the costs to the parties involved, if using ADR, are likely to be significantly less than those incurred by using the court system. The different forms of ADR which are used in the United Kingdom are:
• **Arbitration**: in arbitration an independent third party (the **arbitrator**) is appointed by both parties to make a decision (known as an **award**) which is binding on both parties, though there is a right of appeal against an award to the courts on a point of law. The arbitrator acts in a similar way to a judge, making their decision on the basis of the evidence presented by both parties. The arbitrator is also able to decide on whether the costs connected to the arbitration should be paid by both parties or only one of them. The appointment of the arbitrator is done through an **arbitration agreement** (or if there is no such agreement then the courts will appoint). Arbitration is often used to resolve property-related disputes, and some forms of lease or commercial agreement will have a clause within them agreeing to refer any future disputes to arbitration. In England, Wales, and Northern Ireland, arbitrations are regulated by law through the **Arbitration Act 1996**, whilst in

Scotland they are regulated by the **Arbitration (Scotland) Act 2010**. Arbitration has, historically, often been the default dispute resolution procedure for disagreements relating to agricultural *tenancies in England and Wales; however, legislation has now opened up other alternatives in some cases, including independent expert determination (see below).

- **Adjudication**: method of ADR used in the UK construction and building industry. An independent third-party **adjudicator** is appointed and assesses the evidence which the parties present in order to reach a decision which is binding on both parties unless it is referred to arbitration, or to the courts (or if the parties reach subsequent agreement amongst themselves). Adjudication usually operates to a short timescale, allowing disagreements to be resolved quickly, which may be important in an ongoing construction project.

- **Independent expert determination**: in independent expert determination the **independent expert** is appointed by both parties by contract, and is an expert in their own right. They will make their own investigations and come to a decision known as a **determination** which the contract between the parties will usually specify is binding on both parties. Unless the contract specifies otherwise, the independent expert does not have legal powers to decide how costs should be apportioned between the parties. In the United Kingdom, unlike arbitration there is no *legislation relating to independent expert determinations, though there is some *case law. The process is designed to operate relatively quickly and can be suitable for smaller *valuation disputes.

- **Mediation**: this is a method of ADR which is used in a broad range of types of dispute (both property and non-property related). It is non-binding and is designed to allow the parties to reach an agreement between themselves, rather than having a third party make a decision. The **mediator's** role is, therefore, to reopen communications in order to help the parties reach their own decision. It is usually voluntary, and each party can end the mediation if they wish; therefore, if unsuccessful, it is sometimes followed by the dispute being referred to the courts (though the courts also sometimes encourage mediation during cases to encourage the parties to resolve the issue themselves).

- **Early neutral evaluation**: this is a non-binding form of ADR which is designed to help clarify issues which are of a factual or legal nature before they go forward to arbitration or to the courts. A neutral **evaluator** provides a recommendation of the likely decision if the dispute goes forward to arbitration or the courts, thereby allowing both parties to identify the strengths and weaknesses of their case.

See also CALDERBANK OFFER; EXPERT WITNESS.

amenity woodland Generally, the value of this type of *woodland is in its non-commercial contributions, though there may be some management for timber production. The look and setting of amenity woodland in the landscape is very important, as is its provision of aesthetic qualities, wildlife habitats, and shelter and privacy by screening. It may provide significant opportunities for formal and informal recreational activities such as walking, where key features might include ease of access and large trees.

ammonia A compound of nitrogen and hydrogen with the formula NH_3, ammonia is a colourless pungent gas. Its most common use in agriculture is as a fertilizer in the form of a salt such as *ammonium nitrate or *ammonium sulphate. It is produced naturally during the decay of *organic matter, *slurry, and *farmyard manure, thus increasing the level of ammonia in the atmosphere. Indeed, around 90% of ammonia emissions into the atmosphere comes from farmland. It is a serious pollutant largely responsible for acid rain.

ammonium nitrate A stable salt of *ammonia with the formula NH_3NO_4, the main component of nitrogenous *fertilizers. As a salt it is a white crystalline substance that dissolves easily in water. It is also a constituent of explosives.

ammonium sulphate A stable salt of *ammonia with the formula NH_3SO_4, used predominantly as a *fertilizer for alkaline soils, as when dissociated it forms a mildly acidic ion. As a fertilizer, however, it has a lower concentration of nitrogen compared to *ammonium nitrate. It is also found as a constituent of some pesticides, herbicides, and insecticides, as it aids the binding of cations and improves the efficacy of the active ingredients. It is also used in small quantities as an acidity regulator in processed food products.

anaerobic digestion Process in which *renewable carbon-based materials are digested by microbes in the absence of oxygen in order to produce *renewable energy. A variety of different **feedstocks** can be used for anaerobic digestion, including food wastes, *slurry, and *manure (which may be mixed with crops such as *maize), and *sewage sludge. Anaerobic digestion produces *biogas and a **digestate** (which can be used as a *fertilizer, as it is typically rich in *nitrogen, phosphorus, and *potassium).

The composition of the biogas produced from anaerobic digestion varies depending on the feedstock put into the digester; however, it is predominantly methane and *carbon dioxide and can therefore be combusted on site in order to produce heat and/or electricity (*see also* COMBINED HEAT AND POWER). The biogas can also be cleaned up to remove the other gases, leaving pure methane (known as **biomethane**) which can then be injected into the mains gas grid, or used as a transport *biofuel. *See also* FEED-IN TARIFF; RENEWABLE HEAT INCENTIVE; RENEWABLES OBLIGATION.

anatomy The study of the structure of living organisms. It can apply to both plant and animal life-forms. It is related to physiology, which is the study of the function of each anatomical structure.

ancient monument A construction of a particular or significant historical interest. An ancient monument can range from built stonework to simple earthworks. Many are buried remains and not easily visible in the *landscape. Many ancient monuments are protected by legislation, and permissions would have to be obtained before carrying out any activity that might impact upon these

a

*scheduled monuments. *Agri-environment schemes can be used to encourage less intensive agricultural practices that can benefit and protect both scheduled and unscheduled ancient monuments such as *barrows and *ridge and furrow.

Ancient Semi-Natural Woodland (ASNW) A category of *Ancient Woodland which is broadleaf and comprised of predominantly native *tree and *woody shrubs which has been in existence since at least medieval times, i.e. from before 1600 AD. However, unlike Ancient Woodland it may have been more intensively managed and allowed to regenerate. The long continuity of these *woodlands and their associated and undisturbed soils provide valuable *habitats and associated *biodiversity.

ancient tree *See* VETERAN TREE.

Ancient Woodland (AW) Similar to *Ancient Semi-Natural Woodland, but where there has been continuous *woodland cover since at least medieval times, i.e. from before 1600 AD. The lower levels of disturbance provide high levels of *biodiversity, including some flora restricted to this woodland environment, and potentially significant historical and archaeological values. Ancient Woodlands in England, Scotland, and Wales are identified in the Ancient Woodland Inventory.

(⊕) SEE WEB LINKS
• Website of the Forestry Commission.

anemometry *See* WIND TURBINE.

angling The activity of fishing using a hook or 'angle', hence the derived term. The hook is normally attached to a line and rod. Commonly disaggregated into types of fishing: *game fishing, *coarse fishing, and sea fishing. These all have their associated and specialist organizations, of which an example is the *Angling Trust. *See also* FISHERIES; FLY FISHING; COARSE FISHING.

Angling Trust The representative body for all types of anglers in England. It is a membership organization that is involved in campaigns and legal work to promote and protect fishing.

(⊕) SEE WEB LINKS
• Website of the Angling Trust, outlining its role and providing an enquiry service.

Animal and Plant Health Agency (APHA) An executive agency working on behalf of the *Department for Environment, Food and Rural Affairs (DEFRA), and also on behalf of the *Scottish Government and the *Welsh Government. It merges the former Animal Health and Veterinary Laboratories Agency (AHVLA) with the plant and bee health parts of the *Food and Environment Research Agency. The AHVLA part operates animal disease surveillance labs for either routine testing for disease or for identification of unknown cases. Data are used to inform policy and stakeholders regarding current animal health threats and to implement appropriate action when necessary.

Animal Health and Veterinary Laboratories Agency (AHVLA) An
executive agency working on behalf of the *Department for Environment, Food
and Rural Affairs (DEFRA). It is responsible for safeguarding animal health and
welfare, and public health where relevant, protecting the economy and
promotion of food security through research, surveillance, and inspection. The
AHVLA operates a number of animal disease surveillance labs around the
country, where samples are sent for either routine testing for disease or for
identification of unknown cases. Data are gathered from these centres in order to
inform policy and stakeholders regarding current animal health threats and to
implement appropriate action when necessary. The agency also liaises with
global partners to ensure that cohesive approaches to animal health and disease
threats are undertaken where possible. *See also* DEPARTMENT FOR ENVIRONMENT,
FOOD AND RURAL AFFAIRS (DEFRA).

Animal Health Officer Person who works for the *Animal Health and
Veterinary Laboratories Agency (AHVLA), and whose main role is to support the
actions of the vets employed by the *Department for Environment, Food and
Rural Affairs (DEFRA) in matters concerning animal or public health. Examples
of their duties include inspections of animal housing facilities, promoting welfare
standards, monitoring and enforcing legal obligations of animal-related
industries, and checks at airports and ports over animal and animal by-product
imports.

Animal Transport Certificate Document required by law for the transport
of animals as part of an economic activity that will exceed eight hours' travel.
The form must also be accompanied by a contingency plan, and both sets of
documentation must be retained for six months after the date of travel.
See also ANIMAL WELFARE.

animal welfare The well-being of animals. Comprehensive guidance exists for
what constitutes good animal welfare, and the majority of legislation generated
to ensure acceptable standards of animal welfare stems from the five freedoms
that were originally generated in 1965 by the British Government. Minimum
standards exist for the housing, husbandry, and handling of animals to which
keepers of commercial, pet, domestic, and exotic animals must all comply, with
penalties including prison sentences for those who break the law. Assessments of
animal welfare are centred around physical indicators such as heart rate and
respiration rate, physiological indicators such as compromised health status and
reduced production output, and observable changes in animal behaviour, such
as changes in feeding, social or breeding behaviour and the exhibition of
stereotypes.

anion An atom of an element that has gained an electron and therefore now
carries a negative charge. Anions are usually non-metals, and common elements
that form anions in agriculture are chlorine, iodine, oxygen, and sulphur. As they
carry a negative charge, they are reactive and attract ions of the opposite
(positive) charge—the cations. In this way, compounds are formed through ionic

a

bonding. The balance of anions and cations found in soils is important, as they will dictate the interaction of important nutrients within *soil and soil water, and thus influence plant biochemistry.

annual 1. A plant whose life cycle occurs within a one-year period. This includes germination, flowering, and fruiting. Common agricultural examples of annuals are *wheat and *barley. **2.** Annual rings describe the new wood formed in the trunk of a *tree. These are easily identifiable when cut through, and can indicate the age of a tree. *See also* BIANNUAL; BIENNIAL.

Annual Tax on Enveloped Dwellings (ATED) Tax, operating in the United Kingdom, on high-value residential properties which are owned in some kind of company structure. The tax is paid on an annual basis, with the amount payable being on a sliding scale; the properties which have the highest *capital value trigger the requirement for payment of the highest annual charges. There are a number of situations where relief from ATED can be claimed, meaning that ATED is not paid for the days of the year which qualify; this includes working farmhouses, situations where the property is open to the public on a commercial basis, property developers, property rental businesses, and where employees or partners (*see* PARTNERSHIP) occupy the building. There are also some exemptions from ATED—for some charities and public bodies, and where the property has an exemption from *Inheritance Tax. All of these reliefs and exemptions are subject to specific conditions.

annuity Financial product, offered by insurance companies, which provides those people buying one with a regular source of income, either for a fixed period, or (when used for pension planning) for the rest of their life. In the *United Kingdom, some pension products give those people who contribute to them the option of using part or all of their 'pension pot' (i.e. the fund they have built up through saving into their pension through their working career) to buy an annuity when they wish to retire.

anthelmintic A group of worming drugs used for the treatment of internal gut parasites in animals. Anthelmintics can either target one or two specific species of helminths (gut parasites), or can target a range, in which case they are referred to as broad-spectrum. Anthelmintics are classified into groups according to which active ingredient they contain, and in the case of sheep and cattle, are colour-coded to aid correct selection of drug. There is an increasing level of concern regarding reliance on anthelmintics, as resistance to the active ingredient is becoming more widespread in the populations of gut parasites. With limited scope for the discovery of new active ingredients there is concern that we will lose the ability to effectively treat animals suffering from internal parasite burdens, with the resulting effect on economics, health, and welfare. Alternative methods of controlling internal parasites rely on a combination of strategies where wormers play a part, but grassland management, rotational grazing, and correct worming strategy is as important.

anthers The pollen-producing and bearing component of the stamen which is the male reproductive structure in a plant. The anther is attached to the filament, or stalk, of the stamen, and can usually be found forming a ring surrounding the stigma—the female receptacle for pollen grains—though the reproductive components of plants are extremely diverse in both structure and arrangement.

anthesis The period of flowering starting from the point of the opening of the bud. The flower then expands with the extension of the style to facilitate pollination. The flower may be brightly coloured or emit a fragrance to attract pollinating insects, and the parts of the flower may have contrasting colours to be more visible.

anthrax A generally fatal disease caused by the bacterium *Bacillus anthracis*, affecting humans and other animals. The bacterium produces spores which are highly resistant to environmental conditions, including extremes of temperature, desiccation, and light, such that they can be found in the soils across all continents including the Antarctic, and can lie dormant for many decades. The most common route of infection is through ingestion of contaminated soil, and once inside the animal the spore is reactivated and rapidly multiplies. *B. anthracis* may also gain entry into an animal's body through an open wound or lesion. Anthrax can only be spread through spores, so through direct contact or ingesting an infected body, and spores can be carried on the soles of shoes, on vehicles, on the wind, or by other carriers to infect a new area. Due to a comprehensive vaccination programme, the threat of anthrax infection is rare in the majority of the developed world; however, occasional cases do occur. In recent years the threat from anthrax came more from its use as a biological weapon, and several countries have invested considerably in research into both the use of and containment of such biological warfare.

antibiotic A group of drugs that specifically work against bacteria. Antibiotics can be broad-spectrum, in which case they are active against a range of bacteria species, or very narrow, in which case they only work against one or two identified species.

AONB *See* AREA OF OUTSTANDING NATURAL BEAUTY.

APHA *See* ANIMAL AND PLANT HEALTH AGENCY.

aphid Common term for a group of insect pests that infest crops. They have soft bodies and suck sap from the plants, often secreting a sticky liquid referred to as honeydew. They are sometimes referred to as plant lice, and as the different species have different body colours they are also known as greenfly, whitefly, or blackfly. Infestation by aphids causes distorted growth, and as a result of the honeydew, secondary pests and diseases are common, such as moulds, or they can act as vectors for viruses.

Appaloosa *See* HORSE.

appeal, court *See* COURT.

appeal, planning Mechanism by which an applicant can appeal against a planning decision made by the *local planning authority, including the refusal to grant *planning permission, *listed building consent, or *advertising consent; or to agree a prior approval in relation to *permitted development; or to remove or vary a planning *condition; as well as failure to decide on an application within the required timeframes; or in respect of a decision to take *enforcement action in relation to a planning breach.

Appeal applications in England and Wales are administered at a national level by the *Planning Inspectorate and are dealt with by **Planning Inspectors** who are employed by the Inspectorate. The Planning Inspectorate acts on behalf of the Secretary of State (in England) and Welsh Ministers, both of whom can, in some instances, make the decision directly; these situations are referred to as **recovered appeals**. Appeals in Northern Ireland are made to the *Planning Appeals Commission. In Scotland, appeals where the decision has been made by a planning committee are dealt with centrally by the **Scottish Government Planning and Environmental Appeals Division**, whereas those decided by a planning officer under delegated powers are considered by the local planning authority's **local review body** (which comprises a group of three or more elected members of the planning authority). Scottish Ministers can also recover appeals in some circumstances.

In England and Wales there are three procedures that a planning appeal can follow:

- **Written Representations**, where both the applicant, local planning authority, and other interested parties' representations to the appeal body are submitted in writing (although the Inspector considering the appeal may visit the site).
- An **Appeal Hearing** where the Inspector, having considered the initial evidence presented, discusses the main issues with the applicant, local planning authority, and other interested parties at a hearing. This mechanism is used for more complex cases where written representations would not be sufficient on their own.
- An **Appeal Inquiry**, which is open to the public and where the evidence presented in respect of the case is formally tested through cross-examination, with parties sometimes represented by advocates. Because of the potential costs involved in this process it is used only in a relatively small number of cases, often where there is significant public interest.

The system is broadly parallel in Northern Ireland with the **Commissioners** of the Planning Appeals Commission operating a system of written representations, **Informal Hearings** (equivalent to Appeal Hearings), and **Formal Hearings** (equivalent to Appeal Inquiries).

The Scottish Government Planning and Environmental Appeals Division also operates on a parallel basis, with Scottish Government **Reporters** making the majority of decisions on appeals which can be considered either through Written Representations, an Appeal Hearing, or an Appeal Inquiry.

Appeal decisions are usually binding on the parties. However, if the applicant (or following an appeal, the local planning authority) considers that a decision has been made by the local planning authority or appeal body which is legally incorrect, then in some circumstances this can be challenged through the courts by requesting a *judicial review.

There is no mechanism in England, Wales, Scotland, or Northern Ireland for interested parties (such as members of the public) to appeal in a situation where planning permission has been granted and they do not agree that it should have been. In this situation the only opportunity potentially available to such a party would be, in some circumstances, to challenge the decision within the courts if they consider there was evidence that a legally incorrect decision had been made.

approval of matters specified in conditions application (Scotland)
See PLANNING PERMISSION IN PRINCIPLE (SCOTLAND).

aquaculture The farming of fish, shellfish, molluscs, algae, and aquatic plants. This can take place in coastal areas and inland.

aquatic Descriptor in relation to anything to do with water. More commonly used in conjunction with wildlife *species living in or alongside water, or associated *habitats. Can also be used to describe a category of sporting activities linked to water. *See also* WETLAND.

aquifer *See* WATER.

Arab *See* HORSE.

arable crop A commercially grown plant crop, usually grown as a *monoculture. Arable crops refer explicitly to *cereals, oil crops, bean crops, and root crops rather than soft fruit, vegetables, and salads.

arable farming The practice of producing plant-based crops, usually through *monoculture, for the purposes of food, biofuel, and some biomedical products.

arachnida A class of invertebrates characterized by eight legs, an exoskeleton, jointed legs, and an absence of antennae or wings. It includes spiders, scorpions, mites, ticks, and harvestmen (daddy long-legs). Whilst most arachnids are of little importance to agriculture, the mites and ticks can cause significant problems for livestock, as they are parasitic and can act as vectors of other diseases such as *Lyme disease.

arbitration *See* ALTERNATIVE DISPUTE RESOLUTION.

arboriculture The study of the growth, care, and management of amenity *trees. This is the predominant area of work addressed by **arboriculturalists**, and might include landscape design, planting, and tree surgery.

Area of Great Landscape Value (AGLV) *See* SPECIAL LANDSCAPE AREA.

Area of Outstanding Natural Beauty (AONB) Regions of the countryside in England, Wales, and Northern Ireland that are statutorily designated to ensure that their characteristic *landscape and natural beauty are protected and enhanced. *Natural England, *Natural Resources Wales, and *Northern Ireland Environment Agency have respective oversight and funding responsibilities, and are managed locally by **Joint Advisory Committees** or **Conservation Boards**. The focus and management of each AONB can vary according to local priorities, but they have common objectives comparable to those of *National Parks; namely, to conserve and enhance the natural and cultural heritage; to promote public understanding and enjoyment; and to support the economic and social well-being of the local communities. *See also* AREA OF GREAT LANDSCAPE VALUE; NATIONAL SCENIC AREA.

(⊕) SEE WEB LINKS
• Website of the National Association for AONBs, providing details and links to all the AONBs.

Area of Significant Archaeological Interest (ASAI) *Planning designation used in Northern Ireland for an area of distinctive historic landscape which is likely to include a number of *scheduled monuments and other historic sites. Where such areas have been designated, the local development plan (*see* LOCAL PLAN) will have specific planning policies in relation to them.

Area of Special Control *See* ADVERTISING CONTROLS.

Area of Townscape or Village Character *Planning designation for town or village areas that have a unique identity but may not always warrant designation as a Conservation Area. *Local plans may set out specific *planning policies in respect of these areas.

area plan *See* LOCAL PLAN.

ark Small to medium-sized chicken housing usually of a moveable nature. Often constructed of wood or plastic it offers housing options for groups of birds for small-scale poultry or egg production in backyard farms or small-holders. Also referred to as a coop.

arrears, rent *See* RENT ARREARS; BREACH OF COVENANT.

article 4 direction Order used in the United Kingdom to restrict *permitted development rights in a specific area. The effect of article 4 directions is to require *planning permission to be applied for in relation to some types of project which are covered by permitted development rights in other areas (and which therefore do not normally require an application for planning permission). Article 4 directions are commonly used for *Conservation Areas in order to provide *local planning authorities with a greater degree of control over *development occurring within them.

artificial insemination (AI) The process by which semen is introduced into the female without the need for copulation. Semen is collected from a male through either electro-stimulation or more commonly through the use of a dummy or a mount animal. The male is encouraged to mount the mount animal and then the penis is guided into an artificial vagina that simulates the pressure and temperature the male would experience inside the female. The ejaculate is collected, checked, diluted and, if required, sorted into male or female sperm (sexed semen) and then usually frozen until required. During AI, recipient females are synchronized so that they are in the correct stage of the oestrus cycle to conceive, and the semen is introduced either into the vagina (sheep) or trans-cervically where the catheter goes through the cervix in order to deposit the semen in the uterus. AI's benefits include being able to use semen from animals across the globe, that are deceased, eliminating the need to purchase sire animals, reduced disease risk, and potential reduction in unwanted animals if sexed semen is used.

asbestos Material commonly used in building construction before it was discovered to pose a significant health risk in some circumstances. It was originally considered to be a versatile building material because it has good insulation and fire-resistant properties as well as being relatively lightweight. It was, therefore, very frequently used in applications such as roof and wall sheeting, fire protection, and insulation, and was commonly used, for example, for the construction of agricultural and industrial buildings (though it is also often found in domestic premises). Once the material was in widespread use, however, it became apparent that if fibres from asbestos escape into the air (for example, during manufacture, or in building construction or demolition works, or if it is damaged *in situ*) then they pose a significant risk of causing serious lung diseases (such as mesothelioma and lung cancer) for those who come into contact with them. Asbestos is commonly found in the *United Kingdom in buildings which were built or refurbished before the year 2000, and health and safety laws therefore regulate the ways in which existing buildings containing asbestos must be managed, how any construction and demolition projects which involve asbestos must be carried out, and how it must be disposed of. *See also* HAZARDOUS WASTE.

ash *See* TREE.

ASNW Abbreviation for *Ancient Semi Natural Woodland.

aspect, environmental *See* ENVIRONMENTAL MANAGEMENT SYSTEM.

aspen *See* TREE.

asset Something owned or controlled by an individual, organization, or other entity, and from which they may potentially gain future economic benefits. Assets are usually divided into **tangible assets** and **intangible assets**. Tangible assets are, essentially, things that can be physically touched, and include buildings, land, machinery, trading stock, fixtures and fittings, cash and investments, together with **debtors** (i.e. where money is owed to the asset owner). Intangible assets include patents, copyrights and trademarks, computer software, and trade secrets.

asset of community value Building or land, in England, which is included on the *local authority's **list of assets of community value**, because it is considered to further social well-being or social interests (including cultural, recreational, or sporting) of the local community, and which is then subject to the *community right to bid, which, in the event of the owner proposing a sale, allows community groups to request a delay in the sale of up to six months in order to raise funds to prepare a bid to purchase it. Examples include pubs and village halls.

assignation In Scotland, the transferring of rights in something that is *incorporeal property from one party to another (for example, in the case of a *lease the *lessee may **assign** the rights they have to occupy the property from themselves to another party, thereby having no future interest in the land). The term assignation is broadly equivalent to the term *assignment which is used in the rest of the *United Kingdom. *See also* ALIENATION; SUB-LETTING.

assignment The transferring of rights from one party to another; for example, in the case of a *lease, the *lessee may **assign** the rights they have to occupy the property to another party, thereby having no future interest in the land. Assignment by a lessee contrasts with *sub-letting, after which the original tenant still retains an interest in the land. Some leases contain a clause either expressly allowing assignment by the *tenant (with or without specific permission from the landlord) or forbidding it. The Scottish equivalent of assignment is *assignation. *See also* ALIENATION.

AssocRICS *See* ROYAL INSTITUTION OF CHARTERED SURVEYORS.

assured agricultural occupancy *See* HOUSING ACT 1988.

assured shorthold tenancy (AST) *See* HOUSING ACT 1988.

assured tenancy *See* HOUSING ACT 1988; SCOTTISH RESIDENTIAL TENANCIES.

AST Abbreviation for assured shorthold tenancy. *See* HOUSING ACT 1988.

Asulam A *herbicide used in agriculture, primarily for the control of *bracken, for which it is an effective chemical against that invasive species. It was also used as a herbicide in spinach for human consumption, and was banned by the European Commission in 2011 when research showed that it might be carcinogenic. Since then there has been an annual emergency authorization for its use on *bracken control.

ATED *See* ANNUAL TAX ON ENVELOPED DWELLINGS.

atrazine A herbicide of the triazine group that is used to treat broad-leaf weeds across the USA, Australia, and South America. However, it remains controversial and is banned in the European Union due to its perserverance as a ground-water contaminant. Atrazine is a suspected endocrine-disrupting chemical thought to

affect mammalian reproductive systems, although due to the difficulties in
demonstrating cause and effect the evidence remains minimal.

ATV *See* ALL TERRAIN VEHICLE.

auction The process of offering either goods, rights, or real estate for public sale,
with the person offering the highest public bid being sold the item. Normally, the
property offered for sale can be inspected before the auction, and when real
estate or rights are being sold a pack of legal information relating to it is normally
made available beforehand. The auctioneer normally receives a **commission** for
their work in selling the item (this is a payment based upon a percentage of the
price that the item has sold for, as stated in the auction's terms and conditions).
For smaller items the auction house may also charge the seller a small **lotting
fee** (i.e. a flat-rate payment made per lot entered into the auction which is
normally paid whether the lot sells or not). In addition, for some types of goods
(for example, furniture) as well as the seller paying commission, the buyer may
also pay the auctioneer a **buyer's premium** (again based on a percentage of the
sale price as stated in the auction's terms and conditions).
    The seller may put a **reserve** on the item, which is a minimum price which
they will accept. If the bids do not reach the reserve, then the item is not sold.
Buyers who are not able to attend the auction in person may leave a **commission
bid** with the auctioneer before the auction. This is a price (effectively a maximum
bid) that they are willing to pay for the item, and the auctioneer is then able
to include their bids within the auction. However, commission bids are now
less common, with more auctions now being streamed live on the internet
(allowing online bidders to offer live bids within the auction itself). Alternatively,
people who are not able to attend may be able to speak to auction staff on the
telephone during the auction, with the auction staff passing on their bids to
the auctioneer.

augur Tool used for taking soil samples. The augur in its simplest form consists
of a horizontal handle with a perpendicular rod ending in a screw thread.
When screwed into the ground, a core is removed that remains intact and
contemporary with the layers in the soil from which the sample has been
taken. It allows for soil profiles to be studied and for samples to be taken from
specific layers within the soil.

available water capacity (available water content) The quantity of water
stored in the soil that is available to the plant for growth. Of particular importance
in *arable farming. *See also* FIELD CAPACITY.

Ayrshire cattle A breed of cattle native to Ayr in Scotland, originally bred as a
dual-purpose animal but now almost exclusively used to produce beef animals.

(🌐) **SEE WEB LINKS**
• Website of the Ayrshire Cattle Society of Great Britain and Ireland.

Azotobacter A genus of soil-living, aerobic, and usually motile bacteria that are of particular importance to agriculture because of their ability to fix atmospheric nitrogen. Plants are unable to source nitrogen from the atmosphere and rely on soluble sources of this nutrient. A soil rich in Azotobacter reduces the dependence on the artificial application of nitrogen. Certain members of this genus can be found inhabiting the rhizosphere of leguminous plants, having evolved a symbiotic relationship with this group of plants.

bacon pig A pig specifically farmed to produce bacon as opposed to fresh pork products, or the processed pork market. As a bacon pig it will typically be slaughtered at around 100kg finished weight, the heaviest of the three possible markets. Recently, with improvements in nutrition and breeding it has been possible to finish pigs at heavier weights without their gaining excessive fat, meaning that heavier pigs are also suitable for the other two markets. Consequently, the number of bacon pigs being produced specifically for this market has declined.

bacteria (*sing.* **bacterium**) The major group of microbes found within the prokaryotic organisms. An extremely diverse group of micro-organisms, it is possible to find examples of bacteria living in all ecological niches found on the planet, and they are thought to be one of the oldest forms of life on Earth. Whilst many are of no direct consequence to man or their activities, some are highly beneficial and a few are very damaging (i.e. pathogenic). In all their forms, the study of bacteria has led to many innovations that have directly improved the health and welfare of man and animals.

badger The European badger (*Meles meles*) is a member of the family **Mustelidae** and is native to the British Isles. 'Badger' is a word thought to be perhaps derived from the French *bêcheur*, meaning 'digger', and they live in family groups in underground burrow systems called 'setts'. They are nocturnal and have an omnivorous diet but particularly favour earthworms. They are an animal of significant interest and impact to *livestock farming because they are one of several mammal species that is a reservoir for *bovine tuberculosis (bTB). This can be transmitted to cattle and was once a major human health problem in the UK and in other countries. The measures taken to reduce bTB in cattle have included testing and controlling cattle movements, vaccination, and culling of badgers. These all have a role in reducing and ultimately ridding cattle of bTB, but it is the culling programmes that have caused the most and continuing controversy between *farmers and wildlife and animal rights groups.

badger gate A one-way badger gate, normally made of wood, is part of an exclusion construction placed around a *badger sett to allow passage out but not back in. Operating like a one-way valve, this allows a sett to be closed down and cleared of badgers with minimal harm. Badgers are a *protected species under the **Protection of Badgers Act 1992**, and any such operation would require a government licence.

b

bad husbandry *See* GOOD HUSBANDRY; AGRICULTURAL HOLDINGS ACT 1986.

bag *See* GAME.

bag records *See* GAME.

balance sheet A financial statement, also known as a **statement of financial position**, that shows the *assets and liabilities of an organization, normally at the end of a specified period of trading. A balance sheet for a straightforward *sole trader or *partnership situation essentially shows:

- The value of the assets held by the business (minus any *depreciation accounted for in previous years).
- The amount owed by any **trade debtors** (also known as **trade receivables**) at the date of the accounts (i.e. where goods or services have been sold by the business but payment has not yet been received for them) *less* any bad debts or doubtful debts (i.e. where the business does not expect to be paid money it is owed by others).
- Any **prepayments** (for example, where a business has paid for something for a full year, but only part of that year falls into the accounting period—hence it has prepaid for part of the following accounting period).
- Cash held in the business (either in the bank or physically).
- How the business is financed—the **capital** (money and possibly other assets like land or property) invested in the business by its owners and any **retained earnings** left in the business by its owners (*see* PROFIT AND LOSS ACCOUNT).
- Its current liabilities, including any **accruals** (i.e. service expenses incurred by the business during the accounting period that have not yet been paid for; for example, a business might have used water during the last part of the accounting period but not yet paid for it, so it is shown as an accrual) as well as any **trade creditors** (i.e. amounts owed by the business to other people for items bought as stock by the business but not yet paid for; also known as **trade payables**).
- Any drawings made by the owners of the business.

This then provides an overall financial position for the business at the end of that trading period. *Limited company balance sheets are effectively constructed in the same way except that they will also show a number of other pieces of information (where relevant), including investments that the company has made in the shares of other companies, details of the types of shares that the company has issued, and the current investment of shareholders into the business, any long-term loans or debts of the company, any outstanding Corporation Tax (which is due on the company's profits for the accounting period, but which has not yet been paid), and any **dividends** (shares of the profit which are due to *shareholders) which have been declared, but not yet paid out.

bale A large quantity of usually dried plant material such as *straw, *hay, *wood shavings, or paper that is packed tightly together into a large rectangle (square bales) or cylinder (round bale) for ease of handling, storage, and transport. To maintain its shape, the square bale is tied using nylon string, whereas the round bale is covered in industrial film referred to as bale wrap. Bales can further be wrapped in several coatings of plastic, which ensures that an airtight seal is formed around the material, helping to preserve and protect the material, or in the case of *haylage/*silage production, the plastic contributes towards the conditions necessary for fermentation in order to produce the final product. *See also* BALER.

baler A machine required in order to produce a bale. A baler can produce either square bales or round, and works by collecting the material directly from the field or store, and packing it densely together to eliminate as much air as possible. The baler then completes the process by tying or wrapping the bale appropriately. *See also* BALE.

bank *See* FIELD BOUNDARIES.

BAP *See* BIODIVERSITY ACTION PLAN.

bare land Farmland, without buildings, or a house. *See also* EQUIPPED LAND.

bare licence *See* LICENCE.

bare trust *See* TRUST.

bark The outer protective layer of trees and woody shrubs, including both living and dead plant tissue. Bark is a by-product of forestry operations and is sold for a variety of uses, including bedding, footpaths, and mulching, and in the tanning industry. *See also* RING-BARKING.

barley An arable crop that is one of the three main *cereals grown across the developed world. A member of the *grass family, barley was first domesticated around 8000 BC in Western Asia and Northern Africa; however, it is widely adaptable and is well suited to temperate climates. Barley has been selected over thousands of years of cultivation to improve yield, and to give rise to adaptations in grain content, or plant characteristics, such that the crop can be aimed at specified markets. Barley can be planted either in the spring—referred to as 'spring barley'—or in the autumn—referred to as 'winter barley', but although it copes in cool conditions it is not particularly hardy in cold conditions. The natural genetics of the wild-type barley gives rise to only two fertile rows of spikelets, but natural mutations have caused the appearance of strains of barley with six fertile spikelets. These are referred to respectively as two-row and six-row barley. The two-row barley tends to have a much higher fermentable sugar content than six-row barley, lending itself well to the malting barley market. Malting barley has to be delivered to the processer in a living state. This means that it cannot be subjected to the same degree of drying techniques often employed with other cereal crops to aid storage, as the maltster requires the grain to germinate in order to extract the malt from the grain. The malt is used in a number of products, such as fermented beverages (predominantly beer), non-fermented beverages (e.g. barley water), and food products (e.g. malt loaf). Qualities required in good crops of malting barley include purity of variety, no pre-harvest germination, disease-free, plump, uniform kernels, and a maximum of 13.5% moisture content. Barley can be also be eaten as 'scotch barley' or 'pot barley', which is a de-hulled kernel (the fibrous husk is removed from the kernel), but it is mainly consumed in the form of pearl barley—a de-hulled, steam-processed barley.

Alternatively, barley can be grown as an animal feed product, predominantly using six-row barley varieties that have a comparatively higher protein content compared with malting barley. It amounts to around 50% of the total global barley production. Feed barley has a high starch content that is readily fermented in the rumen, making it a good source of energy for growing animals. For this reason it is often a staple ingredient of beef cattle diets, particularly in the USA, where intensively reared beef animals are often referred to as 'barley beef'. Barley straw can also be a valuable feed ingredient for animals, as it is a cheap and readily available source of fibre. The *straw is also sought after by keepers of ornamental fish, as it contains natural anti-algal products that assist in keeping ponds free of excessive algal growth.

barley yellow dwarf virus (BYDV) A virus that affects economically important crops including oats, wheat, triticale, rice, maize, and barley. It is a single-stranded RNA virus transmitted by aphids, and has a number of subtypes based on serotype and vector. As *aphids feed on the plant, the virus enters the phloem cells, where it then replicates. The symptoms of the disease appear around two weeks after infection, and will vary according to the cultivar of crop and the subtype of virus. Generally, however, leaves become yellow or red, with stunted growth, upright, thickened, stiff leaves, reduced root growth, and reduced, delayed, or inhibited heading, with the consequent impact on yield.

barn An agricultural building used for various purposes. Traditionally quite large with low eaves and steeply pitched roofs, barns are used for the storage of *grain, *hay, and *straw. Often timber-framed on a brick base with a thatched, tiled, or slate roof, there are few doors or other openings, which makes it difficult to convert for residential or commercial use once redundant for agricultural purposes. Barns commonly have central high doors, sometimes opposite each other to allow easy vehicular access for loading and unloading. In some parts of the country they may be built of stone, but modern barns are often built of steel or concrete stanchions with a corrugated metal or asbestos roof.

barren A term referring to a female that is not carrying a foetus despite being serviced. Failure to conceive represents a significant area of financial loss to livestock enterprises, and barren females are often culled from the herd or flock.

barren brome (poverty brome, sterile brome) A species of brome grass that is a common weed of arable crops. Either annual or biennial, the grass can grow up to 90 cm in height, with rough hairy leaves, green to purple in colour.

barrow An archaeological term for a burial mound, constructed of stone, timber, or earth. Long barrows characteristically date from the **Neolithic** period, and round barrows commonly from the **Bronze Age**. They are often cited in highly visible locations within the *landscape. *See also* ANCIENT MONUMENT.

basal area A measure that has two differing ways which are interconnected. **1.** The cross-sectional areas of a tree at breast height of a man (typically measured at 1.3 metres from ground, i.e. the diameter at breast height or *DBH), measured in square metres. The basal area can then be used to estimate the *standing timber volume of the tree. **2.** The cross-sectional area of all trees in a defined stand measured at DBH and quantified as a figure per unit area, normally per hectare or per acre. This figure is a useful determinant of tree density, and is used to inform forest management decisions.

BASC *See* BRITISH ASSOCIATION FOR SHOOTING AND CONSERVATION.

basic loss payment *See* COMPULSORY PURCHASE.

Basic Payment Scheme (BPS) *See* COMMON AGRICULTURAL POLICY.

basic slag A black dusty product similar in particle size and density to cement that is rich in trace elements, particularly phosphates but also calcium, boron, and manganese. It is a by-product of the mining industry and is subject to rigorous controls in its application. It is often a cheaper alternative to artificial fertilizers.

BASIS An independent standards setting and auditing organization for the *pesticide, *fertilizer, and allied industries. It provides initial training towards a professional qualification and thereafter continuing professional development for those advising on pesticide use or responsible for pesticide storage. All *agronomists should be BASIS qualified.

bay *See* HORSE.

BBSRC *See* BIOTECHNOLOGY AND BIOLOGICAL SCIENCES RESEARCH COUNCIL.

BCIS *See* BUILDING COST INFORMATION SERVICE OF ROYAL INSTITUTION OF CHARTERED SURVEYORS.

BDS *See* BRITISH DEER SOCIETY.

beagling Hunting of hare on foot with a pack of beagles; a type of hunting that pre-dates the better known *fox hunting. Beagling takes place in the winter months, with each pack operating from within its own *country. The beagles are a type of small hound bred to hunt the hare, and there are currently sixty-seven packs that hunt throughout England and Wales within the limitations of the Hunting Act 2005. A Master or joint masters are responsible for the running of the hunt. Beagling is represented by the Association of Masters of Harriers and Beagles. *See also* HUNTING; HARE HUNTING.

(⊕) SEE WEB LINKS
• Website of the administrative hub for the Council of Hunting Associations with dedicated site of the Association of Masters of Harriers and Beagles.

b

beak trimming (debeaking) Partial removal of the beak of poultry, particularly layer hens, carried out in order to prevent injury through feather pecking, vent pecking, and cannibalism exhibited by some housed birds. The beak may be permanently removed or some re-growth may be seen; however, methods are continually being improved to make it permanent so that repeat procedures are not required. Usually the procedure is carried out on very young chicks at the same time as they are sexed and vaccinated. Arguments exist that press for improvements in the housing conditions of these birds, as the behaviours it is designed to prevent are signs of stress in housed birds; however, there are significant economic implications for this.

bean A large seed from a member of the legume family grown for either animal or human consumption. In terms of agricultural production, beans produced for human consumption are considered vegetable crops, and the term 'bean' is generally used where the bean is harvested and fed green, rather than dried, in which case it is referred to as a pulse. Animal feed beans ('feed beans') are a good source of protein and often provide a cheaper alternative for protein than imported staples such as soya.

beating The action of driving *game birds through *woodland, gamecrops, or other vegetation to flush the birds to a waiting line of guns. At this point, *sewelling may be hung to encourage the birds to fly upwards. Normally, beating is undertaken by a team of beaters that may also include dogs. Sticks are normally used to 'beat' or tap the trees or vegetation. Where beating is used not to flush the gamebirds but to drive them from one area to another, this is known as blanking.

bedding Material used to provide comfort to housed animals. There exists a wide variety of bedding types, including natural products such as *straw and *wood shavings, and inorganic products including rubber matting and sand. Bedding materials usually share the common properties of absorbing water (to soak up urine and surface water), insulation against extremes of temperature, non-slip, cheap to source, and relatively easy to work with. Ideal bedding materials should also be easy to dispose of, be low in dust, and have good hygiene properties, i.e. they do not easily harbour microbes that could cause disease. Organic bedding such as straw, whilst one of the most common bedding types used, is gradually being replaced with alternatives, mainly due to its poor hygiene properties and bulky nature, making it more difficult to store, handle and dispose of compared to some alternatives. New developments in bedding materials are continually being investigated, such as recycled manure fibres and paper by-products.

beech *See* TREE.

beef Meat product coming from domesticated cattle. Beef can be sourced from heifers, steers, cows, and bulls, though in some countries there are age restrictions on beef entering the human food chain. Meat coming from calves is not classed as beef, but is instead called *veal. Beef is the third most widely consumed meat in the world, and is growing in popularity, particularly in the developing world as disposable incomes rise. Beef as a consumed meat is not culturally acceptable to Hindus.

beetle bank A narrow strip of grass normally 2–4 metres in width set within large arable fields. Its primary function is to provide a *habitat for predatory invertebrates, including beetles, and to give them easier access to pest species such as *aphids within the adjacent crops. Beetle banks can also provide habitats for other wildlife, and can be an option within *agri-environment schemes. The strip is slightly raised to allow for better drainage and to improve overwintering sites in the tussock-forming grasses such as *cocksfoot and **Yorkshire fog**. There is normally a gap of about 25 metres at each end to allow easier cultivation and other machinery operations. *See also* FIELD MARGIN.

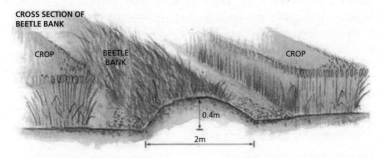

Cross-section of a beetle bank. (Game and Wildllife Conservation Trust.)

beet yellows virus (BYV) A pathogenic virus transmitted by numerous species of aphid, and a particular problem in the production of sugar beet. It causes significant yellowing of the leaves, particularly young, actively growing leaves, followed by thickening and leaves becoming brittle. It is thought to affect production by 2–5% for every week that plants remain infected. Control is mainly through cultivation practices such as avoidance of overwintering crops, physical separation of the seed crop from the beet crop, and controlling aphids.

benchmarking The practice of establishing common standards for key performance indicators (KPIs) for different enterprises that allows an individual farmer to measure their performance against. Benchmarks are often used to illustrate average, bottom 10%, and top 10% of performance for each KPI. The use of benchmarks is well established in some agricultural enterprises such as dairy production, where financial drivers have caused the industry to be very focused on costs of production versus outputs gained; however, it is slower to be utilized in others, such as the sheep industry. This is in part due to the lack of investment, perceived lack of financial incentives, and an absence of supporting technology.

beneficial insect A member of the insect family that is considered to have a positive or necessary influence on the production of a particular agricultural product. Good examples are often found within arable farming as many insects are essential for pollination of flowers, without which seed production would not be possible. Bees are an example of a pollinator without which much of the top fruit production would fail. Beneficial insects can also adopt roles as predators of pest species, such as the lady bug which predates aphids. Beneficial insects are of great significance in organic production which permits only very limited us of chemical controls for pests and diseases. Emphasis is therefore placed on farming practices or interventions that actively promote the presence of beneficial insects in areas of organic production.

beneficiary Person who is entitled to benefit from either a *trust or a will.

best and most versatile agricultural land Term used in national planning policy in England and Wales to identify the highest-quality agricultural land on which development is generally not permitted, given its potential for high-yielding food production. It is generally identified as being those areas of land given the highest classifications in the Agricultural Land Classification system (i.e. grades 1, 2, and 3a). Similar levels of protection from development for the best-quality agricultural land also exist within the planning policies of Wales, Scotland, and Northern Ireland.

betterment *See* COMPULSORY PURCHASE.

BHS *See* BRITISH HORSE SOCIETY.

biannual A plant whose life and reproduction cycle occurs twice within one year.

biennial A plant whose lifecycle occurs over a two-year period. The first year is usually devoted to plant establishment in root and leaf canopy, with the second year devoted to sexual reproduction through the production of flowers or fruiting bodies. Examples of biannual plants important in agriculture include vegetable crops such as onions and cabbages, but mostly the weed species such as thistles. *See also* PERENNIAL; ANNUAL.

Bill (Parliamentary or Assembly) Draft law which is in the process of being considered and voted on by the elected representatives of either the *Houses of Parliament, *National Assembly for Wales, *Scottish Parliament, or *Northern Ireland Assembly. Once a Bill is approved, usually through a series of votes, and is given *Royal Assent (i.e. formal approval by the monarch), it then becomes an *Act of Parliament or *Act of Assembly (i.e. law). A Bill is sometimes preceded by a *White Paper or *Green Paper, and is often subject to consultation with other interested parties, such as interest groups or professional bodies, as well as the general public, before it is formally considered by elected representatives. *See also* DELEGATED LEGISLATION.

(((•))) SEE WEB LINKS
• Website of the United Kingdom Parliament: includes information on how laws are made, and also the progress of legislation currently before Parliament.

bindweed A common *weed of the genus *Convolvulus*. It has a creeping or climbing habit, with trumpet-shaped flowers that are usually white but sometimes edged with pink. Bindweed is so called because as it grows it twists around whatever support it can find, including other upright plant species. As a result it chokes and strangles the supporting plant, stunting growth and preventing optimum cropping. Bindweed is difficult to eradicate due to the deep nature of the extensive root system, its ability to regenerate from root fragments, and the persistence of the seeds.

biocide A general term for something that controls unwanted biological organisms through either chemical or biological pathways. Good examples of biocides from agriculture are *insecticides, *herbicides and *anthelmintics. The use of biocides is overseen by the Health and Safety Executive, which assesses the potential of each biocide to cause harm to the environment, to humans, and to the local ecology, and issues guidelines as to their correct use. *See also* BIOLOGICAL CONTROL.

(((•))) SEE WEB LINKS
• Website of the Health and Safety Executive.

biodiesel A transport fuel for use in diesel engines. In the UK, biodiesel is usually processed from *oilseed rape or other vegetable oils, sometimes after use as cooking oil. These oils can be added to diesel without modification of the engine. Biodiesel may be processed from palm oil, which is one of the reasons for large-scale palm plantations in countries such as Malaysia and Indonesia.

biodiversity Refers to the whole variety of life on Earth, but the established definition was provided at the 1992 United Nations Convention on Biological Diversity, as follows: 'Biological diversity means the variability among living organisms from all sources including, inter alia, terrestrial, marine, and other aquatic ecosystems and the ecological complexes of which they are part: this includes diversity within species, between species, and of ecosystems.' Although technically more limiting, the more common terms such as *wildlife, natural history, or nature conservation continue to be used interchangeably. *See also* BIODIVERSITY ACTION PLAN.

Biodiversity Action Plan This publication is a UK initiative to maintain and enhance biodiversity, and is its response to the 1992 *United Nations Convention on Biological Diversity. Subsequent reports have developed and refined action plans for specified habitats and species. These are known respectively as **Habitat Action Plans** and **Species Action Plans**, but more usually as a Biodiversity

Action Plan (BAP). **Priority Habitats** and **Priority Species** are identified on the basis of a number of criteria. Each BAP includes conservation targets, contributing organizations, and mechanisms for achieving targets. BAP habitats and species are a key tool to focus *agri-environment scheme resources, and can be a material consideration in the planning process.

(⊕) SEE WEB LINKS
• Website of the Joint Nature Conservation Committee: includes all the relevant publications and lists surrounding BAPs in the UK.

biodynamic agriculture Method of farming that is an extension of organic farming and times its farming practice to the rhythms of the Earth. Key tasks such as planting and harvesting are carried out in tune with the astrological calendar, and creating a self-sustaining farm with a high level of ecological diversity is as important a priority for the farm as producing the food products themselves. Of critical importance is nutrient recycling, and close attention is paid to managing manures. They are treated with specific preparations in order to generate valuable composts which are then used on the land.

(⊕) SEE WEB LINKS
• Biodynamic farming organization of the UK.

bioethanol Ethanol produced from plants used for various purposes, including as a transport biofuel. It can be added to petrol at low inclusion rates without modification of the engine and at higher rates if the engine is modified appropriately. In the UK it is processed mainly from wheat and sugar beet but can come from any source of starch such as potatoes. In the USA it is processed from maize, and in Brazil from sugar cane.

biofuels, transport Transport fuels which come from *renewable sources (as opposed to *fossil fuel-based sources such as petrol and diesel, which are not considered renewable because of the length of time they take to form). There are three main types of transport biofuel: *biodiesel, *bioethanol, and *biogas, which can be produced from a variety of different raw materials. Biodiesel (which is sometimes blended at low percentages with conventional diesel) can be produced from materials such as *oil seed rape, and waste vegetable oils. Bioethanol can be produced from crops such as sugar beet and cane, *wheat, and *maize, and can be blended at low percentages with conventional petrol. Biogas is a gaseous form of fuel (biomethane) which can be produced through the *anaerobic digestion of materials such as food wastes, *slurry, and *manure (which may be mixed with crops such as *maize), and *sewage sludge. Next-generation biofuels, which can be produced from **lignin**-rich materials (the woody elements of plants), are also currently being developed. In the *United Kingdom the use of transport biofuels has been promoted by the government through the *Renewable Transport Fuel Obligation.

biogas Gas produced most commonly through the anaerobic digestion of *renewable carbon-based materials (such as food wastes, crops, or *sewage sludge) and comprising mainly methane and *carbon dioxide. Biogas can either be combusted at the point of production to produce heat and/or electricity or can be cleaned up to leave pure methane (known as **biomethane**), which can then be injected into the mains gas grid or used as a transport *biofuel.

biological control The use of a biological agent to bring about the control of pests and diseases. Common examples in agriculture include the use of parasitic nematodes to control slugs or the larvae of damaging insects, ladybirds to control aphids, and specific bacteria to control scale insects. For biological control to be effective, the agent must specifically target the pest species rather than attack the wider species range within the ecosystem, and be able to operate within the environmental parameters of the area where the agent is to be used. Biological control can be imported from other locations to an area where it does not occur naturally. This carries significant risk, as not all variables can be truly explored in test conditions to assess a biological control agent's suitability. Alternatively, naturally occurring biological control agents can be conserved through supportive farming practices, or boosted through artificial release. *See also* BEETLE BANK.

Biological Records Centre (BRC) A focused centre in the UK for recording of terrestrial and freshwater *species. Supported by the *Joint Nature Conservation Committee and other government-funded organizations, it is also heavily reliant on data supplied by the voluntary sectors. It is in turn supported by a network of local and independent biological records centres. The recorded data provide information for scientific research and policy making.

biomass *Renewable carbon-based materials which can be used as fuels for *renewable energy; for example, wood, conventional crops (such as *maize), *energy crops (for example, *miscanthus or *short rotation coppice), food wastes, *slurry and *manure, *sewage sludge, and agricultural crop residues such as *straw. Biomass can be used to generate heat and electricity in a number of ways, including through digestion by microbes in *anaerobic digestion plants, through combustion in *biomass boilers, or in larger-scale power plants. They can also be used to produce *transport biofuels. *See also* COMBINED HEAT AND POWER.

biomass boiler Boiler which uses *biomass as a fuel (typically wood in the form of chips, pellets, or logs). They are considered to produce *renewable energy because the wood fuel is quickly regrown. **Wood chip** is commonly produced from trees which are removed during thinnings of woodland, whereupon the wood is chipped and stored until it is sufficiently dry in order to be burnt in a biomass boiler. **Wood pellets** are generally made from compacted sawdust or untreated waste *wood shavings, such as those from *sawmilling, and because they are very dense they therefore take up less space to store than chips or logs. Biomass boilers are typically used for heat; however, it is possible to find *combined heat and power biomass systems which produce heat and electricity simultaneously. Biomass boilers tend to be larger than conventional boilers and require sufficient space for fuel storage (although a smaller area for pellet storage than for chips or logs). Consequently, they are often best suited to larger rural properties or commercial sites. They can, however, also be used in **district heating systems** (sometimes in urban areas) where the biomass boiler supplies heat to more than one building or dwelling. *See also* RENEWABLE HEAT INCENTIVE.

biosecurity A series of planned measures introduced to a farm or enterprise concerned with food production that minimizes the risk of accidental disease introduction. A risk analysis is carried out and appropriate steps introduced. Common biosecurity measures include the dipping of feet or vehicle wheels in a disinfectant at the point of entry to an enterprise, the washing of hands on entry and exit, the use of protective clothing, and the collation of visitor details. Biosecurity measures are of increasing importance as the movement of animals and agricultural products increases and the issues of traceability become more significant.

biosolids The solids retained following the separation and treatment of domestic sewage. The water fraction is cleaned and filtered and returned to water courses, whilst the solid component, referred to as sludge, is retained. The sludge is then treated, decontaminated, monitored, and dried. Often rich in organic matter, the biosolids are a good source of nutrients for arable crops and can be spread on fields.

Biosphere Reserve There are 651 of this international category of reserve in 120 countries dedicated under a *UNESCO programme, including five in the UK. They are designated for preserving natural resources and for a variety of other objectives including research, training, and demonstration. There is particular focus on integrating low-intensity human activity into protected areas using core and transition areas separated by buffer zones.

Biotechnology and Biological Sciences Research Council One of seven research councils that work together as Research Councils UK. Funded by the Department of Business, Energy and Industrial Strategy, in 2015–16 it invested £473 million in biotechnology and biological sciences research in the UK, supporting 1,600 scientists and 2,000 research students in universities and institutes. For example, it funds research at *Rothamstead Research and the Roslin Institute—two of the premier research stations in the UK at the forefront of worldwide biotechnology research.

biotic Relating to or resulting from living organisms. This includes the organisms themselves and the interaction with other organisms such as predation, infection, and **symbiotic** relationships.

birch *See* TREE.

bird of prey *See* RAPTOR.

Birds Directive A European Council Directive to promote the conservation of wild birds. This is achieved by providing a framework that includes a range of broad wildlife conservation objectives. It also provides the mechanism for designation of *Special Protection Areas (SPAs). The Birds Directive is implemented in each of the EU member states through domestic legislation; for example, in the UK it includes the *Wildlife and Countryside Act 1981 (as amended). *See also* HABITATS DIRECTIVE; SITES OF SPECIAL SCIENTIFIC INTEREST.

bit *See* TACK.

black bean aphid (*Aphis fabae*) A member of the *aphid family, also known as black fly or bean fly, which attacks numerous species of plants but particularly ornamental plants, sugar beet, and leguminous crops. They suck sap and can also act as vectors for viral diseases, thus causing significant damage to plant growth and crop yield. Black bean aphids can overwinter as eggs, so can be a perennial problem. Control measures include crop rotation, *biological control agents such as ladybirds, and the use of *insecticides. *See also* BEAN.

blackgrass A significant weed species of arable crops. Of the genus *Alopecurus* it competes with arable crops for space, nutrients, and light, and can grow in high densities, shedding seeds in large numbers before harvesting, and thus perpetuating the problem for future years. Because of changes in planting and managing arable crops, blackgrass now tends to germinate and grow within them, making it harder to eradicate. It has also developed resistance to a number of herbicides, making it very difficult to control through chemical use. Ideally, blackgrass should be left to germinate in the autumn and then be topped before drilling, which eliminates a large number of viable seeds; however, this does not completely eradicate blackgrass from the land. Use of spring-sown crops, altering drilling distances, and ploughing the land, have also shown benefits in attempting to control this weed. Most control methods aim for minimizing detrimental impact rather than for complete eradication.

blackleg 1. A disease of livestock also called black quarter or quarter ill. It is caused by *Clostridium chauvoei* bacteria, though occasionally other species are isolated from lesions. Animals aged 6–12 months are most susceptible to infection and pick up bacterial spores from contaminated pasture. Usually acute in nature, symptoms include fever, pain, and significant swelling of the affected limb, although animals can commonly succumb to the infection and die before any visible symptoms are observed. The muscle tissue is attacked by the bacterium and causes necrosis with the accompanying production of gas build-up in the infected area. Treatment of the infection is rarely successful, so prevention through vaccination remains the best option. **2.** A disease of brassica crops caused by a fungus called *Leptosphaeria maculans*. Symptoms of the infection include small grey lesions on the plant leaves, lesions on the stems, and root rot, all of which lead to a reduction in crop performance and subsequent yield. As a fungus, the pathogen prefers cool, wet environments, and spores can remain infective in the field for long periods. Integrated crop management is the key to controlling blackleg in brassicas, with rotations, removal of crop stubble (an over-wintering site for the fungus), and the use of fungicides. **3.** A disease of potatoes most commonly caused by *Pectobacterium atrosepticum*, which is a pectolytic bacteria. The bacteria cause chlorosis in the leaves and lesions at the base of the plant's main stem, causing it to wilt and show stunted growth, or in severe cases, rot completely. Consequently, yield is reduced. The source of infection is contaminated potato tubers. Control methods include ensuring that soil is free-draining, as the bacteria favours cool, wet conditions, planting into a warm soil, ensuring contaminated tubers are removed, reducing harvesting damage, and using sterile propagation methods.

black medic *Medicago lupulina*, species of plant also known as yellow clover or yellow trefoil. It is a member of the legume family and is similar in appearance to the clover family with three leaflets and a small yellow flower. It can be an annual, biennial, or short-lived perennial in lifecycle, depending on climatic conditions. It can be sown as part of a clover ley for fodder or hay crops, though it does not appear to withstand grazing particularly well, or can be grown as a green manure, which is a crop sown specifically to help build soil fertility.

black scurf A common name for the infection of potato crops with the fungus *Rhizoctonia solani*. It is a particular problem in crops grown in cool, wet conditions. The fungus is most obvious when seen on the tubers as black or brown 'scabs' on the surface of the potato, which are the resting bodies of the fungus. These, however, cause little harm to the potato, and instead the most damage is caused when the fungus destroys the sprouts emerging from the tuber during the growing season, causing lesions that can be mild to severe. In the worst cases, the affected area becomes sunken and necrotic, severely hampering the plant's ability to grow and to therefore produce good yields. Control measures are of an integrated nature and include crop rotation, planting tubers in warm soil and covered with a minimum of soil to prevent water logging, and the use of fungicide treated seeds.

blacksmith A skilled craftsman who works with metal to make industrial, agricultural, domestic, and decorative items. Most commonly works with wrought iron or steel, and uses bespoke tools to bend and hammer it into shape. A blacksmith operates from a workplace called a forge. *See also* FARRIER.

blackthorn *See* WOODY SHRUB.

blanket bog (bog) *See* WETLAND.

Bleu du Maine A breed of sheep originating in Western France. *See also* SHEEP.

 SEE WEB LINKS
• Bleu du Maine breed society of the UK.

blight *See* COMPULSORY PURCHASE.

blight (disease) A general term used to describe a set of symptoms of disease in plants. The symptoms include chlorosis of the leaves and browning and death of the leaf tissue, followed by death of other parts of the plant. Common blights include tomato blight, potato blight, and fire blight.

bloat A condition in livestock caused by abnormal swelling of the rumen caused by trapped gas. As a by-product of fermentation, gas accumulates in the rumen and is normally expelled through eructation (burping); however, if the entrance to the oesophagus is blocked, the escape pathway is not available. The gas collects in the rumen, stretching the wall and causing significant pain to the animal. If left untreated, bloat can cause the rupture of the rumen wall and death of the animal, which can occur in a matter of hours in some cases. Causes of bloat can be a physical blockage through a large piece of food swallowed without thorough chewing, or through excessive foam produced by the fermentation of readily digestible protein (frothy bloat).

bloodhound *See* DRAG HUNTING.

bloodline Term used to describe the direct relations to an individual animal such that patterns of heredity can be explored. Bloodlines can be used to plan and manage breeding programmes to avoid interbreeding of related animals whilst maintaining and promoting specified desired characteristics.

blowfly Common name for the species of fly of the family *Calliphoridae* that lays its eggs on carrion. They are characterized by a thorax that has an easily recognizable metallic green, blue, or black colouring that has earned these flies their alternative names of bluebottle or greenbottle flies. Certain species of blowfly are of particular interest in livestock production, as they cause myiasis (flystrike) in livestock, especially sheep. Flies lay their eggs in damp areas of the animals' fleece, commonly areas near the anus where the fleece forms 'dags'—lumps of fleece contaminated with faeces or urine. The eggs hatch, and the larvae break through the skin of the animal and feed on the tissues underneath. The resulting open sores are prime targets for secondary bacterial infection that further weakens the animal. Loss of condition, failure to thrive, and potentially death can occur. Control is mainly through management of the sheep to avoid dags, shearing, and chemicals to control the blowflies. *See also* SHEEP.

Blue Book 1. Informal term for *The European Valuation Standards, published by The *European Group of Valuers' Associations. **2.** Informal term for the Royal Institution of Chartered Surveyors' *RICS United Kingdom Residential Real Estate Agency *professional statement.

blue tongue A viral disease of *ungulates, particularly sheep but also cattle. It is carried by biting midges of the *Culicoides* family. The symptoms in sheep include high temperature, reduced appetite, swelling of the tongue and face, excessive salivation, and the characteristic blue colour of the tongue. There is no effective treatment for the disease. It is sometimes fatal, and in highly susceptible animals can have very high mortality rates. Those animals that do recover take a significant length of time to do so, with the subsequent negative effect on enterprise productivity and financial performance. Because of this the disease is notifiable, and where outbreaks occur a compulsory protective vaccine is used together with tight movement restrictions and midge control.

boar Adult male *pig.

BOAT (Byways Open to All Traffic) *See* PUBLIC ACCESS.

bobby calf Purebred male calf from the dairy industry. *See also* DAIRY.

BOD Frequently used in the abbreviated form, Biological Oxygen Demand is a standard method to indirectly assess the amount of organic pollution in water. It is a measure of the amount of oxygen needed for biological degradation of organic waste. Commonly used to monitor sewage outfall in watercourses. Also referred to as Biochemical Oxygen Demand.

bolting Term used in *agronomy and *horticulture to describe the rapid growth associated with plant sexual reproduction. Typified by the production of usually long stems, flowers, and fruiting bodies. These may not necessarily be desirable in commercial crops, as it diverts resources away from other commercially important structures such as roots, tubers, and leaves.

borage (starflower) Commercially cultivated herb grown for its essential seed oil, although the leaves are also edible. Sown in the spring it is harvested in September, requiring a specialist harvesting head to capture the small seed. The optimum borage oil profile is produced when cultivated in the colder climates of the UK such as northern England and Scotland. However, it does not crop well on wet or heavy soils, and wet conditions make harvesting difficult.

Bordeaux mixture A mix of copper sulphate, lime, and water that acts as an antibacterial agent and fungicide for use on vines and top fruit. It is particularly effective against mildews, although continued use over a number of years may give rise to copper toxicity. Suitable for use in organic production. *See also* ORGANIC.

boron An element required for the correct development and functioning of plants. As a micronutrient it is often included in trace amounts in compound fertilizers, and it is necessary for the correct metabolism of protein and sugars, in the process of pollination and the production of seeds, and in cell division. Boron can also be used as a herbicide or pesticide when in large concentrations and in the form of boric acid.

bothy A simple dwelling or cottage with basic accommodation originally for unmarried farm labourers or gardeners. This function no longer exists, and the term is now used mainly to describe a shelter available for anyone to use, often in remote upland areas.

botrytis (grey mould, bunch rot, noble rot) Fungal disease of many plants, but in agriculture it is of particular importance to fruit crops and vineyards.

Where soft fruit is affected it renders the fruit inedible, and is therefore of significant economic impact. Increasing ventilation and lowering humidity help to control the fungus. In viticulture, botrytis can be of benefit or a source of economic loss. As 'noble' rot it draws water out of the grape, causing dehydration and therefore raising the solid components of the grape, which increases the intensity of the final product. However, botrytis can cause significant losses when wounds on the grape allow the penetration of the fungus, causing its destruction.

bovine spongiform encephalopathy (BSE, mad cow disease) A notifiable disease caused by a prion which is a malformed protein. Although present in virtually all body tissues of affected animals, it is most concentrated in nervous tissue and can be transmitted through the consumption of food contaminated with brain, spinal cord, digestive tract, and blood of affected animals. The prion is not destroyed by cooking or freezing. It causes the degeneration of the brain and spinal cord with the subsequent loss of function and control over the body. It has a long incubation period of between thirty months and eight years, and is the reason for the upper slaughter age for cattle destined for the human food chain. In humans the disease is called new variant Creutzfeldt–Jakob disease (new variant CJD). The first confirmed case in cattle was identified in the UK in 1986, and it has been estimated that approximately 400,000 infected cattle entered the human food chain during the 1980s.

bovine tuberculosis (bTB) A chronic disease of cattle and other animals caused by the bacterium *Mycobacterium bovis*. The symptoms may take months or years to appear, and include weakness, loss of appetite, weight loss, fluctuating fever, intermittent hacking cough, diarrhoea, and large prominent lymph nodes. It is a *notifiable disease in the UK, and animals that test positive are slaughtered. The standard method for detection of TB is the tuberculin test, by which a small amount of antigen is injected into the skin, and the immune reaction measured. Definitive diagnosis is made by growing the bacteria in the laboratory, a process that takes at least eight weeks. Other animals that may carry and spread the disease include deer and particularly *badgers, which have been the focus of controversy in the UK where a cull of badgers is part of the Government strategy to eliminate the disease, along with increased *biosecurity. Humans may catch bovine tuberculosis, but the risk is estimated to be very low.

((⊕)) SEE WEB LINKS
• UK government webpage detailing actions and strategies to combat bTB in cattle.

BPEX *See* LEVY BODIES.

BPS (Basic Payment Scheme) *See* COMMON AGRICULTURAL POLICY.

bracken (*Pteridium aquilinum*) A type of fern native to the UK, tolerant of a range of climatic and soil conditions. It can reach heights of over 1.5 metres and is commonly found in *upland areas, *heathland, and *woodland. Bracken is a very competitive plant, and there is significant concern in its ability to encroach onto *grazing land and reduce effective available forage for *livestock. It is poisonous and is carcinogenic, and although it has some limited wildlife value its dramatic spread in many areas has reduced both farming profitability and wildlife interests. *See also* BRACKEN CONTROL.

bracken control A number of techniques have been developed to restore *bracken-infested areas and restore grassland, *heather, or mixed vegetation land cover. Extensive areas could be treated by aerial spraying using herbicides with the active ingredient *asulam, but this is now a banned product. Cutting and rolling can be effective, but the topography of individual areas can make these methods impractical. Ideally, the dead bracken material needs to be removed to allow for other plants to colonize, but the mechanical impact of trampling livestock can also be effective in this regard. The benefits to wildlife and landscape have encouraged financial incentives within *agri-environment schemes and other conservation initiatives to grant aid bracken control programmes. *See also* BRACKEN.

brackish Water that is saline but not as salty as seawater, and although not precisely defined it can be categorized into zones according to the varying salinity levels. Common in places that mix seawater with freshwater, such as estuaries, deltas, and mangrove swamps.

bran The husk of cereal grains separated before or during milling. It should not be confused with the *chaff which is the loose outer seed coating separated during *threshing. Bran is particularly high in dietary fibre and essential fatty acids as well as other nutrients. It is usually from wheat but can come from other *cereals, notably *oats and *rice. It is a food for both humans and animals.

brash 1. A limestone soil noted for its high stone content, as in Cotswold brash. **2.** In *arable *cultivation, brash describes the decaying stems and roots of the previous crop together with any weeds that may be on or near the surface of the soil where *ploughing has not taken place. This brash may impede subsequent cultivation such as *drilling. **3.** In forestry, it describes the small branches and foliage removed when trees are felled. Where trees such as hazel are *coppiced, the brash is often placed over the *stool to deter *deer from grazing the regrowth.

brashing The physical operation of pruning of lower branches up to approximately 2 metres undertaken to improve access and timber quality.

brassica A large genus of plants of the mustard family, including *oilseed rape, *kale, *stubble turnips, and vegetables such as cabbage, cauliflower, broccoli, and Brussels sprouts. They are widely grown in agriculture and horticulture as human and animal feeds, whilst oilseed rape is grown for its vegetable oil content. They cannot be grown too frequently in an arable rotation as they are susceptible to a number of pests diseases such as *clubroot and cabbage flea beetle. There are also some arable weeds such as *charlock.

brassica pod midge (*Dasineura brassicae***)** A *pest of the oilseed rape crop.
It lays its eggs within the seed pods, which turn yellow, ripen prematurely, and
may be twisted and deformed, with localized small swellings. Pods affected by
the midge contain up to fifty white or yellow–white gregarious larvae, and normal
development of seed is prevented, reducing yield. The pest can be controlled
by application of *insecticide but should only be a last resort, with every effort
made to limit the impact on pollinating insects such as bees.

breach of condition notice *See* ENFORCEMENT, PLANNING.

breach of contract Situation where one party has broken, or not complied
with, one or more of the *terms of a *contract. In this situation there are a number
of **remedies** that the party not in breach may have open to them (and the
contract itself may also set out potential remedies). The general principles
relating to situations where there is a breach of contract are similar across the
*United Kingdom, though there are differences in terminology and also in the
detailed application of the law, particularly in terms of the different bodies of
*case law that apply in the different parts of the UK. Approaches that can be
pursued by the party not in breach include:

- *Damages (i.e. *compensation) can be awarded by the *courts. These are paid
  to the innocent party by the party in breach of the contract.
- **Action for an agreed sum** (**action for payment** in Scotland): a court action
  where the party not in breach seeks to recover money which is due under
  the contract (i.e. a debt) from the other party. Because it involves the recovery
  of an amount of money specified in the contract it is different from a claim
  for damages (which is effectively for compensation).
- **Specific performance** (**specific implement** in Scotland): the innocent party
  may apply to the courts for an *order compelling the other party to perform
  an obligation as set out in the contract. Specific performance and specific
  implement are broadly similar; however, specific implement is used in a
  broader range of situations by the courts in Scotland than specific
  implementation is in the courts of England, Wales, and Northern Ireland.
- **Injunction** (**interdict** in Scotland): the courts have the power to grant
  an *injunction to stop one party doing something that is in breach of
  the contract.
- **Set-off** (or **retention** in Scotland) is where the innocent party stops
  performing one of their obligations under the contract (for example, making
  payments) if the other party is not complying with their obligations.
- A party not in breach may also be able to effectively cancel the contract if
  there is a serious or significant breach by the other party, or the other party
  indicates that they are abandoning or not intending to comply with the
  contract (when this is expressed as an intention not to do something in the
  future it is known as an **anticipatory breach**). The innocent party may then
  seek damages from the other party.

See also BREACH OF COVENANT; FRUSTRATION OF CONTRACT.

breach of covenant Situation where one party has broken, or not complied with, a covenant (for example a *tenant or *landlord has broken a *covenant in their *tenancy or *lease). In the case of covenants which have been breached in respect of a tenancy or lease, the landlord or tenant in the *United Kingdom has several possible **remedies**. As a tenancy or lease is, in essence, a *contract, many of these provisions are those of *breach of contract (so will apply, potentially, to all types of contract); however, there are some additional remedies which relate specifically to landlord and tenant situations. Whilst the remedies listed below apply, in general terms, to leases and tenancies, their application can also be varied, or limited, by *statute (particularly in the case of residential tenancies), and not all remedies are available in connection with all types of lease. There are also some differences between remedies available in each of the parts of the *United Kingdom. Remedies available to the landlord in England and Wales (see below for Scotland and Northern Ireland), largely depending on whether the breach relates to non-payment of *rent or some other breach, and also what type of tenancy or lease it is:

- **CRAR (Commercial Rent Areas Recovery)**: this is available for non-payment of *rent in respect of written commercial leases (including *agricultural tenancies), but not residential properties. It replaced the old remedy of *distress in England and Wales in 2014. It involves the landlord employing a certified enforcement agent to seize goods to cover the amount of unpaid rent. However, unlike distress, landlords are required to give notice of their intention to do this, and there are rules about the minimum amount of rent that must be outstanding before CRAR can be used, together with other rules about the procedure.
- **Action for rent**: the landlord may take action to recover rent, as a debt, through the *courts. Depending on the size of the debt, the claim may go through the *County Courts (either through the small-claims process or the court itself), or through the *High Court.
- **Forfeiture** gives the landlord the right to forfeit (effectively, end) the tenancy. It is generally used for non-payment of rent, but can also be used for other breaches of covenant. The landlord is entitled to re-enter the premises and recover the lease (historically, re-entry was done physically, but given modern legal protections of tenants it is now done more normally through court proceedings against the tenant). However, a landlord can **waive** (relinquish) the right to use forfeiture for a breach they become aware of by taking some action which implies the continuation of the tenancy (for example, demanding rent). Forfeiture can also generally only be used if a **right of re-entry** is contained within the lease, and it is generally not effective for residential properties (owing to the *statutory protections of residential tenants under the law, including the *Protection from Eviction Act 1977). If the breach relates to non-payment of rent, then often a formal **rent demand** must be served first. If it relates to another type of breach, then a notice must be served on the tenant before any process of re-entry is undertaken. The notice must specify the breach, and if it is one that can be corrected (i.e. it is a **remediable breach**)

then the notice must require that the tenant does this within a reasonable period of time. If the tenant does so, then the forfeiture proceedings cannot continue any further. When preparing the notice it is for the landlord to decide whether the breach is capable of correction, or whether the damage caused by the breach is not possible to correct (i.e. it is an **irremediable breach**). If the courts consider that the breach is remediable and the landlord has not set out in the notice what the tenant should do to correct it, then the notice is likely to be invalid. There is consequently a body of *case law concerning whether breaches are remediable or irremediable. The tenant can also choose to apply to the courts for **relief from forfeiture**, and if the tenant pays the rent or corrects the relevant breach (for those not relating to rent), the *courts can then effectively stop the process. There are also specific additional procedures if the breach relates to the tenant's *repairing obligations, and for long residential leases. The *Law Commission has recommended replacement of forfeiture in England and Wales with a different procedure due to the complexity of the forfeiture system; however, at the date of writing, forfeiture continues to exist.

- *Damages (i.e. *compensation) can be awarded by the courts for a breach (other than non-payment of rent).
- **Specific performance**: the courts can order the tenant to comply with a covenant if they are reluctant to; however, this is rare.
- **Injunction**: the court has the power to grant an *injunction to order a tenant to stop doing something in order to comply with a covenant (say, for example, the tenant had covenanted not to play music after a certain time but was doing so), but again this is rare.

Tenants also have a similar range of remedies against their landlord for breaches of covenant in England and Wales (though they cannot use forfeiture or CRAR). Tenant's remedies, therefore, include damages, injunction, and specific performance. In some circumstances, if a breach of covenant relates to repairs then the tenant can undertake the repairs and deduct the cost from their rent; however, this can be dangerous if not done carefully, as it may lead to the landlord taking action for *rent arrears. A tenant may also be able to effectively cancel the lease, under general *contract law, if there is a serious or significant breach by the landlord (this option is also potentially open to the landlord in the event of a breach by the tenant; however, it is likely to be covered by forfeiture in any event). The courts also have the ability to appoint a **manager** (usually in the case of a block of flats) to take on the responsibilities of the landlord if they have persistently neglected them and they involve repairs. Additionally, for residential properties the *local authority can take action against the landlord if they consider the property to be unfit for habitation or a significant health and safety risk (*see* HOUSING HEALTH AND SAFETY RATING SYSTEM). Some breaches of covenant (for example, breaches of the tenant's right to quiet enjoyment in a residential property) may also constitute a criminal offence under the *Protection from Eviction Act 1977.

Landlords and tenants in Northern Ireland have a set of remedies similar to those in England and Wales; however, CRAR does not operate in Northern

Ireland, and in general, neither does distress (although goods can sometimes be seized through an order issued by the **Enforcement of Judgments Office**). Similar principles also operate in Scotland, but with some differences in detail and terminology. As in England, Wales, and Northern Ireland, landlords and tenants in Scotland are able to use **specific implement** (broadly similar to specific performance in the other UK courts, but used in a broader range of situations) and **interdict** (the equivalent of an injunction in the other UK courts). They are also able to seek damages through the courts. However, in addition, both landlords and tenants have a number of other options available to them:

- **Action for payment**, which operates through a court action when one party has some obligation to make the other party a payment (for example, rent). This is a general contractual remedy. The court will then issue a court **decree** for debt (i.e. a formal *court order) which can be enforced through **diligence** (in essence, the seizing of property of the person who owes the money). It is also possible for there to be a clause in a lease where the parties have agreed to the lease being **registered for execution** in the **Books of Council and Session** (*see* LAND REGISTER OF SCOTLAND). This, in effect, provides consent for **summary diligence** to be used, without there needing to be a specific application to the court in the event of there being a future outstanding debt as a consequence of a breach of covenant. The process of registering a lease for execution in the Books of Council and Session means that a warrant is given when the lease is registered, allowing for lawful execution of diligence. This is equivalent to a **decree** from *Court of Session (and hence removes the need for the landlord to go to court at a future date to recover unpaid rent if they wish to rely on summary diligence).

- **Retention** allows the innocent party to stop performing one of their covenants if the other party has not been performing an equivalent covenant; for example, a tenant may retain rent if their landlord is not meeting their obligations. However, as in the other parts of the UK where similar approaches may be available, caution needs to be exercised by tenants considering this because it could lead to the landlord taking action for rent arrears.

- *Landlord's hypothec, i.e. retaining the tenant's goods on the leased premises as security against rent. This remedy (which is only available to landlords) has some similarities with the old remedy of distress which operated in England and Wales; however, legal changes in recent years have limited its scope and application, and it is now mainly available only in relation to commercial (but not agricultural) leases.

- **Irritancy**, which is broadly equivalent to forfeiture in other parts of the United Kingdom, and is again only available to the landlord. There are two forms of irritancy: **legal irritancy**, which is provided for in law and is restricted to a situation where the rent has been unpaid for two years (or six months for some kinds of agricultural tenancies); and **conventional irritancy**, where the lease includes specific provisions which allow the landlord to use irritancy; for example, for non-payment of rent over a shorter period, or another type of breach. As with for forfeiture, waiver applies to irritancy. The tenant can also, in some circumstances, **purge** the irritancy (i.e. effectively stop the process) by, for example, paying off rent arrears. As with forfeiture in other parts of the UK, there are specific requirements to serve notice on the tenants if a landlord wishes to rely on irritancy.

- In addition, as under English law, a landlord or tenant may also be able to effectively cancel the lease, under contract law, if there is a serious or significant breach by the other party. However, there are procedural controls over this, and if the landlord wishes to do this the courts are likely consider whether they have acted reasonably.

Where the breach relates to a covenant concerning *freehold land (or its Scottish equivalent, a **real burden**), then the main remedies are likely to be damages, injunction (or interdict in Scotland), or specific performance (specific implement in Scotland).

See also ABATEMENT; FRUSTRATION OF CONTRACT.

bread-making quality The characteristics required of a *wheat variety for bread making. There are quality requirements such as specific weight, moisture, protein content, and Hagberg falling number, an international measure of enzyme activity that relates to good loaf structure. Milling wheat varieties are rated on a scale issued by the National Association of British and Irish Millers, with Grades 1 and 2 being suitable for bread-making, and 3 suitable for biscuits. If the quality fails marginally for bread-making, it may be adequate for cakes or biscuits.

break clause Clause within some *tenancy agreements or *leases which allows either the *landlord or *tenant to serve notice to end the tenancy at a break clause point (i.e. before the expiry of the tenancy or lease).

break crop A secondary crop grown in a *rotation to break the run of the primary crop, usually *cereals. This is done to reduce the impact of *weeds, *pests and *diseases which can build up under continuous cropping. Take-all, for example, is a fungal disease of cereals that builds up in a continuous cereal rotation. The break crop is usually of a different genus, although oats, for example, are sometimes treated as a break from wheat or barley. The most effective break crops are those that are restorative, i.e. they improve the fertility of the soil. Thus *legumes and pulses with their associated *nitrogen fixation are particularly valued.

breed A specific group within a species that show similar characteristics, usually domestic animals that have been selectively bred for specific purposes. Thus, for example, the Friesian breed of cattle is kept primarily for milk production, the Hereford for beef. As a verb, to breed is to mate animals to produce offspring.

Breeding Bird Survey (BBS) The major bird monitoring project in the UK, led by *BTO in collaboration with *JNCC and *RSPB, in which breeding populations of wild birds are surveyed annually using volunteers. Birds are a recognized indicator to measure the environment, and the data collected are a valuable means of assessing short- and long-term trends. The BBS was preceded by the Common Bird Census which ran from 1962 to 2000, and was instigated by concerns over the impact of agricultural practices on farmland bird populations.

Brexit Colloquial term for the departure of the United Kingdom from the European Union following the *referendum decision of 23 June 2016.

bridle *See* TACK.

bridleway *See* PUBLIC ACCESS.

British Association for Shooting and Conservation (BASC) The largest representative body covering quarry shooting, founded in 1908 as the Wildfowlers Association of Great Britain and Ireland. It was renamed to BASC in 1981. It promotes the interests of shooting, particularly in regard to firearms, and also undertakes political, legal, advisory, and research work.

(⊕) SEE WEB LINKS
• BASC website: includes information on the range of its interests and activities.

British Christmas Tree Growers Association (BCTGA) A specialist organization for members in the UK that grow and supply Christmas trees for the market. Its stated aims include the promotion of live Christmas trees produced in a sustainable manner. *See also* CHRISTMAS TREES.

(⊕) SEE WEB LINKS
• BCTGA website: includes information on Codes of Practice, quality standards, and related information.

British Deer Society (BDS) The BDS is a charity that promotes the welfare and management of *deer in the UK. It has more than 6,000 subscribing members, of which the majority have an interest in *deer stalking. One of its most important functions is running training courses, as well as providing information and representing specialist interests to government.

(⊕) SEE WEB LINKS
• BDS website: supplies a range of deer management information.

British Horse Society (BHS) The society is a charity that represents all equine interests and makes representations to government on behalf of all who ride in the UK. It includes within its remit the improvement of horse management and breeding, and promotion of the care and welfare of *horses and ponies. It also takes on a responsibility to protect and improve *Public Rights of Way, particularly in relation to access for horse riders and carriage riders.

(⊕) SEE WEB LINKS
• BHS website: supplies a range of information, advice, and training opportunities.

British Standards Institution (BSI) United Kingdom National Standards body that develops and certifies standards relating to products, services, and systems. These standards set out specifications and requirements relating to products, materials, processes, or services with the objective of ensuring that they are fit for purpose. The BSI develops its own standards (for example, British Standards, or Publicly Available Specifications, which are identifiable by the prefix BS, or PAS, respectively), and owns the BSI Kitemark™. It also works as a certification body, inspecting organizations who wish to demonstrate that they are complying with BSI standards, or those set by the *International Organization for Standardization (ISO). The BSI is the United Kingdom's member of the ISO.

British Trust for Conservation Volunteers (BTCV) *See* Conservation Volunteers.

British Trust for Ornithology (BTO) An independent and charitable organization that combines professional research and data supplied by the general public or citizen science. Their surveys on UK birds are used by government and NGOs. In particular, the BTO undertakes long-term monitoring work, including the Breeding Bird Survey and Wetland Bird Survey, which rely heavily on volunteers.

(((∰))) SEE WEB LINKS
• BTO website, detailing latest surveys and research.

British Waterways and Rivers Trust *See* Canal and River Trust.

British Wool Marketing Board The national body to which all producers with more than four sheep are registered. After the shearing of sheep, the wool is collected, graded, and sold on behalf of producers. Although a commercial enterprise it is non-profit-making, returning all proceeds to farmers after deducting costs, including the promotion of wool. It is the one remaining agricultural commodity board in the UK. *See also* levy bodies.

broadcast To spread seed over a relatively wide area, often by hand, as opposed to placing it in the soil in rows through a seed *drill. A crop may be broadcast into an existing crop before *harvest—for example, *mustard or *stubble turnips into a *cereal crop—when using a seed drill is impractical. The coverage of the seed is random rather than the greater precision produced by a drill. The technique may also apply to the spreading of fertilizer or solid pesticides such as slug pellets.

broadleaf (*pl.* **broadleaves**) Broadleaved tree species have wide leaves as opposed to the narrow needles of conifers. Most broadleaves drop their leaves in winter and in the UK are mainly **native**, i.e. naturally occurring. Timber that broadleaves produce is commonly referred to as *hardwood. *See* also tree.

Broads Grazing Marshes Scheme An early *agri-environment scheme developed jointly in 1985 by the **Broads Authority**, the *Ministry of Agriculture,

Fisheries and Food, and the *Countryside Commission. The scheme was set up to encourage farmers in the Halvergate Marshes area in the county of Norfolk to implement more environmentally friendly farming practices. Its particular significance is that it subsequently developed into the UK's first *Environmentally Sensitive Area Scheme. *See also* DEPARTMENT OF ENVIRONMENT, FOOD AND RURAL AFFAIRS.

broilers A *chicken specifically reared for meat rather than egg production. Breeds have been selectively bred over time to be fast growing and produce the optimum carcass for meat consumption. Most are reared in sheds so that the environment of light, temperature, and humidity can be controlled. There are guidelines for stocking rates and husbandry, but there has been controversy in some countries over the conditions in which the birds are kept. Typically, a broiler chicken is slaughtered at 5–7 weeks old at 2.5 kg in weight.

broody hen A hen *chicken that is trying to incubate eggs. She will stop laying eggs and sit on her nest with or without any eggs beneath her in a psychological state of motherhood. Broody hens are not welcome in an egg production enterprise, but can be very useful when eggs are to be incubated naturally; for example, in the rearing of game birds when large numbers of eggs have been collected and are placed under broody hens for incubation.

brownfield land Land which has, or has had, some form of previous *development. *Planning policy generally favours the redevelopment of brownfield land over *greenfield land; however, this preference does not automatically extend to all types of previously developed sites. For example, the *National Planning Policy Framework in England deliberately excludes land that was, or is, occupied by agricultural or forestry buildings from its definition of previously developed land.

brownfield register *Local planning authority register of *brownfield land sites which are potentially suitable for housing *development. Inclusion of a site on the brownfield register is one of the mechanisms through which a local planning authority can grant *planning permission in principle for housing development.

brown rust A disease of *wheat caused by the fungus *Puccinia triticina*. There are other strains that can infect barley, rye, and triticale. The symptoms are often seen in the autumn on early-sown crops as individual orange-to-brown pustules usually seen on the leaves, though in severe attacks symptoms can be seen on the stem and glumes. In the past, brown rust was not considered to be a major problem in the UK despite early-sown crops generally carrying high levels of brown rust through the winter. However, the occurrence of new virulent strains overcoming varietal resistance in a few key wheat varieties has moved brown rust up the league table of importance. Severe attacks result in a significant loss of green leaf area and hence yield, and infection of the ears will also result in loss of grain quality. Control is maintained by growing resistant varieties, rotation of crops, and fungicide application.

Brundtland Commission *See* SUSTAINABILITY.

Brundtland report *See* SUSTAINABILITY.

brushwood Cut or fallen branches and twigs on the woodland floor that may also include smaller woody vegetation to create a dense undergrowth, a *hedge, or other protective barrier made from cut branches. A brushwood permit or similar can be allocated to local people for the collection of **estovers** or kindling wood. The origin of the term describes birch twigs or other material to make brushes.

BSE *See* BOVINE SPONGIFORM ENCEPHALOPATHY.

BSI *See* BRITISH STANDARDS INSTITUTION.

BTCV *See* CONSERVATION VOLUNTEERS.

BTO *See* BRITISH TRUST FOR ORNITHOLOGY.

buckrake Agricultural tool that consists of horizontal tines sticking out of a frame. It may be mounted on a front loader or on the *three-point hydraulic linkage at the back of a *tractor. It is used to move material such as *bales of hay or straw or to consolidate material such as grass in a *silage clamp. In the latter case it may have a mechanism to push material off the tines.

buckwheat (*Fagopyrum esculentum*) A plant grown for its seed and as a cover crop. Despite its name it is not a *cereal but a member of the polygonum family related to **knotweed**, **sorrel**, and **rhubarb.** However, the seeds resemble the grain of cereals and it thus known as a pseudocereal. It has a short growing season and does not respond to fertilizer, preferring poor fertility. The seeds, containing no gluten, are considered to be a health food for human consumption.

buffer strip Normally a relatively narrow strip of vegetation whose positioning reflects its function as a buffer designed to limit adverse environmental factors. The commonest agricultural buffer strips are sown grass margins around *arable fields that can mitigate loss of *agrochemicals, nutrients, or soil into surrounding non-farmed habitats such as *hedgerows or *watercourses. This can be achieved by slowing water run-off, filtration, and trapping of sediment. Buffer strips can also limit noise and smell disturbance, and provide additional protection from mechanical damage. They can also provide a number of other environmental benefits, and are a popular option within many *agri-environment schemes. *See also* FIELD MARGINS.

building control The system, managed by *local authorities in the United Kingdom (though known as the *building standards system in Scotland), for ensuring that the construction, extension, and in some cases alteration, of buildings complies with *building regulations. This is distinct from *development control, which is the system which ensures that *development complies with any requirements for *planning permission.

building control bodies Bodies that issue *building regulations approvals in England and Wales. Building control bodies (BCBs) can be either *local authority BCBs or private BCBs (known as **approved inspectors**).

Building Cost Information Service of Royal Institution of Chartered Surveyors (BCIS) Provider of cost and price data relating to the construction sector. BCIS data are commonly used by construction industry professionals (for example, *chartered surveyors), as well as government bodies, the insurance sector, and construction contractors. Data relating to construction costs have a variety of uses, including for *development appraisals, producing cost plans for construction projects, assessing building maintenance costs for budgets and *lifecycle costing, and for insurance *reinstatement cost assessments. *See also* ROYAL INSTITUTION OF CHARTERED SURVEYORS.

Building Preservation Notice Temporary notice served in respect of a *listed building which is considered, by the local planning authority (or relevant government minister), to be at risk of imminent damage or destruction, and which has the effect of instantly giving the building the same degree of protection as a listed building. The effect of the notice can last up to six months, during which time consideration can be given by the relevant authorities as to whether the building should be given full listed building status.

building regulations *Regulations in England, Wales, and Northern Ireland which govern the physical design and construction, extension, and in some cases, alteration of, buildings (for Scotland, *see* BUILDING STANDARDS SYSTEM). They are distinct from the *planning system (as this addresses whether the use and appearance of a building, or site, is, in principle, acceptable in general land use terms), as building regulations concern the acceptability of the physical construction (although changes of use of a building, or part of a building, can sometimes trigger a requirement for building regulations approval, as the building regulations may be different for the old and new uses of the building). Whilst, historically, building regulations were primarily concerned with ensuring that buildings met health and safety standards, they now also address other matters such as ensuring minimum levels of energy efficiency.

The system for ensuring compliance with building regulations is known as the *building control system (as opposed to the *development control system, which deals with *planning permission). A building project may require both planning permission and buildings regulations approval.

When building regulations approval is required in England or Wales, this is done through an application to a *building control body (BCB). (A slightly different system operates in Northern Ireland; see later in this entry.) A BCB in England or Wales can be either a *local authority BCB, or an approved private BCB (these are known as **approved inspectors**, who will notify the local authority about the work).

There are three types of application:

1. Full Plans. Under this system, plans showing full construction details are supplied to the BCB, which then assess whether they comply with building regulations. If they do, then a **full plans approval notice** is issued (the equivalent if issued by a private BCB is known as a **plans certificate**). Whilst the work is being carried out, inspections will be undertaken, and on completion, providing the work complies with building regulations, a **completion certificate** (or **final certificate** if issued by a private BCB) can be issued. If there is any disagreement about whether the plans comply with building regulations, then, under the Full Plans system the applicant can apply for a **determination** from the *Ministry of Housing, Communities and Local Government in England, or the *Welsh Government.

2. Building Notice. This process involves a notice or application being submitted to the BCB, rather than a full set of plans. This process is normally only suitable for smaller projects where it is considered that there will be little risk of their not complying with building regulations (as it misses out the plan scrutiny stage). Once work has commenced it is inspected, and on completion, providing the work complies with building regulations, a completion or final certificate can be issued. It is not possible to apply for a determination under the Building Notice procedure.

3. Regularization. This covers a situation where work has already been undertaken without approval, and the application is being made retrospectively to regularize the situation. These types of application can be made only to a local authority BCB. If the BCB is satisfied that the works are compliant (which may involve making alterations), then it will issue a **regularization certificate**.

For some types of work (for example, installing windows), a **competent person scheme** operates in England and Wales. Under this system, tradespeople can apply to belong to the **Competent Persons Register**. They are then able to self-certify that their work complies with building regulations, will notify the local authority that they have undertaken the work, and will then issue a **certificate of compliance**, thereby meaning that an application for building regulations approval is not required.

Local authorities are generally responsible for enforcing the building regulations. In the event of their becoming aware of works which are not compliant, they will usually initially seek to deal with the matter informally in order to try to ensure that the work is brought into compliance. If this is unsuccessful then the local authority has two formal enforcement options available. It can prosecute the person who has carried out the work (for example, a *contractor) through the courts, where they may be subjected to a fine and/or where an **enforcement notice** may be served on the building owner, which may require the owner to either amend, or remove, the non-compliant work. If the owner does not do so, then the local authority may undertake the work and then recoup the costs from the owner. In addition, the absence of the appropriate building regulations certificate may cause the owner of the property difficulties in selling that property, as its absence is likely to become evident during a **local land search enquiry**.

The building control system in Northern Ireland operates in a similar way to that in England and Wales, although applications and oversight are managed only by local authorities. As with England and Wales, three types of application exist (depending on the nature and timing of the work): a full plans application (leading to an approval notice), a building notice application, and a regularization application. Work is inspected, and on satisfactory completion, or compliance, a completion certificate or regularization certificate is issued. There is also a right of appeal to the **Northern Ireland Department of Finance** in respect of building regulations decisions made by local authorities. Failure to comply with building regulations can result in the local authority serving a **building regulations contravention notice** requiring the work to be made good, altered, or removed. In more extreme cases, legal action can also be taken to enforce the regulations.

See also BUILDING STANDARDS SYSTEM.

building standards system System for implementing building regulations within Scotland (*see* BUILDING REGULATIONS for England, Wales, and Northern Ireland). Building regulations (or standards) set out legal requirements in relation to the physical design and construction, extension, and in some cases, alteration of, buildings. They are distinct from the *planning system (as this addresses whether the use and appearance of a building, or site, is in principle acceptable in general land use terms), as building regulations concern the acceptability of the physical construction. Whilst building regulations historically concerned health, welfare, and safety matters, they now also address other matters such as energy efficiency. The local authority can take enforcement action against a building owner if work does not comply with building regulations. If building standards approval is needed for works in Scotland, then this is usually done through the submission of a set of plans to the relevant local authority **building standards verifier**.

If the plans are compliant with building standards, then a **building warrant** will be issued to confirm that the work can commence (it is a legal requirement to have a building warrant before works commence). Once the work has been completed, the property owner or developer submits a **completion certificate** to the local authority, which will then inspect the work, and providing it is satisfied that the work has been done in compliance with building standards, will issue a **notice of acceptance of completion certificate**.

For some types of work it is possible to use an **approved certifier of design** (who is able to self-certify that the design is compliant with building regulations) and/or an **approved certifier of construction** (who can self-certify some types of construction work as compliant with building regulations). Whilst the local authority still needs to issue a building warrant and notice of acceptance of completion certificate, using approved certifiers can speed up the process by reducing, or removing, the need for local authority scrutiny of designs or inspections of work.

bulk tank Large tank for the cooling and storage of *milk until collected by the dairy processor or used on the farm. Milk is pumped from the milking parlour into the bulk tank, which is refrigerated and contains a paddle to circulate the milk.

bull A male bovine animal. A bull is entire, whilst a *bullock or *steer has been castrated. A bull is usually kept for breeding, but some beef animals are not castrated if they are due to be slaughtered at around a year old. A bull may be of a dairy, beef, or dual-purpose breed. A beef bull may be used on dairy cows to make the offspring more suitable for beef. When a bull mounts a cow for the purpose of breeding, this is known as *bulling.

bulling *See* BULL.

bullock *See* BULL; STEER.

bund A retaining bank or wall. Usually built of earth to contain a liquid, it might be a primary retaining structure for a slurry lagoon, waste water, or even a channel for irrigation, especially where the ground is *terraced, or it might be a secondary barrier around a fuel tank or chemical store to prevent any leakage beyond the perimeter.

burning A management tool available and used to restore and enhance grazing land. It is more often used as a last resort to aid restoration of long neglected *grassland back to more productive agricultural use. Burning can remove layers of dead vegetation and stimulate growth, but can adversely impact upon plant and invertebrate diversity and can kill less mobile species such as reptiles. Least 'fire-resistant' plants such as mosses and *lichens can be eliminated if burning is undertaken on a large and frequent scale. Burning is more commonly used to manage *upland areas, including *heather moorland. The burning of heather to stimulate more productive and vigorous growth benefits *livestock and *wildlife, and is carried out on about fifteen-year intervals. Where the primary economic land use is for *grouse, then burning can be at shorter 7–10 year intervals, and in a mosaic pattern to provide a closer mix of different-aged structural heights to provide more intimate nesting and feeding areas.

business property relief (BPR) *See* INHERITANCE TAX.

business rates Local tax paid on non-domestic properties in England and Wales. The rate paid by the business is based on the **rateable value** of that business (which are published on **rating lists**). The rateable value is based on the amount that the property could have been let for a year on the open market (i.e. not the actual amount of rent paid). Whilst most businesses pay business rates, farmers are an exception in this respect in that they will, if they are only undertaking traditional agricultural activities, pay *Council Tax on their farmhouse, but not business rates. If, however, the farmer has undertaken some form of *diversification into other business activities, then they may pay business rates on those diversified activities. There are, in addition, a number of **reliefs** available which may mean that in some situations some other types of businesses and organizations may not have to pay business rates; for example, some small businesses may get relief, as may owners of properties which are temporarily vacant (say between lettings). In Northern Ireland and Scotland, separate business (or non-domestic) rates systems exist, but they both operate on similar principles to those of the system in England and Wales. *See also* CONTRACTOR'S METHOD; PROFITS METHOD.

business tenancies Tenancies of premises which are occupied for business purposes. In England and Wales, *landlord and *tenant law in relation to the occupation of property for business purposes comprises several pieces of *legislation. The key piece of legislation for non-agricultural businesses is the *Landlord and Tenant Act 1954, and alongside it the **Landlord and Tenant Act 1927** (for a discussion of their operation, *see* LANDLORD AND TENANT ACT 1954). Some types of tenancy are, however, specifically excluded from the scope of the 1954 Act. These include *agricultural tenancies which are covered by other *statutes. Other arrangements for businesses are also sometimes found which lie outside the scope of the 1954 Act; for example, *licences and *tenancies at will.

In Scotland there is very little *statutory control of business *leases (though *see* SCOTTISH AGRICULTURAL TENANCIES), with the exception of the tenancies of shops, where there is some statutory intervention.

The position in Northern Ireland is, however, that business tenants generally have *security of tenure with provisions similar to those of the English/Welsh Landlord and Tenant Act 1954. The landlord in Northern Ireland is, therefore, only able to refuse to grant a new tenancy if they can rely on one of the reasons set out in the Northern Irish *legislation. and unlike in England and Wales, it is also not possible to *contract out of the security of tenure provisions.

butterfat The natural fat contained in milk and other dairy products. Chemically, butterfat consists essentially of a mixture of **triglycerides**, particularly those derived from fatty acids such as **palmitic**, **oleic**, **myristic**, and **stearic acids**. The butterfat content of milk is an important component used to describe the quality of milk and determine its price. Typically, the butterfat content of milk is around 4%.

buyer's premium *See* AUCTION.

Byways Open to All Traffic (BOAT) *See* PUBLIC ACCESS.

C3 plant Plants create sugar by *photosynthesis. Carbon dioxide is taken in through stomata or pores in the leaves and combined with water to form glucose and oxygen using sunlight energy captured by chlorophyll. This process takes place in chloroplasts and results in a three-carbon-atom molecule, hence the name. C3 plants have adapted to cool, damp climates as the stomata remain open throughout the day. They include most trees and agricultural crops such as wheat, barley, and soya bean.

C4 plant The process of *photosynthesis is slightly different in C4 plants than in C3 plants, and results in a molecule of four carbon atoms, hence the name. C4 plants have adapted to hot, dry climates, as less water is used in the production of sugar. They include agricultural crops such as maize, sugar cane, sorghum, and millet. As C4 photosynthesis is more efficient than C3, genetic scientists are working to convert C3 plants to use C4 photosynthesis to increase yield, especially where water shortage might be an issue.

CA *See* Countryside Alliance.

CAAV *See* Central Association of Agricultural Valuers.

CAD *See* computer-aided design.

Cadw The *Welsh Government's historic environment service. In addition to managing a collection of properties in state care, Cadw also carries out a range of activities (equivalent to those undertaken by *Historic England) relating to the legal protection of heritage in Wales (for example, managing the process of *listing buildings). Cadw means 'to keep' or 'to protect' in Welsh.

calcareous Mostly or partly composed of calcium carbonate. When applied to *soils, it implies a soil with a high calcium carbonate content over chalk or limestone. Such soils are usually light and well drained with good soil structure, though they may be low in organic matter. They may have a high stone content such as flints or limestone, and may have deficiencies of some trace elements. They are suitable for most arable crops provided that sufficient moisture and nutrients are available.

calcium ammonium nitrate Also known as CAN, this is a granular or prilled fertilizer that contains 27% nitrogen and 8% calcium. A combination of calcium nitrate and *ammonium nitrate which crystallizes into a hydrated double salt.

Universally applied to crops as a source of nitrate, the high lime content considerably improves the fertility of the soil and is particularly useful on acidic soils.

Calderbank offer (Calderbank letter) Letter written by one party offering terms for the settlement of a dispute being considered, at that point, by the law *courts or in an arbitration (*see* ALTERNATIVE DISPUTE RESOLUTION). For example, in a dispute about a *rent review, one party may issue a Calderbank letter to offer to settle the dispute at a certain level of rent without it continuing to proceed through the legal system. In the event of the other party rejecting the offer set out in the Calderbank letter, if that rejecting party then subsequently loses the case or arbitration, the court, or arbitrator, is likely to take the Calderbank letter into consideration when deciding who should bear the costs of the case. By rejecting the offer, the rejecting party may be considered to have unreasonably prolonged the case (thereby incurring additional costs on both sides), so the judge or arbitrator may require them to pay the costs incurred by both sides after the point that the Calderbank offer was made.

calf (*pl.* **calves**) The young of bovine animals, usually to a year old, may be *heifer (female) or *bull (male). They may be reared to adults, either for future breeding stock or for *beef, or slaughtered for meat, known as *veal, when young. The birth of a calf is called calving. As well as other bovine species, such as bison or water buffalo, the term 'calf' is used for the offspring of some other animals, such as camels and red deer.

calf scours Broad, descriptive term for diarrhoea in *calves. There may be a wide range of causes, some non-infectious caused by bad husbandry, and some infectious caused by bacteria, viruses, or protozoa. The diarrhoea causes a loss of body fluids and electrolytes, so treatment includes an electrolyte solution as well as medication for the cause. Examples of organisms causing scours are *E. coli*, Salmonella and Clostridium (bacteria), Rotovirus (virus), and Coccidia (protozoan parasite). *See also* COCCIDIOSIS.

calibration In general terms, calibration means checking the measurement of an instrument against a standard and adjusting the graduations accordingly. In a farming context it might mean ensuring the accuracy of application rates of fertilizer or pesticide through a spreader or sprayer, checking the scale of a weighing machine or the dose rate of a medicine applicator.

calving interval The gestation period of a cow is around 280 days. The calving interval is the length of time between the birth of one calf and the next from the same cow. The shorter the period, the more efficient the herd, but the cow needs time to recover from the pregnancy. The optimum calving period is around 365–375 days.

Cambridge roll A roller is used to flatten ground, to break up clods, or to consolidate a seedbed to ensure good contact between seed and the soil. The roller may be flat with broad or even a single segment, or a Cambridge roll, made of ridged rings around 5–10 cm in width and 50 cm in diameter. The advantage of a Cambridge roll is that the rings can rotate in isolation from each other, which improves cornering ability.

camelid The family *Camelidae* are even-toed ungulate mammals of which there are six living species: the dromedary, Bactrian camel, alpaca, vicuna, llama, and guanaco. The last four are native of South America, particularly the Andes mountain range, and are valued for the fibre of their fleeces. They are herbivores and ruminate but are not true ruminants, with three stomachs rather than four. They have become popular in recent years in the UK as alternative grazing animals with a valuable fleece.

Campaign to Protect Rural England (CPRE) A charitable organization that champions the protection of the England's *countryside through influencing policy and raising awareness. It has more than 60,000 members and 2,000 affiliate bodies, including parish councils and amenity societies operating at national, regional, and county level. It was founded in 1926, and in 2003 changed its name from Council for the Protection of Rural England. Equivalent charitable organizations operate in Scotland and Wales: the **Association for the Protection of Rural Scotland** and the **Campaign for the Protection of Rural Wales**.

((⊕)) SEE WEB LINKS
• CPRE website: includes information on its operations and campaigns.

Campylobacter A genus of bacterium, the most common cause of food poisoning in the UK. Infection is generally mild with gastroenteritis symptoms of diarrhoea and lethargy, but can be more serious, and even fatal in some cases of very young children or the elderly. The bacteria may be found on raw or undercooked meat, especially poultry, unpasteurized milk, and untreated water. Good hygiene in the handling and cooking of meat is the best protection.

Canal and River Trust A charitable organization that champions the protection of over 2,000 miles of rivers and canals in England and Wales. This includes many associated structures, including towpaths, locks, aqueducts, and bridges. Formerly known as **British Waterways**, it changed to the new name in 2012 but continues as a legal entity in Scotland, where an equivalent charity called **Scottish Canals** cares for canals.

((⊕)) SEE WEB LINKS
• Website of the Canal and River Trust, providing a detailed range of recreational activities associated with waterways and volunteering opportunities.

Candlemas *See* QUARTER DAYS.

canker A disease of trees and shrubs which is itself the reaction to a fungal infection of an open injury. The plant's defensive measure is the formation of dead tissue or 'canker' that can cause excessive enlargements.

canola An alternative to *oilseed rape commonly used in parts of the world, notably North America. It describes both the brassica plant and the oil derived from it.

canopy The uppermost 'layer' of a woodland comprising the branches, foliage, and **tree crowns** of the trees themselves. When the tree crowns are in contact with each other, this is described as a **closed canopy**.

CAP *See* COMMON AGRICULTURAL POLICY.

capercaillie (*Tetrao urogallus*) Formerly the largest of *game birds listed as *quarry species in Britain. Found across Northern Europe, but confined in Britain to parts of Scotland. It is not generally present in sufficient numbers for organized shooting, and its population has declined to levels where it is threatened with extinction. It is now a fully protected species.

capital 1. The total amount of *assets of an individual (or other entity) minus their *liabilities. **2.** The amount that the owner(s) of a business have invested in that business (i.e. their assets) minus their liabilities. Essentially, this is the amount they would have the right to withdraw from the business if that business ceased to trade. **3.** In economics, the term 'capital' can be used to describe something which will affect, or provide an opportunity for, economic production. This can include financial capital (money), physical capital (plant and machinery), and also other assets such as *human capital and intellectual capital. *See also* NATURAL CAPITAL.

capital allowance An amount that can be deducted from the profits of a business in the *United Kingdom (to allow for the cost of purchasing *assets such as plant and machinery for use in the business) before calculating the amount of either *Corporation Tax or *Income Tax (depending on the business structure) that is paid. The amount of allowance depends on the type of asset purchased. **Enhanced capital allowances** allow businesses to invest in equipment such as low-energy technologies, water-saving equipment, and some low-carbon vehicles and receive a 100% capital allowance in the year of purchase.

Capital Gains Tax (CGT) Tax payable in the *United Kingdom at the point when a capital *asset is disposed of (usually sold or given to someone else) and based on the increase in its value since it was originally bought or acquired by the party disposing of it. The amount payable is a percentage of the increase in value of that asset since it was acquired (i.e. the 'capital gain'). Normally, the gain is the difference between the value of that asset when it is disposed of, and the cost of the asset when it was first acquired (or for gifts or inheritances, the *market value when first acquired). For assets owned since before April 1982, the market value at 31 March 1982 is used. There are different rates of CGT for residential property and other types of asset (but see 'private residence relief' below).

There are a number of **reliefs** which apply in relation to CGT and which may reduce the amount of tax payable. These include:

- **Annual exemption**: individuals, and most *trusts, have an annual allowance—a specified amount of gain which is allowed before they have to pay CGT.
- **Private residence relief**: gains on the sale of the only, or main, residence are usually exempt from CGT, so none has to be paid if someone is selling their main residential property.
- **Rollover relief**: this defers the payment of CGT (but does not cancel it) if some types of business *asset are disposed of and the receipts are then reinvested in other business assets. Relevant assets can include land and buildings occupied for the purposes of a trade, and fixed plant and machinery.
- **Entrepreneurs' relief**: this applies if all, or part, of a business is disposed of or ceases trading, and means CGT is paid at a lower rate. It does not cover the disposal of individual assets.
- **Gift hold-over relief**: this again defers CGT and can be used if business assets are disposed of in some way which is not considered to be arms' length (i.e. not disposed of in the general market). This type of situation would include something which is given as a gift to someone else.

In addition, CGT is not payable on some items, such as private cars, nor in most circumstances if gifts are made to a spouse or charity. It also does not apply when assets are transferred to other people after someone has died (however, *see* INHERITANCE TAX).

capital investment appraisal *See* PAYBACK PERIOD.

capitalization Process used to convert the income stream from an investment (for example, the rent received on a property investment) into a *capital value. This process underpins the *investment valuation method. In the context of property, the rental value is converted to a capital value by using a **YP (PV of £1 per annum)** multiplier (for further information, *see* DISCOUNTING).

net income × YP (either in perpetuity or for a number of years) = capital value

This is done because what the capital value essentially represents is the right to receive an income (rent) from a property either for all the foreseeable future (in the case of a *freehold) or for a defined period (in the case of a *leasehold). Using the YP figure allows for the variation in the value of money received at different points in time: £1 received in rent now is worth more than £1 rent received in ten years' time because the £1 received now will have had ten years of *compound interest added to it before the £1 received in ten years' time has earned any interest at all. Hence the 'present value' (i.e. today's value) of the rent received in future years is factored into the consideration through the use of the YP figure.

   YP figures are published in tables (such as *Parry's Valuation and Investment Tables and online calculators/computer programmes based on them). The YP selected will reflect the *yield expected in the case of property (or the *interest rate in the case of a cash investment). Generally, the riskier the investment the higher the yield that will be expected (and therefore the higher the potential income from the investment).

In the case of a freehold, as the rent will be available indefinitely, a YP in perpetuity is used. So, for example: the capital value of the freehold of a shop where the *market rent for the shop is £45,000 per year and a suitable yield is 5% can be determined using the following approach:

$$\text{market rent} \times \text{YP in perpetuity @ 5\%} = \text{capital value}$$
$$£45,000 \times 20.0000 = £900,000$$

The YP in perpetuity figure has been obtained from the relevant table in Parry's. The market rent and yield are determined by looking at comparable properties (*see* COMPARATIVE VALUATION METHOD), with appropriate adjustments to the comparables made to reflect the characteristics of the actual property. (In this example it has been assumed that the comparables used for the rent and yields have accounted for the *landlord's expenditure, but this may be somewhat unrealistic, and adjustments may have to be made in reality.) What has, however, been done here is essentially an **investment valuation**.

It is also possible to capitalize the value of a leasehold interest. However, in order for a leasehold to have a capital value there needs to be a **profit rent** in existence, i.e. the leaseholder has to be paying less than market rent to their landlord, with the difference being known as a profit rent (for further explanation of profit rents, *see* INVESTMENT METHOD). In the case of profit rents, the same principles apply as with freehold investment rents, except that a YP is selected for the number of years left until the profit rent is likely to be extinguished (i.e. the point at which the leaseholder's rent is due to be reviewed by the landlord to market rent), rather than in perpetuity. An appropriate yield will also need to be used (again based on comparables), and will need to reflect that it is the leasehold being valued rather than a freehold.

So, for example, the leasehold of a shop is held for a term of 50 years with the lease providing for only a *peppercorn rent. It is let to a sub-lessee at a market rent of £50,000 per annum with a suitable yield for the leasehold interest being 7%. This is a very simplistic example, because the leaseholder is paying only a peppercorn rent and is therefore effectively paying almost no rent to their landlord, so the £50,000 is treated as being entirely profit rent. The capital value of the leasehold interest can be expressed as:

$$\text{profit rent} \times \text{YP 50 years @ 7\%} = \text{capital value of leasehold interest}$$
$$£50,000 \times 13.8007 = £690,035$$
$$\textit{Say } £690,000$$

If it had been the case in this example that the lease instead provided for a rent review to increase the rent to market rent in five years' time, then a YP for five years would have been used instead, and the yield may also have been adjusted to reflect the differing nature of the risk involved.

capital payments These are grants for specific works that meet eligibility criteria that can be combined with the annual management grants that underpin many *Rural Development Programme schemes, including *agri-environment and *woodland payments. Capital agreements can cover, for example, *hedgerow and *field boundary operations—infrastructure that benefits the farmed environment, woodland plans, and woodland creation. There is also a stand-alone capital grant scheme as part of *Countryside Stewardship Scheme.

capital taxes Taxes which are charged on *capital (as opposed to income). These are typically paid on an on-off basis, having been triggered by a specific event, as opposed to ongoing taxes such as *Income Tax in the United Kingdom. Examples of capital taxes in the UK include *Capital Gains Tax, *Inheritance Tax, and *Stamp Duty Land Tax (and its equivalents in Wales and Scotland).

capital value The value of a property, other *asset, or investment.

capping *See* SOIL CAPPING.

carbohydrate A biological molecule made up of carbon, hydrogen, and oxygen atoms. As a food, carbohydrates are the main source of energy for the body and are made up of sugars, starches, and fibre, all of which are essential in a balanced diet. They are broken down by digestion into glucose, which is then transported around the body. There are many sources of carbohydrates, from simple sugars in fruit or milk to more complex starches found in cereals and potatoes.

carbon accounting Process whereby an organization assesses the amount of carbon that it emits through its activities. Organizations that are participating in *carbon trading schemes will be required, as a consequence of their membership of the scheme, to do some form of carbon accounting. Smaller businesses may voluntarily engage in a carbon accounting process, perhaps as part of an *environmental management system. Carbon accounts may include all *scope 1, 2, and 3 emissions or may only include scopes 1 and 2. The term 'carbon' can be used as a shorthand for a larger basket of *greenhouse gases in carbon accounting contexts, and they may incorporate other greenhouse gases in addition to *carbon dioxide (such as nitrous oxides and methane). *See also* CARBON CALCULATOR; CARBON CYCLE; CARBON FOOTPRINT; CARBON SEQUESTRATION; CLIMATE CHANGE; GREENHOUSE GAS.

carbon calculator Method of generating a **carbon inventory** of how much carbon is being produced by an organization or process, perhaps to feed into a *carbon accounting system. Carbon calculators may include all *scope 1, 2, and 3 emissions, or may include, for example, only scopes 1 and 2. They may cover a whole organization or one specific activity. They can vary from relatively simple spreadsheets to more complex models which automatically incorporate a wide range of standard carbon values for different activities. The term 'carbon' in the context of carbon calculators can be used as a shorthand for a larger basket of *greenhouse gases, and these calculators may incorporate other greenhouse gases in addition to *carbon dioxide (such as nitrous oxides and methane). *See also* CARBON FOOTPRINT.

carbon capture and storage Technology where *carbon dioxide produced as a by-product at either industrial sites or *fossil fuel-based energy generation plants is collected and then injected underground into rocks or old oil and gas fields for permanent storage. The process is designed to lock away carbon dioxide underground, therefore removing it from the atmosphere (which is considered desirable given its classification as a *greenhouse gas). The process is still relatively unproven at a commercial scale, though there are a number of government-backed carbon capture and storage (CCS) demonstration projects in existence in different parts of the world. CCS has been criticized by some environmental groups who see it as an unproven technology and a mechanism for justifying the continued use of fossil fuel-based energy generation, rather than increasing the use of *renewable energy technologies.

carbon cycle The process by which the element carbon is cycled in nature. Carbon is found in all organisms, and hence the carbon cycle is fundamental to life.

The carbon cycle happens in a number of stages:

- Carbon is released into the atmosphere (in the form of *carbon dioxide) from **respiration** by animals and plants, and also the **combustion** (burning) of carbon-containing materials (such as wood, *fossil fuels, and so on).
- Plants utilize atmospheric carbon dioxide to produce carbohydrates through the process of **photosynthesis** (releasing oxygen).
- Plants are eaten by animals, continuing the carbon cycle.

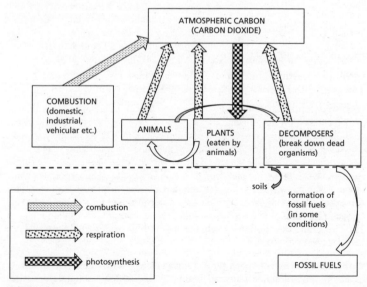

Carbon cycle.

- Animals, plants, and other organisms that die are broken down by **decomposers**, with carbon dioxide being released into the atmosphere. Material from dead organisms tends to accumulate faster than decomposers can break it down completely, contributing to soil formation, and in some conditions, over millions of years, fossil fuels can form.

The oceans and other water bodies also contain significant amounts of dissolved carbon dioxide so also play a role in the cycle, as do organisms living in marine and other water-based environments. *See also* CARBON SEQUESTRATION; NON-RENEWABLE; RENEWABLE.

carbon dioxide Chemical compound of carbon and oxygen (CO_2) which is a central part of the *carbon cycle and is a *greenhouse gas.

carbon footprint The total amount of carbon produced by an individual, organization, object, or process during its lifetime (including, where relevant, manufacture, use, and disposal). The term 'carbon' in the context of carbon footprints is normally used as shorthand for a larger basket of *greenhouse gases in addition to *carbon dioxide (such as nitrous oxides and methane). *See also* CARBON ACCOUNTING; CLIMATE CHANGE; LIFECYCLE COSTING; SCOPE 1, 2 AND 3 EMISSIONS.

carbon monoxide Poisonous, colourless, odourless, tasteless, gas produced when fossil fuel combustion equipment (for example, that burning gas, wood, coal, or oil) is not burning the fuel properly. Faulty boilers or other combustion equipment, as well as blocked or faulty flues, can result in people being exposed to carbon monoxide. Because it cannot be seen, tasted, or smelt, people who are being exposed to it may not be aware that they are, and it can in severe cases cause death. *See also* GAS SAFETY REGULATIONS; HEALTH AND SAFETY EXECUTIVE; HEALTH AND SAFETY EXECUTIVE FOR NORTHERN IRELAND.

carbon offsetting Principle of an individual, business, or organization compensating for their production of *carbon dioxide (or other *greenhouse gases) by undertaking, or paying someone else to undertake, actions which will remove carbon from the atmosphere. For example, an airline may set up a scheme with the objective of offsetting the carbon it produces through its business activities (i.e. flying passengers) by funding the planting of new forests. *See also* CARBON ACCOUNTING; CARBON CYCLE; CARBON SEQUESTRATION.

carbon sequestration The storage of carbon in soils, plant materials (for example, *forestry), the oceans, and rock formations (*see* CARBON CYCLE). *See also* BIOMASS.

carbon trading Process where organizations trade carbon credits (i.e. the rights to emit certain amounts of carbon through their activities). There are various carbon trading systems run by government organizations throughout the world, and in some countries, certain types of larger-scale industrial carbon emitters are required to participate in them. The principle behind them is that organizations are allocated a certain number of **carbon credits** (either initially free, or through purchasing them), each of which allows them to emit a defined amount of carbon. An organization that needs more carbon credits than its initial allocation must then purchase additional carbon credits from other organizations that may no longer need their full allocation (often because they have either started to use, or produce, larger quantities of *renewable energy, or because they have invested in more carbon-efficient manufacturing equipment). The total number of credits within the system is usually capped, and reduces over time (a **cap and trade** system); therefore, overall carbon emissions reduce. The purpose of the schemes is to add a financial cost to businesses for the right to emit carbon, with the aim of incentivizing them to become more carbon-efficient, thereby reducing emissions of *greenhouse gases with the objective of reducing *climate change impacts.

The term 'carbon' in this context is often used as a shorthand for a larger basket of *greenhouse gases, and these schemes may incorporate other greenhouse gases in addition to *carbon dioxide (such as nitrous oxides).

carcass The dead body of an animal, more commonly the body of a slaughtered animal once the offal has been removed. In an abattoir or cutting plant, the carcass of a slaughtered animal is hung up, often for a period of time such as twenty-eight days for beef, to improve the flavour and tenderness as the natural enzymes break down the tissue. The process also allows water in the meat to evaporate, thus concentrating the flavour. After this ageing process, a butcher cuts the carcass into cuts for sale.

care farm A *farm that provides health, social, or educational facilities for a range of less abled or disadvantaged individuals to gain therapeutic benefit from working with animals or plants. They may be people with mental health problems, people suffering from mild to moderate depression, adults and children with learning disabilities, children with autism, those with a drug or alcohol addiction history, disaffected young people, or adults on probation. The farm provides a programme of farming-related activities including animal husbandry, *crop and vegetable production, and woodland management.

(⊕) SEE WEB LINKS
• Social Farms & Gardens website: provides detailed information and identifies individual farms in the UK.

carr A broader habitat description of areas of semi-natural *woody shrubs growing on a waterlogged substrate. Species commonly include *willow and *alder buckthorn. Can typically be a dynamic phase of succession from *swamp or *bog through to eventual climax *woodland. **Carr woodland** or **wet woodland** is an uncommon type of woodland habitat in the UK, and an important habitat for many *wildlife species. It can be dominated by species that might include *birch, *alder, or *willow. *See also* WOODLAND.

carrion The dead and/or decaying flesh of animals. Carrion eaters include the more commonly associated vultures, as well as smaller birds such as crows and magpies. Many carnivores, including the fox and badger, will also eat carrion.

carrot fly A small black-bodied fly of the *psilidae* family, the larvae of which feed on the roots of carrots and related plants, such as parsnips, parsley, celery, and celeriac. Stunting may cause the quality of the crop to be severely reduced or even unmarketable. The adult fly lays eggs in the soil by the crop, so one way to control the pest in a garden is to cover the plants with mesh or fleece to prevent the adult from laying her eggs, or to put out sticky traps. On a field scale, traditionally pyrethroid insecticides have been used, but new chemicals are coming onto the market. A biological approach is also available, as the pest can also be controlled by the use of nematode worms in the soil.

carrying capacity The maximum population of a particular species that can survive indefinitely in its environment in terms of *habitat, food, and water. To that extent, it is a measure of *sustainability. In farming terms it is often used to describe the *stocking rate appropriate for a *pasture or other habitat.

casein The main protein in mammalian milk, making up 80% of the protein in cow's milk; the remainder is the whey. Some people have an intolerance to dairy products, which means that they have an allergic response to lactose, the milk sugar, or the milk proteins, casein or whey.

case law Body of law established through decisions taken in court cases. Law in the United Kingdom is broadly established through two roots: case law and *legislation. Legislation comprises those laws which are subject to scrutiny by elected representatives. These include *Acts of Parliament, or *Acts of Assembly, which are voted on directly by the relevant elected representatives in the *Houses of Parliament, *National Assembly for Wales, *Scottish Parliament, or *Northern Ireland Assembly, and those pieces of delegated legislation, such as statutory instruments, which are issued directly by ministers of the respective governments, under powers given to them by the relevant Act of Parliament or Act of Assembly. It can also incorporate legislation derived from the European Union (EU), although the referendum decision of the United Kingdom on 23 June 2016 to leave the European Union will reduce the influence of European Union law once the United Kingdom leaves.

The second way in which law is established is through decisions of the courts. Whilst legislation generally establishes a set of principles by which the law operates, it is the decisions of the courts which often provide guidance on how legislation is to be applied in practice. The decision of a court in relation to how the law applies in a particular scenario establishes a legal **precedent**, which is a set of principles which can be used by future courts to decide on how the law would apply in the same scenario elsewhere. For some areas of law where legislation could apply in a number of ways on the ground, there is therefore a significant body of case law (i.e. court decisions). An example of this would be the body of case law relating to how the *curtilage of a *listed building is to be interpreted on the ground, and therefore when *listed building consent is required. *See also* COMMON LAW; STATUTE LAW.

casement window *See* APPENDIX 2 (ILLUSTRATED BUILDING TERMS).

cash crop A crop that is grown to be sold for profit. Thus, *cereals grown for sale would be cash crops, but if grown to feed animals on the farm they would not.

cash flow The flow of cash paid into, and out of, an organization. A **statement of cash flow** is a financial statement, showing all the cash that has been received (**receipts**) and paid out (**payments**) by an organization over a period of time. The difference between the receipts and payments for the accounting period (i.e. the net cash flow) is then added, or deducted, from the **opening cash balance** (i.e. the cash available in the business at the beginning of the accounting period) to give the **closing balance** (the cash available at the date of the preparation of the statement of cash flow). This, in effect, shows the cash available to that business at that given point in time, and hence differs from other financial statements, such as the *profit and loss account. For example, in simple terms the statement of cash flow will show the actual amount spent on buying a capital asset in any one year, whereas only a *depreciation figure will be allowed for in the profit and loss account to account for that same purchase. For the purposes of a cash flow, 'cash' is defined as physical cash (coins and notes) as well as instant access deposit accounts (for example, in a bank), plus any overdrafts (bank borrowings) which are repayable on demand. *See also* DISCOUNTED CASH FLOW.

cast Where a *horse becomes trapped on its side against the walls of its *loosebox and is unable to stand upright. They can more often cast at night, and can usually right themselves without assistance. A sheep can also under some circumstances roll over onto its back and not be able to right itself. This is described as a 'cast' sheep.

castration The removal of the testicles of a male animal. This is routinely carried out on farm animals for several reasons: to prevent unwanted pregnancies once the animals reach sexual maturity, to achieve faster growth rates, to reduce aggressive behaviour, and to improve the quality of meat, as that from entirely mature male animals is said to have an unpleasant taste or taint. Animals that are due for slaughter before the age of sexual maturity may not be castrated. One means of castration is to place a rubber ring around the base of scrotum that cuts off the blood supply to the testes and eventually the scrotum falls off.

catch crop A quick-growing crop that is grown at a time between two main crops. It may be a *cash crop—for example, a fast-maturing vegetable in rotation with others—or a green *manure or *cover crop grown between cash crops. For example, it may be sown in autumn to minimize *nutrient *leaching and prevent *soil erosion over winter before another cash crop is planted in spring.

catchment Commonly refers to and describes an area of land that is naturally drained by an individual water body. Thus water may collect and drain from an area of land that might include woodland, hills, and farmland into associated rivers and other watercourses. In an agricultural context, a catchment can be used to focus specific policy measures such as *agri-environment schemes, *Catchment Sensitive Farming schemes, and landscape scale conservation initiatives.

Catchment Sensitive Farming An initiative developed and run by *Natural England in England, and by the equivalent government bodies in the devolved countries within the UK. It focuses on targeted areas that are vulnerable to pollution from agricultural practices, and specifically aims to control diffuse water pollution. This is achieved primarily through training and advice, and appropriate capital items which may be grant aided through *agri-environment schemes such as *Countryside Stewardship Scheme. *See also* WATER FRAMEWORK DIRECTIVE.

cattle Domesticated farm animals of the Bovidae family, ungulate *ruminants. They are kept for milk and meat.

Breeds of dairy and beef cattle in the UK

Dairy	Beef		
Ayrshire	Aberdeen Angus	Galloway/Belted Galloway	North Devon/Ruby Red Devon
Dairy Shorthorn	Beef Shorthorn	Gloucester	Red Poll
Guernsey	Blonde D'Aquitane	Hereford	Simmental
Holstein/Friesian	British White	Highland	Sussex
Jersey	Charolais	Limousin	Wagyu
	Dexter	Lincoln Red	Welsh Black
	English Longhorn	Murray Grey	White Park

SEE WEB LINKS
• UK government website: includes a listing of cattle breeds.

cavity wall insulation Insulation located in the cavity wall of a building (i.e. in the gap between the two layers of block and/or brick which, when a cavity wall-style of construction is used, form the structure of the building). Cavity wall insulation can either be installed whilst the building is being constructed (normally rigid sheets) or, if no insulation was installed in the cavity when the building was being constructed, it can be injected/blown into the cavity subsequently.

CCJ (County Court Judgment) *See* COUNTY COURT.

CDM Regulations *See* CONSTRUCTION (DESIGN AND MANAGEMENT) REGULATIONS.

cedar *See* TREE.

cement Powdery material used in making mortar and concrete, and which causes them to set and harden. There is a variety of different types of cement; however, they typically include a mixture of silicon, aluminium, iron, and calcium compounds. Cement-based mortar is a mixture of cement, sand, and water (sometimes with the addition of lime), and is used for applications such as fixing bricks, blocks, or stones together. (It is also possible to make mortar without cement by using sand and lime, and this is commonly used in renovating traditional buildings.) Other additives can sometimes be used in mortar, such as **plasticizers** which make the mortar more workable. Concrete is also comprised of cement, sand, and water, but also includes small stones (coarse aggregate). Because concrete is mouldable and dries forming a strong, dense material, it is used for applications such as the structural elements of buildings.

Central Association of Agricultural Valuers (CAAV) A specialist professional body representing agricultural and rural valuers and providing advice to government on matters relating to the *countryside, including *Common Agricultural Policy reform, *farm tenancies, and environmental matters. The CAAV was founded in 1910 by representatives of local valuers' organizations, and members (except students) are also required to be members of a local association. Qualified members are known as Fellows and are entitled to use the initials *FAAV. Qualification is normally achieved by passing the Association's own examinations after a period of relevant professional employment. Members typically work as *rural surveyors or valuers in England, Wales, or Scotland, and are often also members of other professional bodies including the *Royal Institution of Chartered Surveyors, the *Institute of Auctioneers and Appraisers in Scotland, and the *Institute of Revenues Rating and Valuation. CAAV is a member of *The European Group of Valuers' Associations and has the right to award Recognized European Valuer status to its members.

(() SEE WEB LINKS
• CAAV website: includes information on the work done by agricultural and rural valuers, and a digest of news and government announcements relating to the countryside.

cereal Edible grain from members of the grass family *gramineae*, such as *wheat or *rice. Rich in carbohydrate, they form the staple diet of most communities across the world and are also used in animal feeds. *See also* ARABLE CROP.

cereal aphids Insects of the *aphidoidea* family, pests of *cereal crops. They suck the sap from the plant, reducing the size and yield of grain when in large numbers on the ears in the weeks before harvest. They may also attack the plant in autumn when the main impact is the spread of viral disease such as *Barley Yellow Dwarf Virus. During feeding they excrete honeydew, which sticks to the crop and provides an ideal environment for *fungi, such as sooty moulds.

cereal beef *Cattle reared for meat fed on a cereal diet. The aim is to prepare the cattle for slaughter more quickly than those on a *grazing regime, usually at twelve to eighteen months of age.

cereal diseases *Cereals are prone to a wide range of *diseases mostly caused by fungi or viruses. They may be controlled by the use of *fungicides or, in the case of viruses, by *insecticides to control the vector.

cereals Members of the grass family *gramineae*, grown for their edible grain. The most common cereals cultivated are *maize, *wheat, *barley, *oats, *rice, rye, sorghum, and millet.

Certificate of Immunity from Listing Certificate, issued in England, Wales, Scotland, or Northern Ireland which confirms that a structure or building named within it will not be *listed for a specified period of time (normally five years).

Certificate of Lawfulness of Existing Use or Development Certificate which can be applied for to confirm whether an existing use of land or buildings, or some form of operational development, or non-compliance with a planning *condition, is lawful in planning terms. The application is normally made to the *local planning authority, who will either issue a certificate confirming lawfulness, or alternatively refuse the application. The provision to apply for such a certificate provides legal certainty in relation to the planning status of an existing use or development, where there may be some element of doubt, particularly in relation to whether *enforcement action can be taken.
See also CERTIFICATE OF LAWFULNESS OF PROPOSED USE OR DEVELOPMENT.

Certificate of Lawfulness of Proposed Use or Development Certificate which can be applied for to confirm whether a proposed use of land or buildings, or some proposed operations to be carried out, will be lawful in planning terms. The application is normally made to the *local planning authority and can be useful if there is some element of doubt as to whether a proposal would either:

• Fall outside the definition of *development (and therefore, not normally require an application for *planning permission), or;
• If it does fall within the definition of *development, clarify whether it is *permitted development (i.e. one of the types of development given blanket planning permission through a permitted development order), or;
• May already have *planning permission following an earlier application.

The local planning authority, having considered the details of the proposal, are able to either grant a certificate or refuse the application, thereby providing a greater degree of legal certainty in relation to the planning status of the proposal.
See also CERTIFICATE OF LAWFULNESS OF EXISTING USE OR DEVELOPMENT.

Certificate of Lawfulness of Proposed Works *See* LISTED BUILDING CONSENT.

cesspit Enclosed tank for collecting *foul water in situations where a property is not connected to the mains drainage system. Sewage (for example, from a bathroom) is drained into an underground chamber or tank where it collects. The tank must regularly be emptied (by a lorry tanker) of both the solid and liquid component of the sewage. Cesspits (also sometimes known as **cesspools**) are only a mechanism for collecting foul water; they do not clean up the water in any way, unlike *septic tank drainage systems or *sewage treatment plants. *See also* REEDBED.

cesspool *See* CESSPIT.

chaff The husks or seed casings of grain separated by winnowing or threshing. Can also include small particles of straw or other waste material from the threshing process.

chain Old unit of measurement, equal to 22 yards (approximately 20 metres), used in land surveying in the United Kingdom. In practical terms, a surveyor would use a linked chain of this length when measuring land (each of the hundred links also being used as a unit of measurement). A piece of land of ten chains in length by one chain in width is one *acre in size. The chain itself is equal to four poles in length (one pole is also known as a rod, or perch). *See also* FURLONG.

chain harrow An agricultural implement that is used to level ground or *scarify grassland. It may be trailed behind a tractor using a drawbar or mounted in a frame on the *three point linkage. The tines are linked loosely as in a chain rather than rigid, allowing the tool to follow the contours of the ground. *See also* HARROW; DISC HARROW.

chalk A soft white porous sedimentary rock composed of calcium carbonate. It underlies the soils of much of southern England, particularly in the downland *landscape. The chalk downs are common in Dorset, Wiltshire, Berkshire, Oxfordshire, Hampshire, Sussex, and Kent. Chalk is permeable but underlain by impermeable substrata so that rainfall soaks into it and forms a reservoir or *aquifer. Where aquifers overflow through springs, the water feeds into chalk stream rivers which are relatively pure and of constant temperature.

chancel repair liability Liability imposed on some landowners in England and Wales to contribute financially towards the costs of repair and maintenance of the chancel (usually the easternmost part) of a local church. This liability applies to land that was historically part of the rectory of a church (i.e. the endowment, often including land, to support the priest of the parish); such land, therefore, does not have to be physically particularly close to the church. There may be multiple properties in a parish that each have chancel repair liability for the same church. Chancel repair liability attaches to the land and, since 2013, has to be recorded at the *Land Registry. It is possible for property owners to obtain insurance to cover a demand for payment made under chancel repair liability. The *Law Commission announced in December 2017 that they would be reviewing the law around chancel repair liability.

change of use A term used in the United Kingdom to describe a situation where a change in the use of a building or land may require *planning permission. Different uses are generally grouped together into **use classes** which are specified in *use classes orders. Changes within the same use class, and which do not involve physical alterations, do not generally require planning permission. In addition, some changes in use between specific use classes may be *permitted development. If a change of use is not covered by either of these two situations, then planning permission is generally required.

charcoal Produced through partial combustion of wood using kilns and by limiting air supplies. Any wood can be **carbonized** by this process, but *broadleaf *trees are preferred, and it provides a market for low-grade timber. Charcoal supplies a range of industrial and domestic uses, including for barbecues.

charge A right affecting land where the land is used as security for the payment of a sum of money (for example, in the case of a *mortgage); in effect, the owner of the charge has the right to the income or proceeds from the sale of the land should the money not be paid as agreed. In England, Wales, and Northern Ireland, charges are recorded against a property's *title at the respective *Land Registry. *See also* LAND CHARGE.

charges register *See* LAND REGISTRY.

charlock A weed with a bright yellow flower, charlock or wild mustard, *Sinapis arvensis*, is a member of *brassica family. At one time considered a troublesome weed of arable crops, especially spring sown, it is easily controlled by herbicides in cereals but is more problematic in *oilseed rape and other brassica crops.

Charolais A breed of *cattle with a white or pale coat, originally from the Burgundy region of Eastern France, kept largely for its meat. *See also* CATTLE.

chartered forester A member of the *Institute of Chartered Foresters (ICF), the UK's organization for professionals working in the *woodland and *forestry sectors. Qualification is by independent assessment, and high standards of practice and professional behaviour are expected and maintained with continuous professional development.

SEE WEB LINKS

• ICF website: includes information on entry requirements, membership benefits, and updated policy and technical materials.

Chartered Institute of Ecology and Environmental Management (CIEEM) A professional membership organization representing and supporting ecologists and environmental managers predominantly in the UK and Ireland. The Institute's objectives include raising the profile of professional ecologists and environmental management, and promoting high standards of practice for the benefit of nature and society. Members are drawn from the public and private sectors, including government, academia, and environmental consultancy. Qualified members can use the nomenclature MCIEEM after their name.

SEE WEB LINKS
• CIEEM website: includes eligibility details and organized events.

chartered surveyor A qualified surveyor who is either a Member or Fellow of the *Royal Institution of Chartered Surveyors. Members are recognized by the letters MRICS after their name, whilst Fellows are entitled to use the letters FRICS. There are several qualification routes, most of which involve degree-level education and a period of employment experience, followed by an RICS assessment. The majority of chartered surveyors have Member status, with Fellowship being given to senior, highly regarded individuals within the profession. Chartered surveyors belong to a professional group within the RICS (which denotes their area of expertise). These groups cover the different types of work done by chartered surveyors, and include arts and antiques, building control (surveyors with expertise in building regulations and construction technology and engineering), building surveying (management and maintenance of buildings), commercial property, *dispute resolution, quantity surveying and construction (management of the construction process), environment, facilities management (management of infrastructure), geomatics (collection and analysis of mapping and spatial data), machinery and business assets (management and valuation of business assets, plant and machinery), management consultancy, minerals and waste management, planning and development, project management (planning and implementing development projects), residential property, *rural surveying (including forestry), and *valuation. In addition to being part of a particular professional group, chartered surveyors may also have additional accreditations in specialist areas of practice; for example, being a *Registered Valuer. They may also hold other professional qualifications; for example, a *rural practice chartered surveyor may also be a Fellow of the *Central Association of Agricultural Valuers.

chattel *See* FIXTURE.

cherry (wild, gean) *See* TREE.

chestnut (horse, sweet) *See* TREE.

Cheviot A hardy breed of *sheep originally from the Cheviot Hills on the Northumberland Scottish border. Often crossed with a Border Leicester or Blue-faced Leicester ram to produce the Scottish Half-bred or Cheviot Mule ewe for lowland prime lamb production. *See also* SHEEP.

chicken (*Gallus gallus domesticus*) The most common form of *poultry in the world kept for egg and meat production. *See also* POULTRY.

chickweed (*Stellaria media*) A member of the stitchwort family, a common *weed of arable land. Annual or over-wintering, it is edible, and is valued for its medicinal properties and used as a cooling herbal remedy.

chipboard Sheet material manufactured using low-value small-diameter *conifer and *broadleaved *roundwood, as well as sawmill offcuts, sawdust, and so on. *See also* PULPWOOD.

chitting The growth of a sprout from a seed, mostly commonly applied to *potatoes. On a garden scale, potatoes may be encouraged to sprout before planting to ensure rapid establishment.

chlormequat An organic compound used as a plant growth regulator, particularly in *cereals. By changing the balance of hormones within the plant, specifically by limiting the production of gibberellin, the stem is shortened and strengthened to prevent the crop from *lodging or falling over under the weight of the ears.

chocolate spot A disease caused by the fungus *Botrytis fabae* or *B. cineria*. Of most importance on field beans, particularly autumn sown, it is a very aggressive disease in humid weather as small brown spots appear on the leaf. If left unchecked, loss of leaves and flowers results in severely reduced yield. There are no resistant varieties of field beans available, but the disease can be treated by the use of fungicide.

CHP *See* COMBINED HEAT AND POWER.

Christmas trees A variety of conifers grown specifically for the Christmas market that include, for example, Norway spruce and Nordman fir. The former is the traditional Christmas tree, whereas the Nordman is becoming increasingly popular and is characterized as 'needlefast'. Both are normally bought in the range of 3–8 feet (90–240 cm). *See also* BRITISH CHRISTMAS TREE GROWERS ASSOCIATION.

churn A container usually made of wood or aluminium holding ten gallons in which milk was collected and transported before the days of bulk storage and transport. Also, the container in which milk or cream is agitated in the production of butter or, as a verb, the act of agitating the milk or cream.

CIEEM *See* CHARTERED INSTITUTE OF ECOLOGY AND ENVIRONMENTAL MANAGEMENT.

CIL *See* COMMUNITY INFRASTRUCTURE LEVY.

circular, planning *See* PLANNING CIRCULAR.

Civil Procedure Rules Rules governing the procedure of some *law courts in England, Wales and Northern Ireland. *See also* EXPERT WITNESS.

CLA *See* COUNTRY LAND AND BUSINESS ASSOCIATION.

cladding Material attached to the outside of the building (sometimes for cosmetic reasons, or to keep water out), but not forming part of the load-bearing structure of the building itself. Cladding made from a variety of different materials can be found, including examples made from tiles, wooden boards (sometimes known as weatherboarding), metal, or plastic-based materials. Whilst cladding (particularly when required for water resistance) is sometimes part of the original design of a building, it can also be applied after the original construction. The impact of cladding on the fire-resistance of buildings is a key consideration in its use.

clamps for silage A bunker usually with a concrete base and concrete panels of railway sleepers for walls in which grass or other crops are placed for *silage. The material is rolled down and covered to exclude air. Animals may be allowed to the silage face to feed for themselves, or the silage may be removed and taken to the animals.

clay The finest *soil particles, less than 0.002 mm, formed from the breakdown of certain minerals such as aluminium **silicates**, **kaolinite**, or **vermiculite**. Because of the size of the particles, they tend to bond together forming a dense material. This limits drainage, so clay soils easily become waterlogged and, when the moisture dries out, form clods that can be difficult to break down. Clay soils are often fertile due to the mineral content, but are difficult to work due to the characteristics mentioned. For these reasons, heavy clay soils may be planted to permanent crops such as grass or trees. For example, clay cap is an area of clay in an otherwise calcareous soil over chalk and, in the past, was planted to trees giving the copses typical of the chalk downland.

clay shooting Shooting at clay pigeons or clay targets using a *shotgun. Clay pigeon shooting has its basis in earlier times when live pigeons were released from traps at competitions. Such competitions were outlawed in 1921, but the English terms used in this sport persist.

clear felling The practice of felling and removing all trees within an area of forest or woodland. It is an alternative to *selective felling approaches where only individual trees, or groups of trees, are removed. The choice of approach is likely to be influenced by the *management objectives for the woodland, which may include *conservation, commercial timber production, amenity uses, or potentially a combination of several objectives. *See also* CONTINUOUS COVER FORESTRY; FELLING LICENCE; SHELTERWOOD; SILVICULTURE.

clear-span building Term commonly used to describe a building where there are no supporting structures extending to the floor (such as pillars) in the interior of the building between the outside walls. *See also* FARM BUILDINGS.

cleavers (*Galium aparine*) An annual herbaceous *weed of the *rubiaceae* family also known by a wide variety of common names such as sticky willy. The surface of the fruits is covered with hooked bristles that catch on clothes and animal fur which spreads the seed. In arable crops the growth habit may drag down the crop, making harvest difficult. It is relatively easy to control by means of a herbicide. It is said to have medicinal properties and is used by herbalists as a lymphatic tonic and diuretic.

climate The prevailing weather conditions in an area over a long period of time. It is distinct from the term **weather**, which describes the atmospheric conditions in a specific place at a particular time. Climate monitoring data most often look at long-term patterns in variables such as temperature, wind, and **precipitation** (rain, snow, sleet, and hail that falls to the ground). *See also* CLIMATE CHANGE.

climate change Term generally used to refer to changes to the earth's *climate which have occurred over a timeframe of many years, usually centuries or longer. The Earth's climate is understood to have changed continually throughout history; however, there is now concern at an international political level that the rate of change has been accelerating in recent decades and that this is the consequence of human activities resulting in increased *greenhouse gas emissions and a resulting change in the *greenhouse effect. These concerns have led to international agreements, such as the *United Nations Framework Convention on Climate Change and the *Kyoto Protocol which bind nations to reduce greenhouse gas emissions.

Climate Change Act 2008 United Kingdom *Act of Parliament which legally binds the *United Kingdom government to reducing *greenhouse gas emissions by at least 80% of 1990 levels by 2050, with the objective of reducing *climate change impacts. This target underpins UK government policies in relation to greenhouse gas emissions and climate change. At the time the Act received *Royal Assent its target was significantly in excess of those which the United Kingdom had already committed to through international agreements, such as the *Kyoto Protocol.

Climate Change Levy Tax added onto electricity, gas, or solid fuels (such as coal) when used by most businesses and public sector organizations in the *United Kingdom (apart from very small energy users) with the purpose of incentivizing them to become more energy efficient, therefore, reducing *greenhouse gas emissions. Up until 2015 renewable electricity was not subject to the CCL and renewable electricity generators were issued with **Levy Exemption Certificates (LECs)**; however, renewable electricity is no longer exempt from the Climate Change Levy and hence LECs are no longer issued.

climax community Can be thought as of a natural community or *ecosystem in equilibrium with the environment. Various stages of ecosystems can develop to a climax community through a series of changes in a process called ecological *succession. Climax communities are normally more stable than the developmental ones, and might include, for example, *habitats such as *Ancient Woodland. The associated habitats of farmland such as *grassland are artificially maintained at a specific stage of succession described as a 'plagioclimax'.

clinometer A simple instrument to measure an 'incline' or angle of slope by sight. It is commonly used in *forestry to help measure the height of a tree, with the operator standing at a measured distance from the base of the tree. *See also* HYPSOMETER.

clipping It is normal practice for a working horse to be clipped during the autumn and winter seasons. Some may also be clipped in the summer months. The coat would otherwise be too heavy and the horse when worked would be likely to sweat, be more prone to chilling, and be more difficult to both wash down and dry. Horses that are clipped would normally need to be rugged to compensate for the loss of hair. Farm livestock, in particular cattle, can also be clipped to varying degrees for welfare reasons or to enhance confirmation for showing. *See also* SHEARING; TWITCH.

close-boarded fence *See* FENCE.

close season (closed season) Any period within the calendar year when the killing of a specified wild animal is prohibited by law. The rules are aimed at protecting individual species for welfare reasons during vulnerable periods (such as reproductive cycles, moulting, or weather extremes), or to promote a sustainable crop. In England and Wales, for example, the Game Act 1831 was the first established that detailed specific dates. *See also* OPEN SEASON; GAME.

clover A genus (*Trifolium*) of annual or perennial herbs. Clovers are legumes which are a group of plants that are characterized by being able to 'fix' (convert) atmospheric nitrogen into nitrogen compounds which are then taken up by the plant. Red clovers and white clovers are of the most agricultural importance, and a balanced mixture of the more productive grasses and clover to aid soil fertility can be an important management goal for farmers, particularly in *organic systems.

clubroot A fungal disease of *brassica species that causes swelling and distortion of the roots and stunted growth. It is caused by a soil-borne organism *Plasmodiophora brassicae*, which produces cysts that can survive in the soil for nine years. There is no fungicide to control the disease, so the only remedy is crop rotation. Some varieties of brassica are less vulnerable to infection.

coarse fishing A description of *fishing for any type of freshwater fish other than *salmon and trout. It is the most popular type of fishing, with around twenty-five species fished within the UK. Carp (*Cyprindae*) are popular with coarse fisherman, and were first introduced into England and Wales in the 1300s and reared by monks as a food source in ponds. There are a number of carp societies often associated with freshwater bodies such as lakes.

Coastal Change Management Area Coastal area identified by a *local planning authority in England as being at risk of physical changes (for example, through coastal erosion) and, therefore, where *development may be restricted. A similar approach to coastal planning is also found in Wales, Scotland, and Northern Ireland, although the specific term is not always used. *See also* PLANNING DESIGNATION.

cob 1. The central, cylindrical, woody part of the *corn or *maize ear to which the grains, or kernels, are attached. Corn on the cob: corn cooked and eaten from the cob. **2.** A mixture of compressed *clay and *straw used, especially formerly, for building walls. **3.** A general type of *horse that is usually characterized as stout and well built with strong bones, ideal for everyday riding activities.

coccidiosis A parasitic *disease of the intestinal tract of mammals and birds caused by a protozoan organism. The primary symptom is diarrhoea, which may become bloody in severe cases. The disease is spread by oocysts through faeces or other contaminated tissue and may be controlled by adding a coccidiostat to feed or treated by a sulfonamide antibiotic. The disease may be chronic with symptoms showing only when the animal is under stress. Cattle, sheep, dogs, and poultry are vulnerable.

cocksfoot A tall perennial grass, *Dactylis glomerata*, with somewhat coarse foliage that forms dense tussocks if unchecked by mowing or grazing. The seed heads are stout, one-sided, and clumped, resembling a cock's foot, hence its name. It grows wild in a range of habitats with deep roots that protect it from drought, but it also thrives in wet conditions. In a cultivated grass seed mixture it is potentially a very palatable and productive grass for both grazing and hay, but is quick-growing and competitive, becoming coarse with age, so needs to be managed carefully. *See also* GRASS (GRASSLAND).

Code for Leasing Business Premises in England and Wales Code setting out a set of voluntary good practice principles and guidance for *landlords and *tenants in relation to the negotiation of business *leases in England and Wales. It was drawn up by a group comprising representatives of landlords, tenants, and the government, and has been endorsed by a range of organizations including the *Royal Institution of Chartered Surveyors (RICS), **The Law Society of England and Wales**, the **Association of British Insurers**, and the **Confederation of British Industry**. It also includes model (example) *heads of terms for business leases. Although the current Code is voluntary, the RICS was, at the time of writing, consulting on a proposed update, including some mandatory (compulsory) requirements.

Code of Good Agricultural Practice A 2009 publication (with the full name, **Protecting our Water, Soil and Air: A Code of Good Agricultural Practice for Farmers, Growers and Land Managers**) produced by the UK government *Department for Environment, Food and Rural Affairs (DEFRA) and designed to provide farmers, growers, and land managers in England with practical guidance on 'good agricultural practice'. In addition to including practical interpretation of legislation, it also suggests actions that can be taken by businesses 'to protect and enhance the quality of water, soil, and air' and notes where this may also offer potential cost savings for individual businesses. It covers the avoidance of soil erosion and diffuse pollution, runoff from farm buildings and hard standing, disposal of waste, plant nutrition, husbandry techniques, management plans, *cross-compliance regulations, and *agri-environment schemes. It also covers good practice in farming coming from other sources such as legislation, tenancy agreements, and assurance schemes.

((())) SEE WEB LINKS

• Website of the Department for Environment, Food and Rural Affairs (DEFRA): details government policies and provides information on related topics, with links to relevant publications.

code of practice Document setting out principles, standards, and good practice in relation to a specific, often procedural, activity. Codes of practice are produced by a variety of different types of organization, including government authorities and professional bodies, and their significance or status varies depending upon the publishing organization. In the context of the *Royal Institution of Chartered Surveyors (RICS), the term 'code of practice' has a specific meaning and is one of the five categories of professional guidance produced by the RICS: namely, *international standards, *professional statements, codes of practice, *guidance notes, and *information papers. Under the bye-laws and regulations of the RICS, its members must comply with RICS international standards and professional statements. Codes of practice and guidance notes provide advice and 'best practice' guidance to practitioners (often in relation to procedures), with some codes of practice being mandatory. Information papers provide information and explanations to RICS members on various matters relating to surveying (although do not recommend or advise on professional procedures). Although the different types of guidance have slightly different statuses, all practitioners are expected to familiarize themselves with new or updated professional guidance, and the RICS reiterates that if it is alleged that a surveyor has been professionally negligent, it is likely that the courts will take into account the professional guidance produced by the RICS in deciding whether or not the surveyor has acted with reasonable competence.

colic A general term that is commonly applied to abdominal pain suffered by *horses. There are five different types of colic, caused by a variety of factors including changes in diet, excess gas in the intestines, or high worm burdens. Its severity can also range widely from mild up to fatal cases.

collecting yard The area adjacent to the milking parlour where *dairy cows are collected prior to milking.

colostrum Milk produced by the mammary glands immediately before and after parturition which is rich in nutrients and antibodies. It is important that newly born animals receive colostrum and thus a good start in life.

coloured horse/pony The commonest forms of coloured *horse or pony are those described as piebald or skewbald. The coat of a skewbald is covered in large white patches and any other colour except black. The coat of a piebald consists of large white and black patches. In both types, the colour of the mane will correspond to the colour on the neck.

combinable crops Arable crops that are cut using a *combine harvester, such as *cereals (*wheat, *barley, *oats) and oilseeds (*oilseed rape).

combination boiler A type of boiler consisting of a water heater and central heating boiler in combination, where, rather than water being heated and held in a hot water tank (as with a **conventional boiler**), it is heated on demand, so that when hot water is required the boiler heats cold water directly from the cold water system as it passes through the boiler. They are sometimes referred to as **combi boilers**. *See also* CONDENSING BOILER; WET HEATING SYSTEM.

combined heat and power (CHP) A type of energy generation system in which both heat and electricity are produced simultaneously. CHP is generally considered to be a more efficient way to use fuel than producing electricity alone, as it utilizes the heat which is potentially wasted in other electricity generation systems. CHP plants can utilize either *renewable fuels or *fossil fuels.

combine harvester Usually known simply as a combine, an agricultural implement, usually self-propelled, that cuts, threshes, and cleans a *combinable crop in one operation. *See also* FARM MACHINERY.

Commercial Rent Arrears Recovery (CRAR) *See* BREACH OF COVENANT.

commercial tenancies *See* BUSINESS TENANCIES.

commission *See* AUCTION.

commission bid *See* AUCTION.

commodities Raw materials that can be traded in **commodity markets**; for example, *grain, tea, cocoa, *wool, and cotton (sometimes known as **soft commodities**), and base metals and other solid materials (sometimes known as **hard commodities**). Many of these commodities are traded internationally. Commodity markets, known as **commodity exchanges**, often trade **actuals** (i.e. the actual goods), **futures**, and **options**.

Actuals can be traded as **spot goods** (i.e. available for immediate delivery and traded at a **spot price**) or on **forward contracts** (where there is an agreement to buy or sell a fixed quantity of something for delivery at a defined date in the future for a fixed price). Forward contracts allow both the buyers and sellers a degree of certainty about the price they will pay, or receive, in the future. However, both parties are also accepting that the spot price may go either up or down significantly at a future date in a direction which is not in their favour (and hence that the buyer may have been able to buy at a cheaper price had they not committed to the forward contract, or the seller may have been able to sell at a higher price).

Futures markets operate through a mechanism of **futures contracts**. Futures contracts effectively operate on a very similar basis to forward contracts (with an agreement to buy something at a fixed price for delivery at a date in the future); however, unlike forward contracts, future contracts can, in theory, be effectively cancelled out through the mechanism of a bought contract being cancelled out by an equivalent sale contract, so often no physical goods change hands in the futures market. A futures contract may be traded many times before the agreed delivery date. The price of futures fluctuate relative to the anticipated availability of that commodity in the future. So, for example, if it is anticipated that there may be a shortage of grain in the future (perhaps because there are concerns that poor weather forecasts may mean that harvests are predicted to be poor) then the futures price will increase, with the opposite happening if it is anticipated that grain *yields will be higher in the coming months.

An option is an agreement to buy a futures contract at a price (known as an **exercise price** or **strike price**) at a future date. However, the potential purchaser is not compelled to buy the futures contract specified in the option so will usually only do so if it is likely to make them a profit. If they decide not to buy, then they will only lose the **option money or premium** (i.e. the money they paid to set up the option initially). In this situation, whilst the seller will not have the benefit of the sale, they will still retain the premium.

Through buying and selling futures and options, alongside physical products, via **brokers**, commodity producers and buyers are able to **hedge** (protect) themselves, to some degree, against future fluctuations in the market. **Speculators** also operate in the futures and options markets, but they are purely interested in making money from buying and selling futures and options at the right time, rather than in the purchase or sale of the commodities themselves.

Common Agricultural Policy (CAP) Objectives for this policy of the then European Economic Community, predecessor to the *European Union, were to increase agricultural productivity, to improve earnings in agriculture, to stabilize markets, to guarantee regular food supplies, and to ensure reasonable food prices. Funding to implement these policies is provided centrally out of EU funds, and the measures involved direct payments to farmers and payments for a range of rural development schemes. These are respectively referred to as **Pillar 1** and **Pillar 2** payments. Measures also include market support in the form of intervention and export subsides, import protection including tariffs and quotas, and regulation.

Rural development which included a significant proportion of funding directed at *agri-environment schemes was not originally part of the CAP. It was developed as part of the MacSharry reform, of the CAP in 1992. The reforms of Agenda 2000 in that year recognized rural development as the second pillar of the CAP. The Fischler reforms followed with further development of the CAP that included significant *decoupling of payments. The evolution of the CAP has been required in order that it remains compliant with the *World Trade Organization, and it remains the largest component of the EU budget.

Payments to individual farmers were administered through the *Integrated Administration and Control System, and subsequently through the **Single Payment**, and, at the time of writing, are made through the **Basic Payment Scheme**.

Common Bird Census *See* Breeding Bird Survey.

common (common land) Land that is owned privately, or very often by a statutory organization such as a local council, or by a charitable institution such as the *National Trust, but over which certain people or *commoners have specific rights. Most often these rights are linked with the ownership or tenure of particular properties. Such rights are described as **common rights**, which might include, for example, the right to graze *livestock, to collect *firewood, or to fish on the common land. Key related legislation includes the definition and registration of common land in the UK with the **Commons Registration Act 1965**, and the subsequent *Countryside and Rights of Way Act 2000 allowed for formal public rights of access.

(⊕) SEE WEB LINKS

• Government website detailing rights and responsibilities pertaining to common land, with links to registers of common land and village greens.

commoner Can be used simply to describe a person who does not belong to the nobility, but more frequently used to describe a person who has a right in or over *common land held jointly with another or others. Commoners would normally occupy land or property connected to the common land to which certain specified privileges or rights are attached. These may include rights of *pasture, the right to graze cattle, sheep, ponies, and donkeys, *pannage, the right to turn out pigs to feed, and *estover, the right to collect fuel wood. For example, the occupier of a particular cottage may be permitted to graze ten cattle, six sheep, and two ponies or donkeys, but the numbers permitted for their neighbours could be different.

commonhold Legal mechanism in England and Wales which allows the owners of individual **units** in a group property (say a block of flats, but could also be individual units in a business park) to jointly own the common parts of that property instead of the more traditional model of having a single *freeholder responsible for the common parts and multiple long *leaseholders each owning the leasehold of their individual unit. The purpose of commonhold (which was introduced by the **Commonhold and Leasehold Reform Act 2002**) is to avoid a situation of having a single freeholder whose interests may not be the same as those of the leaseholders (although there are also other mechanisms which can achieve this end; *see* LEASEHOLD REFORM). Commonhold can only be used for property where the *freehold has been registered at the *Land Registry. Under a situation of commonhold, each unit-holder has two commonhold interests—the interest in their own unit and a collective interest—together with the other unit-holders, in a **commonhold association** which owns the common parts of the property. The situation is similar in practical terms to a series of freehold estates which are linked to each other, because unlike leases, commonholds have no limit on their duration and they are not, therefore, *wasting assets.

The commonhold association will be a **company limited by guarantee** (*see* LIMITED COMPANY) and will be responsible for maintaining the common parts of the building and overseeing the individual units. It will have to have two documents: a **memorandum of association** and a **community statement** setting out the principles by which it operates, what it is responsible for, who the members are, and what their responsibilities are. The community statement can therefore be regarded as similar to what would have been a lease under the traditional system, although there is only one community statement for the whole property, rather than a series of individual leases. When a unit is bought, the new owner will also buy into the association and become a member (the members controlling the association through a general meeting). The directors of the association can either be members or other independent people (potentially professionals), and the association must produce an annual statement of the income required from the unit-holders for management of the common parts of the property.

If the occupiers of a property with group occupation wish to change to a situation of commonhold then they must persuade the freeholder to apply to register the property as commonhold (although they may, themselves, already collectively own the freehold). Doing this effectively extinguishes the existing leaseholds as well as the current freehold, establishing a freehold estate in commonhold land. Consequently, all parties involved need to agree to this, and to date there has been very little take-up of commonhold. In December 2017 the *Law Commission announced its intention to review commonhold and long-leasehold law.

common law A body of law established through custom and the decisions of the *courts (*case law). It is contrasted from *statute law, which is law established through the formal approval of a **legislature** (i.e. an assembly or parliament). In the United Kingdom, the law comprises both statute and case law. Today's English common law (for Scottish law, see below) can arguably be regarded as a combination of common law and **equity**, though they have differing origins. Common law stems from a system originally put in place by the Normans to unify the laws of England (and remove local variations), and which was administered by the common law courts. Alongside this evolved rules of equity which were, until the late nineteenth century, administered by a separate court, the **Court of Chancery**, which was notorious for the complexity and length of the cases heard there. Equity originally evolved from petitions which were made directly to the king (often from those who considered that the common law courts had not provided them with a suitable remedy for their case). As the number of these petitions rose, they were then subsequently made to the incumbent lord chancellor, and were decided, not on the basis of common law, but on the basis of what was just and equitable (i.e. 'equity'). A more rigid set of principles and remedies (such as *injunctions) which were not available in the common law courts then developed as the Court of Chancery was recognized as a court. The Court of Chancery also recognized rights such as *mortgages and *trusts, which the common law courts did not. By the late nineteenth century it was, however, thought that it was ineffective to continue to operate a system where the same issue could be heard in two courts: the common law courts and the Court of Chancery. *Legislation was therefore introduced in the 1870s in England and Wales, and Ireland, to combine both systems with administration by a single court. However, there are still some areas of law where the law differentiates between the two historic systems—for example, in land law, where both **equitable** and **legal rights** can exist—and the **Chancery Division of the High Court** still hears cases concerning equity.

Scottish law has developed from a combination of Roman law (which forms the basis of much law in mainland Europe today) and English law, and arguably encompasses principles of equity; however, equity has never been a separate body of law in Scotland, nor had its own separate courts. *See also* LEGISLATION.

community council Elected body of councillors in England, Scotland, or Wales, being part of the first level of local government and serving a small geographic area. *See also* LOCAL AUTHORITY; PARISH COUNCIL; TOWN COUNCIL.

Community Forest A series of designated areas launched by the *Countryside Commission and the *Forestry Commission in 1990 to promote areas of woodland and recreational access around twelve major urban centres. The aims included the promotion of *multifunctional woodland within a variety of landscapes and land-ownerships. Each Community Forest is a partnership between local authorities and the Forestry Commission and *Natural England. They work with and support community involvement to improve and enjoy the local environment.

(⊕) SEE WEB LINKS
• The website includes signposting to individual Community Forests and links to further and localized information.

Community Infrastructure Levy (CIL) System in place in England and Wales, where a charge is made by the *local planning authority (on a per square metre basis) in relation to certain types of development in order to contribute towards infrastructure within the area where the development is taking place. The CIL can fund a range of types of infrastructure, including health and social care facilities, schools, cultural and recreational facilities, and transport infrastructure. It cannot, however, be used to fund *affordable housing, which is provided for instead through *section 106 planning obligation agreements. It is more generally distinct from planning *conditions and section 106 planning obligations in that it is designed to fund infrastructure within the locality, whereas planning conditions and obligations relate specifically to making the development of a particular site acceptable in planning terms. Local planning authorities who wish to impose CIL charges on development must produce a **CIL Charging Schedule**, which is subject to consultation and public examination by an independent examiner, and which sets out the charges which will be imposed for each type of development included in the schedule, so that these are publicly available to developers. The *Scottish Government is also considering introducing a form of infrastructure levy in Scotland.

community rights Set of rights, applying in England, which were introduced by the 2010–15 Conservative/Liberal Democrat government under the *localism agenda, through the **Localism Act 2011**, and which are designed to allow local communities to have greater influence over *planning, and the provision of services in their locality. The introduction of the *neighbourhood planning system (incorporating neighbourhood plans, *community right to build orders and *neighbourhood development orders) forms a significant element, and allows authorized community organizations in England (typically *town councils and *parish councils in rural areas) to draw up *planning policies and *permitted development rights for their locality, which, subject to approval in a local *referendum, become a formal part of the planning and *development control system. In addition, the *community right to challenge allows community groups to bid to run some publicly delivered services. The *community right to bid allows community groups to request a delay of up to six months in the sale of a community facility in order to prepare a bid for its purchase. *See also* COMMUNITY RIGHT TO BUY (SCOTLAND).

community right to bid A right introduced under the *localism agenda, through the **Localism Act 2011**, which allows community groups in England to request a delay of up to six months in the sale of a community facility in order to prepare a bid for its purchase. In order for this right to apply, the community facility has to be included on the local authority's list of *assets of community value. Buildings or land can be nominated by community groups or *parish councils for inclusion on the list of assets of community value if their principal use is considered to further, or has recently furthered, the social well-being or social interests (including cultural, recreational, and sporting) of the local community, and it is likely that this will continue in future. The list may include assets such as pubs and village halls. Once included on this list, the owner is then required to inform the *local authority if they intend to sell the asset, and, should a community group be interested in preparing a bid to purchase it, the local authority can order that the sale is delayed for a period of up to six months (known as a **moratorium**) to allow the community group time to raise funds. Although this right potentially delays the date of sale, at the end of the moratorium the owner is free to decide who to sell the asset to, so even if the community group's bid is the highest, the owner may choose to sell to someone else. *See also* COMMUNITY RIGHT TO BUY (SCOTLAND).

community right to build order Type of *permitted development order which sits within the *neighbourhood planning regime in England, and grants *planning permission for development specified in the order. They differ from *neighbourhood development orders in that they can be prepared by community organizations.

community right to buy (Scotland) A right, introduced under the Land Reform (Scotland) Act 2003, for bodies which represent communities in Scotland, to register an area of land and/or buildings with which they have a connection, with the effect that, if the landowner wishes to sell them, then the community group has the first right of refusal. Once the Scottish Government has approved the establishment of a **community body**, that body can apply to add an area of land and/or buildings to the **Register of Community Interests in Land**. If the landowner wishes to sell the land/buildings and the community body expresses an interest in buying it, then the *Scottish Government appoints an independent valuer to carry out a *valuation of it. Public support for the purchase must then be demonstrated through a ballot, and the potential purchase must also meet other legal requirements, including that it is in the public interest and that the future use of the land/buildings will be *sustainable development. If the Scottish Government is satisfied that the purchase meets the relevant requirements, and the community body can raise funds, it normally has eight months (or longer by agreement with the landowner) to conclude the purchase.

community right to challenge One of the *community rights introduced in the **Localism Act 2011**. This right allows a set of relevant bodies, such as *parish councils, or voluntary or community organizations, to submit expressions of interest to deliver some services (of a public nature) on behalf of some public authorities in England such as county and district councils. The right relates only to the delivery of services, not the function of *statutory decision making. It is an example of *decentralization and forms one element of the delivery of the 2010-elected coalition, and subsequent Conservative governments' *localism philosophy.

compaction A condition found in soil when it has become compressed. It is usually caused by livestock grazing, *cultivations, or farm machinery travelling on land that is too wet. When caused by cultivations, the compacted area can be described as a cultivation pan or *plough pan. Compaction reduces the ability of the soil to absorb water, and, depending upon the depth at which the soil is compacted, can result in the soil becoming waterlogged or, in water, *ponding at the surface thereby increasing the likelihood of *soil wash or *runoff. In addition, it can also restrict root development and reduce soil aeration, potentially affecting plant growth. *See also* POACHING.

comparable *See* COMPARATIVE VALUATION METHOD.

comparative valuation method One of a number of market-based approaches used in establishing property values. When the method is used to estimate *market values of properties with *vacant possession, it normally involves analysing evidence of the sale prices of **comparables** (similar properties to the subject property) in order to establish a value for the relevant property. In doing this, differences between the comparables and the subject property, as well any relevant considerations, such as market factors and the timing of sales, are taken into account. If a property is not being valued with vacant possession (for example, it is subject to an existing *lease), then the comparative valuation method may not be suitable for estimating market value, as it may not be possible to find enough appropriate comparables (given the added complexity of the potential variations in *lease terms between properties). This is more likely to be the case if the particular type of property in question is not normally sold subject to an existing lease, and, in this type of circumstance, alternative *valuation methods are likely to be used.

The comparative method can also be used in establishing *market rents. Essentially, the same approach is used; however, agreed rents, rather than sale prices, of comparables are analysed, and the terms of occupation are also an important factor in the assessment. *See also* VALUATION.

compartment A permanent division of a *forest or *woodland into a numbered block normally based on hard boundaries such as roads, tracks, *rides or natural features. Broadly analogous to a field within a farm. These can be divided further into sub-compartments which reflect a broadly similar **stand** of trees in relation to characteristics such as age, species, condition, and so on, and is the basic unit of forest management.

compensation A sum of money paid by one party to another, usually in recognition of some kind of loss, injury, or damage.

completion of sale *See* CONVEYANCING.

compost Organic matter or waste such as plant material that has been allowed to break down or rot into humus over a period of weeks or months. The process of decomposition may be encouraged by the addition of water and organisms or by the periodic turning of the decaying material. It can then be applied to the *soil as an *organic *fertilizer or soil conditioner. It is an important feature of *organic farming when *inorganic or artificial fertilizer cannot be used.

compound fertilizer *Fertilizer that contains the three main nutrients required for crop growth: nitrogen, potash, and phosphate. It may be a mixture of the individual ingredients, but is usually turned into granular or prilled form to ensure the individual nutrients do not separate. Some specialist compound fertilizers may contain other nutrients such as magnesium, boron, manganese, or sulphur.

compounding *See* INTEREST RATE.

compound interest *See* INTEREST RATE.

compulsory acquisition *See* COMPULSORY PURCHASE.

compulsory purchase Provision whereby either a local or national government body, or a private-sector organization, can compulsorily acquire land (including, potentially, buildings) under powers granted to them under the law, often with payment of compensation to the owner (and, in some circumstances, to *tenants, *leaseholders, and others affected by a compulsory purchase project but without having had land directly taken). It is distinct from the ordinary sale of assets on the open market because it is compulsory, so the owner does not have the option to refuse. Whilst compulsory purchase powers are often used by government bodies, they are also available to be used, in some circumstances, by privatized organizations, such as water or electricity companies where it is considered to be in the public interest (and hence is provided for within *legislation). Whilst compulsory purchase is often associated with the acquisition of land for infrastructure projects (such as roads or railways), it can also sometimes be used in other circumstances, such as when the owner of a *listed building has not been ensuring it is maintained properly and it is considered to be at significant risk (although compensation provisions and procedures in this type of situation have some differences from the more standard infrastructure-type acquisition which is discussed below).

The body that has the power to compulsorily acquire property is known as the **acquiring authority** and to exercise their rights in England or Wales they must have a confirmed **Compulsory Purchase Order (CPO)** (or a *Development Consent Order). (See the end of this entry for the compulsory purchase process in Scotland and Northern Ireland.) The acquiring authority may have the power to use the compulsory purchase process as a consequence of general compulsory purchase legislation, or it may be necessary for a specific piece of legislation to be passed. The acquiring authority's powers will, however, be subject to approval by an approving authority. The approving authority will be specified in legislation and is normally a Government Minister.

There are a number of stages in the process of developing a confirmed CPO:

- Formulation/information gathering: at this stage the acquiring authority is likely to be carrying out initial investigations to determine the land that they require for the project. This may include carrying out surveys (the acquiring authority may have legal powers to enter land to do this). They may also negotiate with landowners to try and purchase the property by agreement (rather than having to rely on compulsory purchase powers).

- Resolution: at this point the acquiring authority will formally decide to use compulsory purchase powers and will make a resolution which will specify the land they wish to acquire and the reason.

- **Referencing**: at this stage the acquiring authority will gather information on all those with a legal interest in the land and/or who are occupying it (including *freeholders, leaseholders, tenants and other occupiers). To do this they will often serve **requisition for information notices** on known owners and occupiers asking for details of all owners and occupiers.

- Making the compulsory purchase order (CPO): the acquiring authority will, at this stage, make the CPO which will specify the land to be acquired, the purpose, and the legislation on which the acquiring authority is relying. A schedule will also detail all of those people known to be affected by the CPO (i.e. those who are likely to be entitled to claim compensation).

- Notices and objections: at this stage the acquiring authority is required to serve various notices on those who will be affected by the CPO as well as advertising its existence (through mechanisms such as newspaper notices). The purpose of this stage is to allow objections to be submitted to the approving authority (i.e. Government Minister). If objections are received which cannot be resolved through negotiations between objectors and the acquiring authority, and the Minister considers they need to be considered further (the circumstances for this are set out in law), then either a **public inquiry** will be arranged or they will be reviewed through a **written representations** procedure. A public inquiry will involve both the acquiring authority, and objectors presenting, and potentially being cross-examined on, their cases in front of an Inspector. The written representations procedure will address the same matters, but in writing, and in both cases the Inspector will make a recommendation to the Minister.

- Confirmation, rejection, or modification of the CPO: at this stage the Minister will decide whether to confirm the CPO. They may do this in the form originally presented by the acquiring authority, or may decide to amend or reject it. Once a CPO has been confirmed, the decision must then be publicized by the acquiring authority and notice served on those parties known to be affected. At this stage the decision to either confirm, modify, or reject the CPO may still be challenged through the *courts.

Once the acquiring authority has a confirmed CPO, there are a number of different ways in which they acquire the land:

- By agreement with landowner: at this stage although they have a CPO they may still reach voluntary agreement with the landowner (i.e. not rely on their compulsory purchase powers).
- **Notice to Treat** followed by **Notice of Entry**: the acquiring authority will serve a Notice to Treat which will state that the acquiring authority is willing to negotiate (or 'treat') to acquire the land and inviting the party on whom the notice is served to submit a claim for compensation. The next stage is for the acquiring authority to serve a Notice of Entry which will specify the date of entry that the acquiring authority intends to take possession of the land (at which point they can carry out works in relation to the main purpose of the CPO, although legal ownership does not pass until compensation is agreed and the land has been through the *conveyancing process).
- **General Vesting Declaration (GVD)**: the acquiring authority may use a GVD instead of Notice to Treat/Notice of Entry. This process also operates through a series of notices; however, the main difference is that it culminates on a **vesting date** when the acquiring authority both takes possession of the land and ownership of the land is vested to the acquiring authority.
- Short tenancy: the acquiring authority may seek to acquire a short tenancy.
- In response to a **Blight Notice**: blight is a concept which sits within planning law and addresses a situation where a property is difficult to sell (is 'blighted') because some type of planning proposal implies that it may be compulsorily purchased in future. The law provides for some parties (predominantly *owner-occupiers) to serve a Blight Notice on the party that is likely to acquire the land in order to accelerate the acquisition process. In the case of a compulsory purchase situation a Blight Notice would be served on the acquiring authority requiring them to buy the property at a value which is unaffected by the compulsory purchase project, rather than rely on the uncertainty of waiting for the full CPO process to take effect.

In most cases, compensation is paid to the owner, and in some circumstances, tenants, leaseholders, and other occupiers of the property affected by the CPO scheme. This can include groups of people who have had land taken, and may also include those who have not had land taken directly but who may still be affected by the scheme (for example, because their property is affected by noise coming from a new road even if it does not directly pass over their land). The amount of compensation that is paid is governed by legislation as well as *case law (together known as the **Compensation Code**). The framework is complex, but there are several aspects which can, potentially, be compensated for by the acquiring authority (known as **heads of claim**), although the person or business affected has a duty to minimize their losses (for example, by obtaining a number of quotes from any *contractor):

- Value of the property taken: this is generally the market value of the property taken, not taking into account any increase or decrease in its value as a result of the scheme.
- **Disturbance**: this is designed to cover reasonable costs or losses incurred as a consequence of having to leave the property (for example, removal company fees, or in the case of a farm, extra time taken to access land due to road closures), so is mainly available to the occupiers of the property affected.
- **Severance**: this is designed to address a situation (not unusually found in relation to farms) where, as a result of a CPO project, a property has been divided into blocks (say by a road running through it) and this has resulted in a reduction in the value of retained land, perhaps because one piece of land has been left which is not possible to access by vehicle.
- **Injurious affection**: this is compensation to cover the loss in value of any land that is retained, as a result of the project construction works themselves, or by the subsequent use of the new project (perhaps by a new railway passing close to a property).
- **Home loss payment**: in some circumstances, if a house is being compulsorily purchased an additional payment may be made to cover the inconvenience and distress caused by having to leave a home.
- **Basic and occupier's loss payment**: these are payments to reflect the inconvenience and disruption of being subject to a CPO, in situations where a home loss payment does not apply (for example, for businesses). The basic loss payment is paid to a freeholder or leaseholder whilst the occupier's loss payment goes to the occupier. Therefore an owner-occupier may receive both payments.
- Professional fees: this is designed to cover the reasonable costs of professional advice which someone affected by a CPO may require (such as, for example, from a *chartered surveyor).

In some circumstances, a CPO project can actually increase the value of rest of the land which is retained by the landowner (for example, if a new road improves access to a neighbouring *development site which is owned by one of the parties who are entitled to compensation). This is known as **betterment**, and the acquiring authority can seek to offset the increased value due to betterment against the compensation paid to the landowner.

The acquiring authority may undertake, or pay for, voluntary **accommodation works** usually on retained land, such as, for example, putting in a new bridge to access an area of land which has been cut off by a new railway line. The purpose of these works from the acquiring authority's perspective is that they are likely to reduce the amount of compensation payable for severance/injurious affection and/or disturbance, as they reduce the impacts on the landowner's retained land.

The general compulsory purchase process and provisions for compensation in Scotland follow the same principles as in England and Wales. There are, however, no provisions for basic and occupier's loss payments in Scotland, although at the time of writing, Scotland still maintains a **farm loss payment** (which is a payment specifically available to farmers, in some circumstances, to recognize that having to move to different land may result in a short-term fall in profits). England, Wales, and Northern Ireland have replaced their farm loss payment provisions with the basic and occupier's loss payment system which is available to a wider range of businesses.

In Northern Ireland, the compulsory purchase process also operates on similar principles to that in England and Wales, and the Northern Irish Compensation Code facilitates compensation under the same broad heads of claim as in England and Wales. There are, however, some procedural differences in the acquisition process itself. Rather than operating on the basis of a CPO, in Northern Ireland the acquiring authority must have a **Vesting Order**. To do this, the intention of the acquiring authority to make such an order is publicized (and those with an interest in land which is potentially affected are notified). Following this, there is a period during which objections can be made, and a local inquiry may be organized if there are such objections which are not withdrawn. If the local inquiry supports the compulsory purchase, then the vesting order can be made and the order (and its supporting map) are publicized. Following a further period when the order can be challenged, the vesting order becomes operative. At the operative date, ownership of the land passes to the acquiring authority.

Compulsory Purchase Order (CPO) *See* COMPULSORY PURCHASE.

computer-aided design (CAD) Computer technology used by architects, planners, and engineers (as well as others, such as artists) in order to assist in design. CAD technology enables the production of drawings and three-dimensional forms, and may be used, for example, to produce a series of virtual walk-throughs for a new building.

conacre agreement Longstanding form of short contractual agreement operating in Ireland whereby someone grants a farmer access to an area of land to plant, cultivate, and harvest a crop.

concentrate feed Animal feed made up of grains and other nutrients. *Ruminant animals eat *forage such as *grass, *hay, or *silage, but may also be fed concentrate feed to increase the nutritional value of the diet. Used particularly for high performance, for example high-yielding *dairy cows, or to finish animals for meat more quickly. Young animals may be given *creep feed, a concentrate feed designed to help them grow more quickly and from which adults are excluded. Concentrate feed is the primary feed of non-ruminant animals such as *pigs and *poultry.

concrete *See* CEMENT.

condensing boiler A type of boiler where, in order to extract the maximum amount of energy from the fuel, heat is extracted from the exhaust gases that would normally go out through the flue, thereby causing the water in the gases to condense. The term describes the technology, rather than the type of boiler,

which can include *combination boilers, or other types of system. *See also* WET
HEATING SYSTEM.

condition A measure of the environmental quality of *Sites of Special Scientific
Interest, using six reportable condition categories: favourable, unfavourable, no
change, unfavourable declining, part destroyed, and destroyed. These
assessments are used to monitor the quality of individual SSSIs and national
biodiversity programmes over time. This information is normally available to the
general public.

conditional exemption (relief for heritage assets) *See* INHERITANCE TAX.

conditions, planning Conditions attached to the grant of *planning
permission or *listed building consent. The type and nature of conditions can
vary widely depending on the proposed development; however, the underlying
principle is that they should be necessary in order for the development to
become acceptable in planning terms, i.e. that without them, permission or
consent would not be grantable. Typically, conditions relating to the date by
which the development must have commenced are attached to most
permissions. Other examples of conditions which may be used include
requirements for samples of materials to be approved by the *local planning
authority prior to construction, or restrictions on the times of day that potentially
noisy activities can be carried out on the site. Conditions are set out on the
*decision notice which is issued in relation to the application. Whilst conditions
cannot generally positively require the payment of money or other
considerations in relation to the development, they can prohibit development
until a *section 106 planning obligation agreement (or equivalent in Scotland or
Northern Ireland) is entered into (which may require the payment of a financial
contribution towards infrastructure).

confirmation (Scotland) *See* PROBATE.

conformation The shape and structure of the body of an animal and thus a
measure of the effectiveness of the body for its purpose. For example, a meat
animal that has large muscle blocks that make up the primary cuts of the *carcass
may be said to have 'good conformation'. Similarly, a *dairy cow has anatomical
characteristics that may aid the production of milk. *See also* ANATOMY.

conifer An evergreen *tree of the order *coniferales* characterized by modified
thin leaves (needles) that, unlike *deciduous trees, are not dropped in winter (the
larch *Larix* is an exception to this characteristic). They are **gymnosperms**, i.e.
they produce fruit in the form of cones. They are favoured for commercial timber
production as they are faster growing and ideal for supplying paper making and
building construction markets. They are also favoured as a nurse crop. All
conifers grown in the UK, with the exception of the Scots pine, are *non-native
species. They are less favourable to the environment, and generally forestry
policies militate against non-native conifers, particularly in *Ancient Woodland,

*Ancient Semi-Natural Woodland, and protected areas. *See also* TREE; SOFTWOODS; CONNECTIVITY.

connectivity *See* ECOLOGICAL NETWORKS.

Connemara pony *See* HORSE.

conservation The protection and careful management of our cultural and *natural heritage, and that can refer to specific sectors or be applied more generally. A pragmatic division of the more common derivatives are presented as follows: **1.** Conservation of nature describes the protection of *biodiversity and the *natural environment, and in particular protected areas such as *Sites of Special Scientific Interest and *protected species, especially from the impacts of human activity. **2.** Conservation of the built environment is concerned with important structures and buildings that have a particular interest from an architectural, archaeological, or other human interest. The protection of these interests from human and natural impacts such as climate is the focus of legislative and regulatory measures, such as *Ancient Monument, *Conservation Area, and *listed building consent. *See also* ECOLOGY; HERITAGE ASSET.

Conservation Area A*planning designation used in the United Kingdom for areas it is considered desirable to protect as a result of their special architectural or historic interest. The *local planning authority has additional powers to control changes within Conservation Areas. These include requirements to obtain either *planning permission (England) or *Conservation Area Consent (Wales, Scotland, and Northern Ireland) for the demolition of the whole or part of most buildings, and to give notice, in most cases, of proposed works to trees, plus changes to some *permitted development rights and *advertising controls. In addition, proposed developments are required to exhibit high standards of design and, generally, to make a positive contribution to the character of the area. *See also* ARTICLE 4 DIRECTION.

Conservation Area Consent A form of permission required from the *local planning authority in order to demolish most types of building if they are located within a *Conservation Area. Conservation Area Consent is required in Wales, Scotland, and Northern Ireland (and until 2013 was also required in England). Whilst the requirement to obtain permission to demolish buildings in Conservation Areas continues in England, rather than this being a separate consent, it is now merged into the *planning permission system; therefore, in England planning permission is now required to demolish buildings in Conservation Areas (rather than Conservation Area Consent). *Listed buildings within Conservation Areas are protected under the *listed building consent regime.

Conservation Board The establishment of Conservation Boards was enabled by the *Countryside and Rights of Way Act, which allowed for the creation of boards to oversee large *Areas of Outstanding Natural Beauty; in particular, those that ovelapped county boundaries and were not easily managed by a single local authority. Conservation Boards have a management structure independent of local authorities and overseen by a Board comprising members drawn from local statutory representation and appointees from the *Department for Environment, Food and Rural Affairs (DEFRA). The Board has a statutory duty to publish a Management Plan, and to promote its implementation by a wide range of organizations, including local authorities, government agencies, community groups, and landowners.

conservation headland The area of land between the *arable crop edge and the first *tramline, which is usually 6 metres wide to accord with the boom width of the *sprayer. This area of arable crop is treated more sympathetically with selective *pesticides to control the more aggressive weed species and diseases. The aim is to tolerate a level of broadleaved weed density and associated insect populations to promote *biodiversity benefits, with a yield reduction of up to 10%. The origins of conservation headlands were to improve the breeding success of grey *partridge. *See also* FIELD MARGIN.

conservation officer Official of a *local planning authority who has a specialism in, and responsibility for, planning matters relating to *heritage assets (including *listed buildings and *Conservation Areas). Conservation officers will be consulted by *planning officers on applications for *planning permission or *listed building consent involving *heritage assets.

Conservation Volunteers, The (TCV) A volunteering charity that promotes a range of environmental *conservation tasks and projects. It aims to help local communities and to benefit participating volunteers through training and general well-being. It was formerly known as the **British Trust for Conservation Volunteers (BTCV)**.

(((⊕))) SEE WEB LINKS
• Website detailing a range of types of volunteering opportunities and local activities.

consignment note Paper-based, or electronic, record which must accompany all movements of *hazardous waste in the United Kingdom, as defined in the *hazardous waste regulations. A consignment note is required to set out information including what is in the waste, what risks it presents, addresses and times when the waste leaves one site and arrives at another, and who has been involved in moving the waste. Copies of the consignment note must be retained by all parties involved, as set out in the hazardous waste regulations. The system is designed to ensure that the movement of hazardous waste is fully traceable. A separate system, using *waste transfer notes, exists for certain types of non-hazardous waste transfer. *See also* SPECIAL WASTE.

Construction (Design and Management) Regulations Health and safety *regulations, commonly referred to as CDM Regulations, which aim to reduce health and safety risks within construction by placing obligations upon all those involved in any project (including those people for whom the work is being carried out, anyone involved in design within the project, any *contractors, plus workers) to assess those risks and plan for, communicate, and manage them in such a way as to minimize them. The same CDM Regulations apply throughout Great Britain, with Northern Ireland having its own.

constructive dismissal Situation where an employee has resigned from an organization but argues that they have been left with no alternative but to resign because of the actions or omissions of their employer, and therefore argues that their employer has, in effect, dismissed them indirectly and, hence, they (as the former employee) should be entitled to compensation. Disputes over constructive dismissal are usually heard at **employment tribunals** (*see* COURT, LAW).

consultation *See* LOCAL PLAN; PLANNING OFFICER; STATUTORY CONSULTEE.

Consumer Prices Index (CPI) Measure of *inflation in the *United Kingdom, published by the government's **Office for National Statistics**, which is based on the costs of a basket of commonly purchased household goods and services. There are two variants of the CPI: the CPI itself, which does not include housing costs, and the CPIH (*Consumer Prices Index Including Housing Costs), which does include owner occupier's housing costs and *Council Tax. The CPIH is now the main measure of inflation used by the government, having largely replaced the CPI and the older *Retail Prices Index.

Consumer Prices Index Including Housing Costs (CPIH) Measure of *inflation in the *United Kingdom, published by the government's **Office for National Statistics**, which is based on the costs of a basket of commonly purchased household goods and services, including owner occupier housing costs and *Council Tax. There are actually two variants of the Consumer Prices Index: the *Consumer Prices Index (CPI), which does not include housing costs, and the Consumer Prices Index Including Housing Costs, which does. The CPIH is now the main measure of inflation used by the government, having replaced the CPI and the older *Retail Prices Index.

consumer unit Unit located at the point where the mains electricity supply coming into a building (most commonly a residential property) is split into a number of circuits to go around it, and usually comprising a main switch which can be used to manually switch off the electricity supply, and a series of **circuit breakers** which are designed to automatically switch off if there is a problem with the electricity system in the building. Their main function is to ensure that the electricity supply is switched off if the circuit into which they are wired is overloaded. In addition to circuit breakers, **residual current devices** are now being used which offer increased levels of safety protection due to their higher levels of sensitivity to damage to electricity circuitry.

Consumer units have replaced the previous system of **fuse boxes**. These comprised a series of fuses which would break in the event of a circuit being overloaded, thereby switching off the electricity. With a fuse box, once the overloading issue was resolved the broken fuses would then need to be replaced.

contaminated land Land which has substances in, on, or under it which are either causing, or are likely to cause, significant harm to people, property, or *protected species, or which are polluting waters. The legal definition of contaminated land and the way it is addressed through the law, as it applies in England, Wales, and Scotland, are set out in the Environmental Protection Act 1990, as amended, which also goes on to address contamination by radioactive substances.

There is a wide range of substances which can possibly contaminate land, including oils and tars, chemical substances, and *asbestos.

The law sets out the principles regarding who is responsible for cleaning up land contamination, and these generally follow the *polluter pays principle (i.e. the person or persons who caused the contamination, or allowed it to happen, are responsible for cleaning it up). This can be either one party, or sometimes several. It is not always possible, however, to identify who was originally responsible for the contamination (for example, if it occurred a long time before it was identified). In this instance, the liability for cleaning up the contamination can fall to the current landowner or occupier (even if they had had no involvement in the original contamination). Cleaning up contaminated land is a highly specialized task, and the costs of doing so can be extremely high. Therefore, land contamination can potentially present a significant liability for landowners, and therefore forms a key consideration in both the *valuation, purchase, and sale of land.

*Local authorities have a duty to maintain a register of contaminated land, and where it is considered to present a risk, they seek to ensure that it is cleaned up (or **remediated**) in three main ways. Firstly, they may reach a voluntary agreement with the responsible party that the responsible party will clean up the contamination voluntarily. Secondly, if the land is the subject of a planning application (which can often be a trigger for the identification of contamination) then the local authority may require the land to be cleaned up as a *condition of the *planning permission.

The third approach is usually used only if neither of the first two approaches have worked nor are relevant. In this instance, the local authority can serve a **remediation notice** on the responsible party instructing them to clean up the contaminated land. If the party does not clean up the site (it is an offence not to do so) then the local authority (or potentially the *Environment Agency (England), *Natural Resources Wales, or *Scottish Environment Protection Agency) may arrange for the site to be cleaned up, and will then pursue the responsible persons through the *courts to seek to recover the costs of the clean-up.

Whilst Northern Ireland has its own legislative regime for managing contaminated land, in practice it operates on the same principles as those within the rest of the *United Kingdom.

contingent valuation method Method used to place a value on goods and services which do not have a market (and hence have no *market value in the conventional sense). This emerging approach is increasingly used in government policy making in order to try to place a financial value on natural assets. In essence the contingent valuation method operates by placing a set of statements concerning hypothetical situations in front of a representative sample of members of the public (or other target population) and asking them to state how much they would pay for particular environmental services (either to enhance them, or avoid their loss). *See also* NATURAL CAPITAL.

continuous cover forestry Essentially lower-impact management of forests and woodlands that avoids *clear felling and replanting, e.g. by *selective felling, *coppicing, and so on. This system is more environmentally beneficial and is encouraged in the *United Kingdom Forestry Standards, but is required in forests and woodlands managed under the United Kingdom Woodland Assurance Standard.

continuous stocking A *grazing regime where *ruminant animals are kept at low *stocking rate for an extended period of time rather than moved to pastures new. *See also* GRAZING; STRIP GRAZING.

contract In English law, a legally binding agreement between two parties that one party will do something for another party in return for a consideration (usually some type of payment). If one of the two parties fails to comply with what was agreed, then the other party is entitled to challenge them through the civil *courts (commonly known as suing them), in order to ensure that what was originally agreed is delivered, or that *compensation is paid. In Scottish law, contracts do not necessarily require a consideration in order to be enforceable through the courts.

contract farming Farming arrangement where a **farmer** (the *landowner) and a **contractor** sign an arms'-length legally binding *contract with each other; in effect, to undertake the farming of an area. As opposed to *contracting (where the contractor is only undertaking one defined task), in contract farming, the contractor commonly agrees to undertake all of the husbandry tasks required for the farming operation (sometimes known as a ***stubble-to-stubble arrangement**). In return for undertaking these tasks, the contractor is paid a **fixed fee (or basic fee)** by the farmer.

In a contract farming arrangement situation the farmer will usually open a bank account (often referred to as a **No. 2 Account**) which, although in the farmer's name, will be operated in accordance with the terms of the contract. All income from the enterprise is paid into this account, and expenditure relating to the contract farming activities, say feed, seeds, *fertilizers, and sprays, is paid out from this account. Assuming that there are sufficient funds available as a result of operating the enterprise, the farmer will then take a **farmer's basic return**. This is likely to be at a similar level to a *farm business tenancy rent, but carries more risk than a rent, as if the contract farming enterprise does not generate sufficient funds then the farmer will not be able to take their basic return. (The contractor's fixed fee, which is designed to cover their costs, is paid before the farmer's basic return, so, if after the contractor's fee is paid there are not then sufficient funds left, the farmer will not receive their basic return.)

At the end of the farming year, the **divisible surplus** (if any) is calculated. This is done using the following formula:

Income sources from the contract farming enterprise
- Direct costs of enterprise (including contractor's fixed fee and farmer's basic return)
= Divisible surplus (divided between the farmer and the contractor)

Once calculated, the divisible surplus (if any) is then divided between the farmer and the contractor, as per the terms of the contract farming agreement, with the contractor usually taking the majority of the surplus. (Generally, the higher the farmer's basic return, the greater a percentage of the divisible surplus that will go to the contractor, with the opposite applying if the contractor has a higher fixed fee.) *See also* SHARE FARMING.

contracting The carrying out of defined task by one independent party (the **contractor**) for another party, in return for payment. Examples on a farm may include using a contractor to carry out hedge cutting or tree felling. The party for whom the contractor is carrying out the work is the **client**. *Health and safety *legislation sets out a number of responsibilities for both contractors and their clients when undertaking work, and where there may be more than one contractor working on the same project (for example, a construction project), sets out specific responsibilities for all of the parties involved. *See also* DIVERSIFICATION.

contractor *See* CONTRACTING.

contractor's method Valuation method (also known as the **contractor's test**) which is used, in the *United Kingdom, for determining rateable values for *business rates purposes (and occasionally for *compulsory purchase) of non-residential properties for which there is no obvious market but where it is assumed they will continue to be used for the current use (for example, oil refineries or universities). It uses an approach where the value of the site is added to the construction costs of an equivalent building (with some allowances) to give a *capital value of the land and buildings which is then decapitalized (*see* CAPITALIZATION) to give an annual rental value (the rateable value; *see* BUSINESS RATES). The contractor's method operates on essentially the same principles as the **depreciated replacement cost method (DRC)**, although the DRC is usually used to produce a capital value used for accounting purposes, so does not include the last decapitalization step, and has some variations in the method.

In essence, both the contractor's and DRC methods operate on the following basis:

value of site
+ cost of buildings allowing for depreciation and obsolescence
= value of land and buildings (which, if using the contractor's method, is then decapitalized, and adjusted for any specific features of the property not already accounted for).

Because this kind of property is likely to have been subject to some wear and tear and may not be as adaptable as a new building might be, an allowance for *obsolescence and **depreciation** is made (though in this context depreciation refers only to an allowance for obsolescence rather than the more specific meaning in an accounting context; *see* DEPRECIATION for more information).

The major differences between the contractor's and the DRC methods are the purpose for which they are used and also the costs used. The contractor's method when used for ratings purposes uses an **estimated replacement cost** for buildings (plus any other rateable buildings and structures) which is based on costs issued by the *Valuation Office Agency (VOA) (England and Wales), the *Scottish Assessors Association, and Northern Ireland *Land and Property Services, as at the last date when rateable values were assessed. The VOA (and its equivalents) also provide guidelines about determining adjustments for obsolescence, and the decapitalization rate is also issued by each of the respective governments. When DRC is used, this is done with reference to market evidence of site values and costs, usually at today's date. This contrasts with the historic government data used for ratings valuations using the contractor's method.

Both the contractor's and DRC methods are subject to potential criticism because the values produced by them are not well tested in the market and they are also based on an assumption, which is not usually accepted, that cost and value are the same thing (the value usually being the amount that the market will pay for something, which is not the same as the cost of it; for example, if a sweetshop is set up, the 'cost' of the sweets to manufacture are not the same as the 'value' of them when they are sold to the customers). That said, in this instance, if one of these types of property is sold, a replacement usually needs to be built so that the relationship between cost and value in this instance is closer than for most types of property. Therefore, for property where there is no market (because they are rarely let or sold) both are accepted valuation methods.

contract out To exclude, by written agreement between the parties, some form of default *statutory provision from a *contract (more particularly often a *lease or tenancy agreement); it is commonly used in reference to contracting out of the *security of tenure provisions under the *Landlord and Tenant Act 1954, but can be applied more widely. The opportunity to contract out of default statutory provisions under *landlord and *tenant law in the *United Kingdom is relatively restricted, and many of the statutory provisions set out in the various pieces of *legislation cannot be contracted out of (even by agreement between the parties).

Contracts for Difference (CfD) *United Kingdom low carbon energy support system designed to incentivize large-scale generators to produce low-carbon electricity (including from *renewable energy sources). The Contracts for Difference (CfD) scheme was introduced in 2014 and has replaced the *Renewables Obligation as the main mechanism for supporting larger-scale renewable electricity generation. Under the scheme, contracts are offered to generators using specified low-carbon technologies. These contracts ensure that during the term of the contract the generator receives a stable price for their electricity (known as the **strike price**) by guaranteeing that if the average price of electricity on the markets (the **reference price**) falls to underneath the strike price a top-up payment is made. If the strike price increases to higher than the reference price then the top-up payment is paid back.

The scheme is designed to provide generators with a predictable return on the money they have invested in new-generation equipment by guaranteeing a stable price for the electricity they are generating. It also allows the government to support selected technologies by only offering contracts to generators using the technologies in which the government wishes to see an expansion in their use. The CfD scheme operates alongside the *Feed-in Tariff scheme, which supports smaller-scale renewable electricity generators in England, Wales, and Scotland (though this is expected to close in spring 2019).

contractual licence *See* LICENCE.

contractual obligation *See* OBLIGATION.

contributory negligence *See* NEGLIGENCE; DAMAGES.

controlled traffic farming *See* TRAMLINE.

controlled waste Term used in the *United Kingdom to refer to forms of *waste which are subject to *legislation regarding how they can be dealt with.

Control of Pollution (Silage, Slurry and Agricultural Fuel Oil) Regulations *See* SSAFO REGULATIONS.

Control of Substances Hazardous to Health Regulations *Health and safety *regulations, commonly referred to as **COSHH Regulations**, which require employers to control substances that are hazardous to health. The requirement to undertake appropriate *risk assessments to assess and control related risks is central to the Regulations.

(⊕) SEE WEB LINKS

- Website of the Health and Safety Executive: provides information and guidance on health and safety regulations in Great Britain.
- Website of the Health and Safety Executive for Northern Ireland: provides information and guidance on health and safety regulations in Northern Ireland.

convention, United Nations A formal, and usually significant, international legal agreement or treaty made between member states at the *United Nations; for example, the *United Nations Framework Convention on Climate Change.

conventional agriculture Farming using traditional husbandry as opposed to *organic agriculture. The term is slightly confusing as conventional agriculture involves the use of artificial fertilizers and pesticides which have only been available in relatively recent times, whereas prior to this time, all agriculture was organic.

conventional boiler *See* COMBINATION BOILER.

conversion rate/ratio In *agriculture the term conversion rate or ratio is usually applied to the ratio of the amount of food fed to an animal relative to output such as weight gain or milk. It is commonly applied to the production of pigs or poultry where the conversion ration might be 3:1, which means that 3 kg of feed is required for 1 kg of weight gain. For ruminants the rate is higher, perhaps 6:1, which is why some vegetarians argue that this is a very inefficient way to produce food. More widely, conversion rate is simply the rate at which one commodity is converted into another; for example, the exchange rate of currencies. Thus the payment to UK farmers by the EU under the CAP depends upon the conversion rate of euros to sterling.

conveyancing The process of creating, extinguishing, and transferring the ownership of interests in land. The transfer of ownerships in land (for example, as a consequence of sale, gift, or by will) is also referred to as **disposition**. Although there are many commonalities in the process in the different parts of the *United Kingdom, there are some differences (most notably between Scotland and the rest of the UK). At the time of writing, much of the system in the United Kingdom is still paper-based; however, there are longstanding proposals to establish a fully electronic system for conveyancing (known as **e-conveyancing**), but this is progressing at varying speeds in the different parts of the UK.

- England and Wales. The process of purchase and sale of a property takes place in a number of stages:
  - Initial pre-contract negotiations. The potential purchaser makes an offer to the seller (often '**subject to** *contract'), who then accepts it; however, the parties are not, at this stage, contractually bound, so either side can withdraw subsequently. This stage of property purchase is not binding, in order to allow the potential purchaser (or a professional instructed by them) to carry out initial investigations; for example, building surveys, or enquiries into the *title of the property. However, the lack of a binding contract means that the phenomena of **gazumping** and **gazundering** are possible. Gazumping tends to happen in property markets where sale prices are increasing, and describes a situation where after the seller has accepted an offer, a second potential purchaser then comes along and offers more money than the first and this higher offer is then accepted by the seller. This potentially leaves the first purchaser in a situation where they may have spent money (for example, on surveyor's or solicitor's fees) but are no longer in a position to buy the property. Gazundering is essentially the opposite and happens in a market where property prices are falling. In this situation the potential purchaser, after making an initial offer, then subsequently reduces it or withdraws from the sale.
  - Exchange of contract. A contract is prepared setting out in writing all of the terms of the agreement and it is then **exchanged** (i.e. there are two copies of the contract, one of which is signed by the purchaser and one by the seller with each sending the one they have signed to the other party). There are specific requirements in law in relation to contracts for the transfer of *interests in land.

After exchange of contract the two parties are bound by it; if either party then withdraws it then becomes a *breach of contract.

- ○ **Completion** of the sale. This is the last stage of the process and is essentially the transfer of formal legal title to the purchaser after the passing over of the purchase monies. It usually happens a few weeks after the exchange of contracts, but can happen on the same day. It is done by *deed. For **unregistered land** (i.e. that not already registered at the *Land Registry) the title deeds are passed to the purchaser (including the transfer deed) and, in most cases, the law now requires that the land is registered at the Land Registry; if this is the situation then the purchaser must then apply for **first registration**. For land which is already registered at the Land Registry then the system is slightly difficult because there are, in effect, no title deeds. Instead, once purchase monies are received, a Land Registry form (which takes effect as a deed) is completed and submitted to the Land Registry (although solicitors are able to use an online system at this stage). The transfer is not finalized, however, until the purchaser is registered in the **Proprietorship Register** (*see* LAND REGISTRY), and as this may take several weeks, this gives rise to a **registration gap** (although this would be removed if a fully electronic system of conveyancing is introduced).

- Northern Ireland. The system of buying and selling land operates on similar principles to England and Wales, with compulsory first registration often being triggered by transfers of ownership. For unregistered land, deeds are, however, often already held in the **Registry of Deeds** (*see* LAND REGISTRY), although this does not guarantee title.

- Scotland. Although many of the same principles apply to Scotland there are some practical differences between Scotland and the rest of the UK. Unlike in the rest of the UK, solicitors often also act as estate agents, and there are also some variations in the mechanics of the process; for example, in the case of the sale of a residential property:

  - ○ The seller of most existing residential properties must provide a **Home Report** pack. This contains a **Single Survey** which must be prepared by a *chartered surveyor (including an assessment of the condition and value of the property), an energy report with an *Energy Performance Certificate, and a property questionnaire which provides general information such as the *Council Tax band, and any issues that might have affected the property previously (for example, storm damage), and so on. (Although the Home Report contains a survey, the purchaser may subsequently arrange their own survey, although this is fairly rare.)

  - ○ A potential purchaser initially notes their interest with the seller's agent (although this does not bind them to make an offer). The next step is that the potential purchaser then makes an offer (the seller may set a closing date for the submission of best offers by potential purchasers when they think there is sufficient interest, or the property may be advertised at a fixed price).

  - ○ The seller will then decide which offer (if any) to accept, and that acceptance will be set out in writing (usually called a **qualified acceptance** unless there are no matters for further agreement, in which case it is a straightforward acceptance).

○ After a qualified acceptance has been made, the purchaser's and seller's solicitors will then enter an ongoing process of written negotiations about the terms of the sale (for example, the date of entry); the letters are each known as a missive and the process of exchanging them is referred to as **missives**. Once all of these points have been agreed in writing then a contract will have been formed (often referred to as **concluded missives**). At this stage, the contract is binding on the parties.

○ **Settlement** (the point at which the purchaser gains possession, usually on the date of entry, and essentially equivalent to completion) happens following the transfer of the purchase monies and the delivery of a deed (referred to as a **disposition**), and there is usually a requirement to register the transfer at the *Land Register of Scotland for ownership to pass to the purchaser.

○ In Scotland, the person transferring the land (for example, the seller) gives a grant of **warrandice** (either expressly within the deed, or it is implied automatically in law). Warrandice is essentially a personal guarantee that the title is good (this is separate from the **Keeper's Warranty**, *see* TITLE). Court action can be taken for a breach of warrandice.

See also ALIENATION; ASSIGNATION (SCOTLAND); ASSIGNMENT; LOCAL LAND CHARGE; LAND CHARGE; STAMP DUTY LAND TAX.

copper A metallic element that is an essential requirement for life. *Livestock need copper in the diet and it may need to be added if not available naturally. However, copper is also toxic, so too much in the diet can lead to a build-up in the liver, causing liver failure. Copper is also essential for plant growth and copper salts may be added to soil if there is a deficiency. In *organic farming where artificial chemicals are prohibited, copper sulphate has traditionally been used as a fungicide. *See also* BORDEAUX MIXTURE.

coppice (coppicing) Any type of *broadleaf woodland that describes where tree growth from individual *stools are cut at ground level, regrow, and the cycle repeated, normally at regular intervals between five and twenty years. The **underwood** from this system was commonly used to supply small-diameter timber for pre-industrialization uses, but there is now an increasing interest and use of coppice to supply *fuelwood. The operation of **coppicing** is the physical management of this type of woodland. **Coppice with standards** describes where a small number of *standard trees are allowed to grow on to produce larger timber. *See also* SHORT ROTATION COPPICE; BIOMASS.

copse A small and often detached area of *woodland, group of *trees, *thicket, or *grove that is isolated from other woodland. An earlier function was the provision of timber for hurdles, fuel wood, or other farm functions, but more commonly associated with game shooting and hunting activities, and for landscape value and as a wildlife *habitat. *See also* COVERT; SPINNEY.

cord A measure normally for a stack of small-diameter broadleaf timber or **cordwood**, destined for the pulp or *woodfuel markets. The cord traditionally measures about 2.4 × 1.2 × 1.2 metres, typically varying in weight between 1 and 2 tonnes.

core strategy *See* LOCAL PLAN.

CORGI *See* GAS SAFE REGISTER.

corn In the UK, a generic term for any *cereal, notably *wheat, *barley, *oats, or rye. In North America, the word applies strictly to *maize.

cornice Decorative moulding, internally immediately below a ceiling, or externally around the top of the walls of a building. *See also* APPENDIX 2 (ILLUSTRATED BUILDING TERMS).

corporate social responsibility (CSR) Principle by which businesses and organizations take account of the wider social and environmental impacts of their activities, often reporting publicly on these aspects of their business in addition to more traditional economic measures. There is some variability in the definitions of corporate social responsibility between different sectors and countries, although there are some social responsibility standards, such as that produced by the *International Organization for Standardization. However, underpinning principles are that the business or organization operates in an ethical and transparent way underpinned by the objective of contributing to *sustainable development (i.e. balancing economic, environmental, and social considerations).

Corporation Tax Tax paid on the profits of doing business in the *United Kingdom by companies operating in the UK and some clubs and associations. The tax is not paid by all types of business; for example, in *partnerships, each of the partners pays *Income Tax on their share of the profits of the partnership.

corporeal hereditament Under English law, an item of property which exists physically; for example, land and buildings. *See also* INCORPOREAL HEREDITAMENT; HEREDITAMENT.

corporeal property Term used in Scottish law for physical items of property (say a piece of land, or a painting), and the rights of ownership of it. **Incorporeal property** is all other property which is not physical (say the right of an author to their copyright). *See also* CORPOREAL HEREDITAMENT; HERITABLE PROPERTY.

Corsican pine *See* TREE.

corvid (*Corvida*) A family of medium-to-large birds distributed across the world, of which eight species breed in the UK. The commonest among these include the crow, rook, magpie, and jackdaw. The crow and magpie are considered pest species, and can in particular predate wild bird and *game bird eggs. A favoured method of control is often by using Larsen traps, which are a live trap device that uses a living crow or magpie as a lure to trap another.

COSHH Regulations *See* CONTROL OF SUBSTANCES HAZARDOUS TO HEALTH REGULATIONS.

cotyledon The first embryonic leaves of a flowering plant as it emerges from the soil. The number of cotyledons that a plant produces, one or two, is a characteristic used in categorizing angiosperm flowering plants. *See also* DICOTYLEDON.

couch (*Elymus repens*, formerly *Agropyron repens*) A quick-growing invasive perennial grass *weed. It flowers and produces seed, but the main means of propagation is vegetative by *rhizome or underground stem. This allows rapid spread and makes it very difficult to control. The most effective *herbicide is *glyphosate.

council housing *See* AFFORDABLE HOUSING.

Council Tax Local tax paid by the owners and occupiers of domestic property in England, Scotland, and Wales. The rate paid is based on the value of each domestic property, with properties being banded into groups. Council Tax is, in the main, paid by domestic households, as opposed to *business rates which are paid for property which is occupied for non-domestic uses. Farms are, however, an exception in this respect in that farmers will, if they are only undertaking traditional agricultural activities, pay Council Tax on their farmhouse, but not business rates. If, however, the farmer has undertaken some form of *diversification into other business activities, then they may pay business rates on those activities. In Northern Ireland, domestic rates are paid rather than Council Tax.

counter-notice Notice served by the recipient of a notice on the party who originally served that notice. For example, a *landlord may serve notice on their *tenant, who may then, in some circumstances, serve a *counter-notice on their landlord (usually to challenge, in some way, the original notice). The provision to do this in a landlord and tenant situation is often provided for under the law, and is therefore often subject to specific timescales.

country 1. A general description of land that is predominantly used for *farming, *forestry, or in its natural or *semi-natural condition. It is not built up, such as towns and cities, but would encompass, for example, *farm buildings and villages. **2.** A country is also the description applied to the area within the boundaries in which each of the UK's *hunts operate. *See also* RURAL.

Country Land and Business Association (CLA) A membership organization representing the interests of landowners and rural businesses in England and Wales. Support is provided by influencing government at EU, national, and regional levels, and through an advice network with a particular focus on legal and tax issues. It is still commonly referred to as the CLA, reflecting its earlier title of Country Landowners Association. *See also* SCOTTISH LANDOWNERS' FEDERATION.

countryside *See* RURAL.

Countryside Agency *See* NATURAL ENGLAND.

Countryside Alliance (CA) An organization that was formed in 1987 and numbers around 100,000 members. It was formed from an amalgamation of bodies including the British Field Sports Society. It concentrates more on the political aspects of campaigns that promote all *field sports, but it has been the lead voice and is at the forefront of campaigning particularly in relation to *fox hunting and other live quarry hunting sports. It has broadened its interests to include a range of rural issues including food and farming, and threats to rural services and housing.

Countryside and Rights of Way Act (CROW) A major piece of *countryside legislation (England and Wales) that added to and amended the *Wildlife and Countryside Act 1981 and contains measures to improve public access to the open countryside and registered *commons (*see also* OPEN ACCESS). This was the most high-profile and innovative part of the legislation, but it also strengthened the protection for *Sites of Special Scientific Interest, provided a basis for the conservation of *biodiversity, and provided for improved management of *Areas of Outstanding Natural Beauty. Some equivalent legislation in relation to Scotland is addressed in the **National Parks (Scotland) Act 2000**.

Countryside Code A guidance note aimed at walkers and other users of *public access to encourage the protection and enjoyment of the *countryside. Useful and relevant advice is included for farmers and landowners. This code is published by *Natural England, but there are equivalent advisory notes available from various statutory and non-statutory countryside organizations.

Countryside Council for Wales *See* NATURAL RESOURCES WALES.

Countryside Stewardship Scheme The main *agri-environment scheme available in England that provides financial incentives for farmers and landowners to undertake environmentally sensitive land management practices. Not to be confused with an earlier scheme under the same title that ran from 1991 to 2004. It was developed from, and combines, the *Environmental Stewardship Scheme, the *Woodland Grant Scheme, and the *Catchment Sensitive Farming scheme. There are three main elements to the scheme: Higher Tier, Mid Tier, and Capital grants. The activities that the scheme promotes include conserving and restoration of wildlife *habitats, flood risk management, *woodland creation and management, reduction of water pollution from agriculture, protection of *landscape character, protection of historic features, and encouraging access for education. *See also* NATURAL ENGLAND.

⊕ SEE WEB LINKS
• Government website with an introduction and full details of scheme and application procedures.

county council *See* LOCAL AUTHORITY.

County Court Courts in England, Wales, and Northern Ireland that hear the majority of civil (as opposed to criminal) cases in front of a judge. Appeals against decisions made in the County Court are generally heard in the *High Court. Matters dealt with by the County Court include situations where one person wishes to claim money back from another person or business (i.e. for non-payment of debt). This type of claim is sometimes known as taking someone to a **small claims court**. If the Court agrees that such a debt exists and should be repaid, it will issue a **County Court Judgment (CCJ)** against the person who owed the debt, ordering them to pay it. County Courts are also able to issue *injunctions which either forbid another party from doing something or instruct them to undertake a positive form of action to address a problem. *See also* COURT, LAW; SHERIFF COURTS.

County Court Judgment (CCJ) *See* COUNTY COURT.

coupe Normally used to describe a defined area of woodland or a **felling coupe** that is to be *clear felled or managed under *continuous cover forestry.

Court of Appeal Court in England, Wales, and Northern Ireland hearing civil appeals from the High Court (it also hears criminal appeals). The Court of Appeal of England and Wales sits in London and Cardiff, with the Northern Irish Court of Appeal normally sitting in Belfast. In some circumstances, Court of Appeal decisions can be further appealed to the **UK Supreme Court**. *See also* COURT OF SESSION; SHERIFF COURTS.

Court of Session The highest civil court in Scotland. *Judicial reviews in Scotland are considered by this court, which sits in Edinburgh. The Court of Session is divided into an **Outer House** and **Inner House**. Appeals against decisions made by the Outer House can be appealed to the Inner House, and in turn, on to the **UK Supreme Court**. *See also* SHERIFF COURTS; HIGH COURT; SCOTTISH LAND COURT.

court order Direction issued by a *court ordering, or authorizing, a party or parties to take some form of action, or to cease an action. *See also* INJUNCTION.

court, law Body with the authority to make decisions in relation to the implementation of the law. The court system in the *United Kingdom divides into the criminal courts which deal with matters involving crimes, and civil courts which deal with other non-criminal matters (for example, disputes between parties where one party may consider the other owes them money, or has not delivered what they agreed to in a *contract).

The civil courts in the United Kingdom operate in a hierarchy, with matters generally being heard first in one of the lower courts. If one of the parties disagrees with the initial decision then the decision can usually be appealed up through the hierarchy of courts. There are some differences in the hierarchy of courts in different parts of the United Kingdom, as shown in the Figure, but the **UK Supreme Court** is common to all parts of the United Kingdom as the highest UK civil court. In some cases, decisions can be appealed beyond the UK Supreme

United Kingdom Civil Court Hierarchy

	England and Wales	Northern Ireland	Scotland
Higher Courts	Court of Justice of the European Union UK Supreme Court		
	England and Wales *Court of Appeal	Northern Ireland Court of Appeal	*Court of Session
↑	England and Wales *High Court	Northern Ireland High Court	*Sheriff Appeal Court
Lower Courts	*County Court (although in some cases civil matters can be decided in a Magistrates' Court, which is a lower court)	County Court (although in some cases civil matters can be decided in a Magistrates' Court, which is a lower court)	*Sheriff Court

Court to the **Court of Justice of the European Union**; however, following the United Kingdom's referendum decision of 23 June 2016 to leave the European Union, the UK Supreme Court may, in future, be the highest court to which decisions can be appealed.

In addition to the courts, legal matters within the United Kingdom can be decided by tribunals. Tribunals are specialist legal panels which are usually less formal and less complicated than a court process.

The tribunal structure in England and Wales (and in some cases in Northern Ireland and Scotland) is divided into First-Tier and Upper Tribunals. An appeal against a decision of a First-Tier tribunal is generally heard in an Upper Tribunal. Further appeals can, in some circumstances, be made to the Court of Appeal (England, Wales, and Northern Ireland) or Court of Session (Scotland).

The First-Tier Tribunals are divided into seven chambers, including a *First-Tier Tribunal Property Chamber and Tax Chamber. The Upper Tribunal has four Chambers, including a Tax and Chancery Chamber and a **Lands Chamber**. The jurisdiction of each chamber in both the upper and lower tiers varies substantially in different parts of the United Kingdom. In addition to the First-Tier and Upper Tier tribunals there is a parallel system of Employment Tribunals.

covenant 1. An obligation, or promise, established by *deed in English law. Covenants can exist in relation to *freehold land and also in *leases (these are each discussed here as separate entries, as are the equivalent **real burdens** and **obligations** under Scottish law). **2.** An obligation, under English law, which requires one party to either do, or not do, something in order to benefit the freehold land of another party and which is agreed through a deed. The person whose land has the **benefit** of the covenant is known as the **covenantee**, and the person who has the **burden** of the covenant is known as the **covenantor**.

Freehold covenants be divided into **positive covenants** and **negative covenants**. In the case of a positive covenant, the covenantor has to do something; for example, repair and maintain a fence (so the covenantor will often have to spend money). A negative covenant (also known as a **restrictive covenant**) will require the covenantor not to do something; for example, not to carry out any building work on the covenantor's land. English law surrounding covenants is complex, with *case law having extensive influence; for example, whilst some covenants can bind future purchasers of land, others (for example, most positive covenants) often cannot. Under Scottish law, the broad equivalent to a freehold covenant is a **real burden**. **Negative burdens** are similar to negative covenants, whilst the equivalent to positive covenants are known as **affirmative burdens**. **3.** In the context of a *tenancy or lease, an obligation on either a *landlord or *tenant. Strictly speaking, covenants must be made by deed, so the term 'obligation' may be better used in the context of leases and tenancies, as not all tenancies must be made by deed (and in Scotland the term **obligation** is always used). However, the term 'covenant' is commonly used in the rest of the *United Kingdom. In the context of leases and tenancies, covenants are divided into **express covenants**—i.e. those specifically included in a written document (more correctly often **express obligations**)—and **implied covenants**—i.e. those not specifically included within a written agreement but assumed to exist under the law (often correctly **implied obligations**). Under English and Scottish law a series of implied obligations exist for leases; however, these can also be overridden for certain types of lease (for example, residential and *agricultural tenancies) where *statute establishes specific implied obligations in relation to that particular type of tenancy in the relevant part of the United Kingdom.

General implied landlord's obligations in England and Wales include allowing the tenant '**quiet enjoyment**'(this means not doing anything to disturb the tenant's enjoyment of the property or allowing a third party to do so on the landlord's behalf); that the landlord will not '**derogate from their grant**' (i.e. not initially granting the tenant the right to do something, but then by some other action making it difficult for them to exercise that right); and that the premises are fit for the purpose for which they have been let (for example, they are habitable if it is a residential property). If the property is residential then the requirements of the **Landlord and Tenant Act 1985** are also implied into the tenancy; these include obligations on the landlord (for a tenancy of less than seven years) to keep in repair the structure and exterior of the property as well as aspects such as the water supply system, heating, and sanitation. Express obligations may include items such as repair, insurance and maintenance obligations.

The tenant will also have implied obligations under law. In England and Wales these include paying the *rent; paying any taxes for which the landlord is not responsible (for example, *business rates or *Council Tax); allowing the landlord to enter the property at an agreed time in order to exercise their responsibilities, for example, for maintenance; and not committing **waste** (by damaging, or changing the nature of the property by, for example, knocking down walls). Depending on the type and length of tenancy there can be an implied covenant not to commit either **voluntary waste** (as above) or **permissive waste** (allowing damage to occur through not acting when it is clear that a problem needs dealing with, say a water leak), or both. Express obligations on the tenant commonly include a restriction on the tenant's ability to sub-let, assign (*see* ASSIGNMENT) or otherwise part with possession (a 'catch-all' to cover other potential situations in which the tenant might part with possession). These restrictions can be either **absolute** (where a tenant is not allowed to assign or sub-let, and so on) or **qualified** (where they would have to get the landlord's consent, although the law specifies that this cannot be unreasonably withheld). (Restrictions on assignment and *sub-letting are also implied into some residential tenancies, but are usually also included as express obligations for the avoidance of doubt.)

Similar obligations to England and Wales also exist in Northern Ireland, but there are some differences in detail, particularly relating to the assignment of leases.

In Scotland, similar implied obligations exist for landlords and tenants (albeit that the terminology used for some is different from other parts of the United Kingdom). Implied obligations on the landlord in Scotland include the landlord giving the tenant possession of the property and not interfering with that possession (the principle of derogation from the grant also operates in Scotland, though in Scotland there is no directly equivalent right of the tenant to quiet enjoyment). The landlord in Scotland is also obliged to ensure that the property is reasonably fit for the purpose for which it was let, and also to carry out any repairs and maintenance for which they have an obligation under the law (which can differ depending on the type of tenancy). This is notwithstanding that the repairing obligations of both the landlord and tenant (where relevant) may also be set out as express obligations in the agreement between them.

In Scotland the tenant has an obligation to use the property (i.e. take possession of it), to pay the rent when due, and to take reasonable care of the property. They must also not '**invert the possession**' by making unauthorized changes to the property or not using it for the purposes for which it was originally let (and thereby not complying with any '**use clause**'). In addition, if the *landlord's hypothec is available to the landlord (i.e. a provision in some commercial tenancies where the tenant's *moveable property in the leased premises can act as security for the rent), then the tenant has an obligation to keep (i.e. **plenish**) sufficient goods in the property to act as security for that rent. *See also* BREACH OF COVENANT; REPAIRING OBLIGATION. **4.** An informal description of the ability of a potential commercial tenant to comply with the obligations of the lease (for example, whether their business is secure enough to pay the rent for the duration of the lease, and so on). A tenant is said to be a 'good covenant' (or vice versa), and the strength of the covenant is likely to influence the *yield used in a valuation.

cover crop A *crop grown primarily to cover the *soil, to reduce *soil erosion, to minimize the *leaching of *nutrients, or to improve *soil structure and the *organic matter. In this context, also known as a green manure, it is sown after the harvest of a *cash crop and is usually left over winter to be replaced in the spring by another *cash crop. As such, it is also a *catch crop. The term is also used to describe a crop grown as wildlife habitat, mostly usually for farmland or *game birds. Such crops may be eligible for a grant from an *agri-environment scheme, and would usually be a mixture of species such as *kale, *cereal, *oilseed and others.

covering The impregnation of a female animal by a male or by *artificial insemination.

covert A small area of *woodland that is or has been created and managed to provide a sheltered *habitat for *game birds and other hunted quarry such as foxes. Normally characterized by thick undergrowth that provides 'cover' for the wildlife. *See also* COPSE; SPINNEY.

cow The female of the bovine species or *cattle. The cow may be kept to produce milk, known as a *dairy cow, or for *beef. The cow gives birth to a *calf (*pl.* calves), which in turn may be kept for milk or beef. Calves of dairy cows are usually removed from the cow at an early age so that the cow is milked rather than suckled by the calf. Cows kept for beef production are known as *sucklers or single-suckle cows. The female is known as a *heifer until it has produced its first calf.

CPI *See* CONSUMER PRICES INDEX.

CPIH *See* CONSUMER PRICES INDEX INCLUDING HOUSING COSTS.

CPO (Compulsory Purchase Order) *See* COMPULSORY PURCHASE.

CPRE *See* CAMPAIGN FOR THE PROTECTION OF RURAL ENGLAND.

CRAR Commercial Rent Arrears Recovery. *See* BREACH OF COVENANT.

cream A *dairy product processed from *milk by separating the *butterfat from the skimmed milk.

creep feeding The feeding of young animals, especially *lambs, *calves or *piglets with a *concentrate feed is such a way that the adult, such as the *ewe, *cow, or *sow cannot access the feed. This may be done by means of a covered *trough with bars or other barrier.

creeping bent (*Agrostis stolonifera*) A quick-growing invasive perennial grass *weed. It flowers and produces seed, but the main means of propagation is vegetative by a stolon (creeping) horizontal plant stem or runner that takes root at points along its length to form new plants. This allows rapid spread and makes it very difficult to control. The most effective *herbicide is *glyphosate.

creeping thistle (*Cirsium arvense*) The most common thistle, an aggressive weed of pasture. It is spread by wind-blown seed and by a *rhizome or underground stem which can grow 6–12 metres per year. It is one of the five pernicious *weeds specified by the *Weeds Act 1959, the others being *ragwort, spear thistle, curled dock, and broad-leaved dock. Control is by *herbicide, pasture management or by digging it up by the roots.

crib biting The action whereby a *horse grabs hold of the *manger (or crib) and sucks air into its stomach. More often, parts of the stable, gates, or fencing are chewed and can cause extensive damage if not protected. This behaviour is a mechanism for coping with stress, and can cause indigestion and poor health.

CRM *See* CUSTOMER RELATIONSHIP MARKETING.

croft An agricultural holding (typically of around 5 hectares in size) mainly in the Highlands and Islands of Scotland, normally held in a tenancy, and which is governed by specific legal duties on the **crofters**, including a duty to be resident on, or within 32 km of, their croft, and not to neglect their croft, but instead to cultivate it for some kind of purposeful use. Crofters are required to seek permission from the Crofting Commission for a range of different things, including dividing or enlarging a croft, or letting all or part of a croft.

crofter A person who occupies and works a croft. Traditionally, all crofters were tenants, paying rent to a landlord, usually a large estate, but successive legislation of land reform in Scotland has given the crofter the right to buy the land under certain circumstances.

Crofting Commission (Scotland) Board of nine commissioners, three of whom are appointed by Scottish Ministers and six of whom are elected by crofters, to set policies and to decide on a range of application matters relating to *crofts.

crop Any plant grown for food, fuel, or other purpose; for example *cereals, vegetables, or fruit. *See also* CROP PROTECTION; CROP ROTATION; CROP SPRAYER; CATCH CROP; COVER CROP.

Common arable crops

Cereals	Oilseed crops	Pulses
Wheat	Oilseed rape	Combining peas
Durum wheat	Sunflower	Field beans
Barley		Lentils
Oats		Navy beans
Maize		Soya beans
Millet		Vining peas
Rice		
Rye		
Sorghum		
Triticale		

crop protection The science and practice of protecting *crops against *weeds, *diseases, and *pests, including the formulation and application of *pesticides, specifically *herbicides, *fungicides, and *insecticides.

crop rotation The growing of different crops in succession on a piece of land to maintain fertility and to avoid the build-up of *weeds, *pests, and *diseases. This is particularly important in *organic farming where *inorganic fertilizers and*pesticides are not used. *Fallow may be included in a rotation to give the soil a complete rest for a year. The original Norfolk Four-Course Rotation introduced by Viscount 'Turnip' Townshend in the early eighteenth century was *wheat, *clover, with or without *ryegrass, *oats, or *barley, and turnips. The second and last were grazed by livestock, whilst the clover encouraged the fixation of nitrogen, thus improving fertility.

crop sprayer An agricultural implement used to apply chemicals including *pesticides, specifically *herbicides, *fungicides, and *insecticides, and liquid *fertilizer. It may be self-propelled, trailed behind a tractor, or mounted on the *three-point linkage. Composed of a tank to hold the liquid and a boom with nozzles through which the chemical is applied, *calibration of the sprayer is important to ensure that application rates are accurate.

crossbreed The offspring of two parent organisms of different *breeds, varieties or *species. Such cross-breeding may bring *hybrid vigour in the offspring and will combine characteristics from both parents. For example, *ewes in lowland flocks, usually crossbred to bring hardiness and mothering qualities, may then be crossed with a *ram of a breed that has meat characteristics such as *conformation.

cross-compliance A regulatory mechanism to encourage compliance with a range of basic environmental, animal and plant health, and animal welfare regulations by recipients of the EU agricultural support payments, *agri-environment and *woodland grant schemes. It also includes compliance with *Good Agricultural and Environmental Conditions. Although varying in detail and implementation, in 2005 it was further developed across all countries of the EU, and is now a common term within the *Common Agricultural Policy industry. However, its use to describe a condition policy linked to eligibility for an agricultural government support scheme originated in the USA in the 1970s. *See also* GREENING MEASURES.

cross-country A course to test the skill and stamina of the *horse and its rider, commonly over a route through *pasture and *woodland. Larger courses would involve natural obstacles such as ditches and logs, as well as a variety of fences. *See also* EVENTING.

CROW Act *See* COUNTRYSIDE AND RIGHTS OF WAY ACT.

Crown Land Land owned by the monarchy or the state. In the *United Kingdom the term Crown Land is used to refer to land owned by the monarch as a consequence of their position as monarch, plus private property owned by the monarch, and, in some cases, government land. Crown Lands include **The Crown Estate**, which is land owned by the monarch for the period of their reign (but not as their private property). The Crown Estate owns around half of the foreshore around the United Kingdom coastline, together with significant land holdings in both urban and rural areas of the country. Crown Land is often subject to separate *legislation from land in other types of ownership.

crown rust A fungal disease caused by *Puccinia coronate* that amongst *cereals infects only *oats. Symptoms are similar to *brown rust of *wheat and *barley. Orange brown pustules appear on the leaf, and the oat panicle can also become infected. The disease is favoured by high temperatures (20–25° C), so epidemics usually occur in June–July, when yield may be reduced by 10–20%. Several grasses can also be infected, notably *ryegrass, but there is no cross-infection with *oats.

CSF *See* CATCHMENT SENSITIVE FARMING.

cubicle A form of housing, particularly for dairy cows, in which each cow has an area with a barrier down each side, open at the foot and sometimes a feed passage at the head. Free-standing cubicle housing has been the most common method of housing dairy cattle in the UK since the 1960s.

cud Food that is returned to the mouth from the *rumen of a *ruminant animal for further chewing. Cellulose, a major component of forage, is very difficult to digest, which is why ruminant animals have four stomachs with a symbiotic relationship with microbes that aid digestion.

cull To select from a group of animals the weak, aged, or infertile, or those otherwise lacking beneficial characteristics. These are then likely to be slaughtered. Thus a cull cow or ewe is one that has reached the end of its useful breeding life and is slaughtered immediately or maybe kept until improved body condition adds value for meat. The term is also applied to the killing of wild animals such as deer to reduce numbers and to achieve a sustainable population, or a cull of badgers to reduce the spread of *bovine tuberculosis.

cultivar A plant variety that has been produced by selective breeding.

cultivate To prepare land for the sowing of crops. There is a wide variety of *cultivation machinery, including the *plough, *harrow, *disc harrow, cultivator, and roller. The soil may be inverted by ploughing or simply moved by means of discs or tines to break up clods and create a *seedbed which may be levelled or firmed down.

cultural services The non-material benefits that people obtain from *ecosystems through spiritual enrichment, cognitive development, reflection, recreation, and aesthetic experience, including, for example, knowledge systems, social relations, and aesthetic values. They may also be seen as 'cultural benefits', as they directly relate to changes in human welfare. *See also* ECOSYSTEM SERVICES.

culture 1. A descriptive noun of bacteria grown in the laboratory, or as a verb to describe growth of bacteria in a culture medium. **2.** Relating to ideas, identities, and customs, and to social behaviour and norms of individuals, groups, and societies. **Cultural heritage** can equate to, and describe ways of, living and thinking that have existed for a long time in a society. *See also* CULTURAL SERVICES.

culvert *See* DRAIN.

Curry Report A 2002 report to the United Kingdom government written by the Policy Commission on the Future of Farming and Food, and officially titled **Farming and Food: A Sustainable Future**. The Policy Commission on the Future of Farming and Food, which was chaired by **Sir Donald Curry**, was set up by the government in 2001 with a remit to consider the future of the farming and food sectors in England. The report had a significant influence on the government's 2002 Strategy for Sustainable Food and Farming. One of the key recommendations in the report was that the previous system of *agri-environment schemes be reformed and replaced with a new tiered *environmental stewardship system. As part of this it was suggested that a new and easily accessible 'broad and shallow' entry level scheme tier be established. The aim of this recommendation was to increase the area of land in agri-environment schemes, and it was a proposal that was subsequently adopted by the government.

curtilage The area of, and around, a building or structure which can in itself include other buildings or structures which are in some way related to the main building. There is an extensive body of *case law where the courts have assessed the factors that determine the extent of the curtilage of a building. A range of aspects has been considered relevant, including age, layout, history, ownership, function, and the degree to which a building, structure, or piece of land is related to the main building; however, the picture is complex, and it is often not easy to determine the situation on the ground. Defining curtilage on the ground can be important in a range of property-related situations; for example, if it is necessary to define the extent of a dwelling house in relation to tax relief, or if a question arises in relation to the extent of a property passed via a *conveyance. It is also particularly significant in relation to *listed buildings, as the requirement for *listed building consent can apply to buildings or structures within the curtilage of a listed building, even if they are not specifically referred to within the listed building entry. *See also* SETTING.

customer relationship management *See* CUSTOMER RELATIONSHIP MARKETING.

customer relationship marketing (CRM) A strategic approach to sales and marketing which is focused on building long-term customer relationships with a view to retaining existing customers, rather than focusing only on immediate sales. Implementing a CRM process requires an organization to be focused around the customer, with sales operations being fully integrated into marketing strategies and business planning. Other aspects may include the setting up of teams to manage key accounts and relationships, and using customer relationship management software systems where all staff are able to see all the interactions an organization has had with a customer. *See also* MARKETING MIX.

cutter bar The bar of a *mower, *combine harvester, or similar implement with triangular guards through which a reciprocal knife runs.

cypress (Lawson) *See* TREE.

DA (Disadvantaged Area) *See* LESS FAVOURED AREA.

dagging The *shearing or clipping of the wool around the rear end of sheep, especially *ewes, to remove fleece contaminated with faeces or to prevent such contamination. This helps to prevent fly strike in summer. It is common before lambing and/or *tupping to allow clear access to the genital tract.

dairy Pertaining to milk. Thus a *dairy cow is a female of the bovine species that is kept for milk production, and a dairy farmer one who keeps cows for milk production. The building or premises where milk is produced, stored, or processed.

DairyCo *See* LEVY BODIES.

dairy cow The female of the bovine species that is kept for milk production. *See also* CATTLE.

dairy products Products that come from a *dairy or are processed from milk; for example, cream, butter, cheese, yoghurt, and ice-cream. Virtually all milk is processed, at least by *pasteurization, and the most popular liquid milk sold in the UK is semi-skimmed. Around half of milk produced is sold as liquid milk, and the rest is processed into dairy products as listed above in a plant run by a dairy or dairy processor. This adds value to the raw milk, so some dairy farmers diversify into processing their milk to improve their income.

dam 1. A general term that can be applied to the female parent of a domestic animal for example a sheep or horse. **2.** A wall, barrier, or bund to control water flow, to raise its level and to create an artificial lake or reservoir. This man-made supply of water can be used to maintain water storage for multifunctional uses, including irrigation and electricity generation.

damages Sum of money paid, following a *court action, for example, as *compensation for a *breach of contract, or a *tort such as *negligence, *nuisance, or *trespass (or the Scottish equivalent of a tort, a **delict**). Damages are designed to compensate for losses incurred, and in principle put the party claiming them in the same position that they would have been had the breach of contract, or the tort/delict, not occurred. The courts will, however, expect that the party claiming the damages has taken reasonable steps to mitigate (minimize) the loss. Damages can also potentially be reduced in a situation of **contributory negligence** (i.e. where it is considered that the person claiming the damages

themselves contributed to the situation which has triggered the damages).
Liquidated damages are a pre-agreed sum of money specified in a *contract
which will be paid in the event of a breach (unless the courts consider that the
sum is beyond that of the loss), whereas **unliquidated damages** are a sum
decided by the courts.

damping off A disease of seedlings caused by a range of *fungi or similar
organisms particularly common in wet and cool conditions. Infection occurs
before or after germination and causes the stem and roots to weaken or rot at or
below the soil surface. The plant will then droop and may die. There is no cure
once the plant is infected, but the condition may be prevented by good soil
hygiene or the use of fungicidal seed dressing.

DARD *See* DEPARTMENT OF AGRICULTURE, ENVIRONMENT AND RURAL AFFAIRS.

daylighting A term that can be used to describe targeted and small-scale
woodland *felling operations designed to allow sunlight to encourage ground
*flora and *woody shrub growth.

dbh *See* DIAMETER AT BREAST HEIGHT.

DCF *See* DISCOUNTED CASH FLOW.

DDT Dichloro-diphenyl-trichloroethane, an organochlorine *insecticide, was
developed as the first of the modern synthetic insecticides in the 1940s. It became
widely used to combat malaria, typhus, and the other insect-borne human
diseases, and was also effective for insect control in crop and livestock
production. However, resistance amongst insect pests grew, limiting the
effectiveness, and serious impact on human and animal health was identified,
including reproduction and a carcinogenic effect. It was banned by various
countries in 1970, by the USA in 1972, but not in the UK until 1984. The
cumulative long-lasting nature of the chemical means that it still exists in the
environment. It became a *cause celebre* when highlighted by Rachel Carson in
her book *Silent Spring*, published in 1962.

Deadly nightshade (*Atropa belladonna*) Also known as belladonna, it is a
perennial herbaceous plant of the tomato family *solanaceae*, native to Europe,
north Africa, and western Asia. Both the foliage and berries are extremely toxic,
and the plant is notorious as a poison. Atropine is the principal ingredient of eye
drops as it causes the pupils to dilate, and it was widely used in the Middle Ages
as a cosmetic to make eyes larger and more beautiful, hence the name
belladonna.

deadweight The weight of an animal or carcass after slaughter. An animal may
be sold for meat either by *liveweight or deadweight. If the former, the animal is
weighed before slaughter and payment made on a liveweight basis. Once
slaughtered, depending upon the dressing specification, material is removed,
such as the head and skin, leaving the carcass. The weight of this carcass is the
deadweight on which payment is made. The difference between liveweight and

deadweight is called the *killing out percentage. This is around 45–55% for ruminant animals such as *cattle and *sheep, and around 75% for *pigs.

dead wood Dead standing *trees or timber left on the ground to decay that provides important *habitats for invertebrates, especially beetle *fauna. Commercially managed woodland can generally be poor in dead wood, but this can be increased by leaving poorer quality timber and stumps to rot on the ground. *See also* WOODLAND; COPSE; SPINNEY.

decarbonization A term used within the energy sector to describe the process of moving away from energy generation sources that increase *carbon dioxide levels in the atmosphere (for example, conventional *fossil fuel generation) to other systems that do not have this effect on overall atmospheric carbon dioxide levels, such as *renewable energy, nuclear power, and potentially *carbon capture and storage. *See also* DECENTRALIZED ENERGY.

decentralization The process by which a central government redistributes some of its powers to smaller authorities. Decentralization is a key element in the delivery of *localism.

decentralized energy A term used to describe an energy generation system where the energy needs of a population are met by multiple small generation plants located near to the end users (typically *renewable energy sites), rather than through a more conventional centralized energy structure which relies on a small number of large-scale generation plants (such as conventional *fossil fuel-based power stations), which feed into the national distribution grid and each meet the needs of large numbers of users. The process of transitioning from a power-generation structure which is centralized to a more decentralized one presents challenges for electricity distribution networks, as it often requires the installation of additional capacity in local networks to transmit the electricity generated at large numbers of new locations to nearby users. However, it also has potential benefits in terms of reducing the amount of energy that can be lost through transmission (thereby improving overall efficiency) and also potentially improving security of supply, as the system is less vulnerable to failures in a small number of centralized generation sites. *See also* DECARBONIZATION.

deciduous More commonly used to describe a *woodland or *forest that is comprised predominantly of *trees and *woody shrubs that drop their leaves in the winter period and regrow new leaves in the spring. Can also describe a single tree or woody shrub. The period of dormancy when the plant has no leaves is an adaptation to dealing with extremes of temperature. *See also* CONIFER; EVERGREEN.

decision notice Notice, issued in relation to an application for *planning permission or *listed building consent, which confirms whether permission has been granted or refused. In the event of permission being granted, the decision notice will list any *conditions which must be complied with in relation to the permission. If the application has been refused, the decision notice will detail the reasons for refusal. *See also* PLANNING OFFICER.

decompose The breaking down of organic matter into constituent parts by the action of microorganisms such as *bacteria and *fungi. An essential process in nature, as it allows nutrients to be recycled. It is the basis of the creation of *compost.

decoupling 1. The removal of the link between *subsidy and production in the mechanisms of support payments of the *Common Agricultural Policy. In earlier CAP regulations before decoupling, CAP subsidies were paid depending upon the area of *arable crop grown or the number of *livestock kept, known as a **headage** payment. Subsequent to various reforms of the CAP, that link was gradually removed with direct payments being made regardless of the area of crop grown or livestock kept. An overall effect has been to move the agricultural sector towards more market demand production. **2.** Term used to describe the removal of a link between a tractor or similar vehicle and attached equipment such as a trailer or machine, usually, for example, a hydraulic pipe or *power take-off.

Dedication Scheme An early forerunner of the *Woodland Grant Scheme, aimed at landowners and farmers to encourage them to manage their woodlands more effectively, especially for timber production. It was introduced in 1946 and effectively closed in 1974, and was superseded by a number of subsequent woodland schemes.

deed Formal written legal document, which meets the requirements of a deed (as set out in the relevant law). The law on the execution of legal documents varies slightly in the different parts of the *United Kingdom, but the general principles are that a deed must be in writing, must state that it is intended to be a deed, and be signed, witnessed, and delivered. Whilst there are some circumstances in which a deed can be used optionally, the law sets out some areas where a deed must be used. The formalities required for a deed to be executed have varied over time (for example, the requirement for a seal to be applied to a deed was removed in England and Wales in 1990), and increasing moves towards the use of electronic *conveyancing will replace the use of paper deeds in property transfers.

Deed of Grant of Easement Written legal agreement creating an *easement in England, Wales, and Northern Ireland. A *Deed of Servitude is the equivalent in Scotland.

Deed of Servitude Written legal agreement creating the Scottish equivalent of an *easement.

deemed consent *See* ADVERTISING CONTROLS.

deep litter Animal housing system in which bedding, such as straw or sawdust, is repeatedly spread on the floor of an animal building. As the bedding becomes fouled by urination and defecation, more material is added, building up the layers, thus forming *farmyard manure.

deer Six species of native or naturalized deer exist in Britain. The truly native deer include the red deer (*cervus elaphus*), which is the largest and is more associated with the open hill country of Scotland and some parts of England. The roe (*capreolus capreolus*) is smaller and more widespread. The fallow deer (*Dama dama*) was probably reintroduced in Norman times, and was a favoured species that particularly adapted to the confines of a deer park which might be associated with a rural mansion. The sika (*cervus nippon*), muntjac (*muntiacus reevesi*), and Chinese water deer (*hydropotes inermis*) all derived from captive stock, and established breeding populations in the south of England and spread across Britain. *See also* BRITISH DEER SOCIETY; DEER STALKING; STAG HUNTING.

deer stalking The hunting and shooting of *deer, more commonly organized and let out on a commercial basis. Stalking can be an important part of the local rural economy and culture, particularly in parts of Scotland. It can also play a significant part in the management and control of deer populations. *See also* BRITISH DEER SOCIETY.

deficiency In simple terms, the lack or shortage of something. In animal and plant nutrition it is used to indicate the shortage of an essential nutrient, such as a trace element. It is also used to describe the symptoms caused by the shortage. For example, an animal might be suffering from magnesium deficiency.

defoliate To remove the leaves of a plant. This may be done by disease, thus limiting the productivity of the crop, but also may be done deliberately when the crop is ripe by use of a chemical to aid harvesting. *See also* DESSICANT.

deforestation The removal of trees and subsequent loss of forest cover, usually as a result of human activities such as growing crops or grazing livestock, building developments, and unsustainable extraction of timber. Can also occur by natural climate change. The reverse of *afforestation.

DEFRA *See* DEPARTMENT FOR ENVIRONMENT, FOOD AND RURAL AFFAIRS.

dehorn To remove the horns of an animal to make it easier to handle and to reduce the risk of injury to other animals. Some breeds naturally have no horns, whilst others have been selectively bred so that they no longer have horns, in which case they are described as poll or polled.

delegated legislation Laws (also known as **secondary** or **subordinate legislation**) made directly by government ministers in the *United Kingdom, under powers given to them through an *Act of Parliament or *Act of Assembly, and which are not usually subject to direct vote by elected representatives. The over-arching principles of laws are usually set out in an Act of Parliament or Assembly, with delegated legislation usually setting out the detail, which can be

in the form of **Statutory Instruments (SI)** or **Rules** or **Codes of Practice**.
See also REGULATION.

delegated powers Provisions under which a *local planning authority can
delegate decision making powers in relation to planning applications to
*planning officers, rather than the *local planning authority planning committee
(comprising elected councillors). This is done in order to assist in the efficiency
of the decision-making system. Applications decided under delegated powers
tend to be those considered to be uncontroversial, whereas bigger, or potentially
controversial, decisions are usually made by the planning committee. In this
event, the planning officer will normally provide advice and a recommended
decision, but the committee is not bound by this.

delict *See* TORT.

denature To change a protein or enzyme by external force, usually heat or
chemical such as acid or alkali. When meat is cooked, the heat denatures the
protein, breaking down the complex structure and making it more digestible.

denitrification A natural process whereby *nitrate in soil is broken down by
*bacteria, releasing nitrogen into the atmosphere. It occurs principally when soil
is wet and waterlogged to the extent where pores in the soil are filled with water,
thus limiting oxygen. The bacteria then use the oxygen in nitrate for respiration.
In growing crops this is a loss of a vital nutrient, but in other cases it may be
beneficial in, for example, reducing the nitrate content of drinking water.

Department for Communities and Local Government Government
department now known as the *Ministry of Housing, Communities and Local
Government.

Department of Agriculture, Environment and Rural Affairs The
devolved Northern Ireland government department that has responsibility for
agriculture, fisheries, forestry, and the environment. *See also* NORTHERN IRELAND
ENVIRONMENT AGENCY.

Department for Environment, Food and Rural Affairs (DEFRA) The UK
government department that has responsibility for the environment, agriculture
and food production, forestry, and fisheries. It operates in cooperation with the
devolved UK governments that take devolved responsibilities for these matters.
A number of organizations are answerable to DEFRA, including those applying
across the UK, such as the *Joint Nature Conservation Committee and *Rural
Payments Agency, as well as many that apply to England alone, including, for
example, *Natural England. DEFRA takes the lead in dealing with the *Common
Agricultural Policy and other EU-related matters.

deposit 1. Sum of money collected when something is rented, to cover potential
damage or other financial losses, and returned at the end of the period of rental if
there are no such losses or damage. They are used extensively in the *United
Kingdom in relation to residential tenancies where a deposit is typically

requested from the *tenant at the beginning of the *tenancy in order to cover the potential costs of any damage (in excess of *fair wear and tear) or unpaid *rent at the end of the tenancy. If there is no damage and the tenant does not owe any rent, then the deposit is returned in full to the tenant when the tenancy ends. Under the law, in the UK, the *landlords of most new residential tenancies must now protect the deposit in a *deposit protection scheme. **2.** Advance payment (usually a proportion of the whole of an amount that will eventually be due) made by one party to a *contract to the other party to guarantee that the contract will be carried out; for example, someone may pay part of the cost of a new car as a deposit when they order it, with the remaining money being paid at the point when the car has been manufactured and is collected. **3.** Amount of money placed in a bank or building society account, usually to earn *interest. **4.** A naturally formed layer of minerals underground.

deposit protection scheme One of a number of deposit schemes applying to some *residential tenancies in the *United Kingdom. Since 6 April 2007, all *deposits taken by landlords and agents for *assured shorthold tenancies in England and Wales have had to be protected by such a scheme. This is designed to ensure that tenants receive their deposits back at the end of their *tenancy. Each scheme also provides an *alternative dispute resolution service which helps resolve disputes between landlords and tenants over deposits, without recourse to the courts. There are two main types of scheme: **custodial** and **insurance-based** schemes. Custodial schemes hold the deposit, and at the end of the tenancy pay it back to the person who is entitled to it. In insurance-based schemes, the landlord or agent holds the deposit but pays an insurance fee to the scheme. If, at the end of the tenancy, the landlord or agent does not pay the tenant the money to which they are entitled, the insurance-based scheme will pay the tenant and will then claim the money back from the landlord or agent. Deposit protection requirements were introduced in Scotland in the Tenancy Deposit Schemes (Scotland) Regulations 2011 and apply to all those types of tenancy which are included in the *landlord registration provisions. Therefore, landlords in Scotland who are required to register, and who take a deposit from a tenant, are required to comply with the regulations and protect the deposit in an approved scheme. Since 2013, private landlords in Northern Ireland have also had to protect tenants' deposits in a government-approved scheme.

depreciated replacement cost Method of valuing property that is rarely, or never, sold in the market. *See also* CONTRACTOR'S METHOD.

depreciation A function, used in accounting, which allows for the cost of a *wasting asset to be distributed across a number of years' financial statements, on the assumption that that asset will be used by the business for more than one accounting period and will itself also fall in value as it becomes older. For example, a piece of machinery bought in one year is likely to be used by that organization for several years before either being scrapped or sold (normally at a lower value than it was bought for originally); therefore, it would be unrealistic for the initial cost of that machinery to be shown as a cost to the business for only one year, as this would mean that the *profits of that business would be artificially low in the year of purchase, but then artificially high in later years (potentially meaning that insufficient funds would be retained in the business to replace that piece of machinery at the end of its life). There are two main ways of showing depreciation: **straight line depreciation** and **reducing balance**. In straight line depreciation an equal charge is made every year for the useful lifetime of the asset. In reducing balance depreciation, any depreciation charges made in previous accounting periods are taken away from the original cost of the asset before a percentage depreciation rate for that accounting period is then taken away. This means that the depreciation charge is usually much higher in the earlier life of the asset, which makes it more suitable for assets such as cars, which tend to depreciate more in the first years after they are bought new than as they get older. *See also* CONTRACTOR'S METHOD.

Derris An organic insecticide, also known as rotenone, derived from the roots of derris, a climbing leguminous plant of Southeast Asia and the Pacific Islands. It was widely used in liquid or powder form, especially in gardens to control *aphids and other insect pests, until banned in Europe in 2009.

desalinization The removal of salt from seawater by forcing it through filters or by distillation, boiling off the water leaving the salt behind. It some areas of the world where fresh water is very limited, it may be the only method of ensuring adequate supplies, but the process uses significant amounts of energy. Saline soils have a high salt content that limits growth of crops. Most occur in arid or semi-desert areas where there is insufficient rainfall to leach the salt from the soil. This is becoming an increasing problem due to climate change, with millions of hectares affected around the world. Irrigation can exacerbate the problem by adding salts in the irrigation water. In these areas the desalination of soils is achieved by application of gypsum (calcium sulphate dihydrate).

desiccant A chemical that is hygroscopic, i.e. it removes water. In farming, a desiccant is used to artificially ripen a crop immediately before harvest to remove any green plant matter, weeds, or unripe crop, to ease harvesting. *Glyphosate is commonly used for this purpose.

design and access statement Supporting statement required with some *planning applications, and applications for *listed building consent, in the *United Kingdom which sets out how design and access principles have been considered by the applicant within the proposed *development. The precise aspects that must be covered in a design and access statement differ between application types, and are set out within *legislation; however, they normally include information about how the site is to be accessed by its users, in addition to information about how local design characteristics, and the context of the development (for example, common building materials used in the local area), have been reflected in the design of the development. In Wales, in 2017 the requirement for a design and access statement to support applications for listed building consent was replaced by a requirement for a *heritage impact statement.

designations Most commonly used to refer to an area or feature that has been recognized for its environmental worth and merits a degree of protection. The designation will be with a defined and mapped boundary, and can be supported by legislation such as with *Special Sites of Scientific Interest or *National Parks. Some designated areas have a lower level of protection, for example from development, such as *Local Wildlife Sites and *Ancient Woodland. Areas of countryside can also be designated to reflect areas of particular environmental importance to help target delivery of *agri-environment schemes.

detailed planning permission Type of *planning permission granted in Scotland when full details of the application have been provided at the point of application. It is equivalent to *full planning permission in England, Wales, and Northern Ireland, and may also be granted subject to planning *conditions. *See also* PLANNING PERMISSION IN PRINCIPLE (SCOTLAND).

developer contribution *See* SECTION 106 PLANNING OBLIGATION AGREEMENT.

developer's profit *See* DEVELOPMENT APPRAISAL; RESIDUAL VALUATION.

development Term, as defined in *United Kingdom planning *legislation, for a change to land or buildings (either physical or in terms of *change of use) which requires the grant of *planning permission. Development is defined within the **Town and Country Planning Act 1990, as amended** (England and Wales), the **Town and Country Planning (Scotland) Act 1997, as amended**, and the **Planning Act (Northern Ireland) 2011, as amended** as: 'The carrying out of building, engineering, mining or other operations, in, on, over or under land, or the making of any material change in the use of any buildings or other land'. (In Scotland there is an additional provision relating to the operation of marine fish farms.)

Anything covered by the definition requires planning permission; however, the legislation specifically excludes some things from the definition of development, and therefore from the requirement for planning permission. The use of land for the purposes of agriculture or forestry is an example of this (although the erection and alteration of most agricultural buildings is defined as development, and hence requires planning permission). Most internal works to buildings are also not included within the definition of development.

A system of *permitted development also operates in the United Kingdom. This grants blanket planning permissions (either at a national or local level) for certain types of development subject to their meeting certain criteria and conditions. Whilst this means that a developer does not have to submit a specific application for planning permission if something is covered by permitted development rights, the change itself is still included within the definition of development.

For changes which are either not excluded from the definition of development in legislation, or which are not permitted development, an application for planning permission must be submitted.

The planning permission regime (as defined by whether something is included within the definition of development) is one of a number of consent systems in operation in relation to changes to buildings or land. Examples include *listed building consent, *scheduled monument consent, and *building regulations regimes. The requirements for permission under each system vary, and therefore it is possible for the same change to require permission under only one system, or possibly under multiple regimes in parallel. *See also* PLANNING REFORM WALES.

development appraisal The process of assessing whether a potential *development is likely to be financially viable, when all costs, and the risks of the project, are taken into account. This type of process is undertaken to try to determine whether it is worthwhile undertaking a potential development (particularly when obtaining a *profit is the main motivator behind it), and can also be used as a tool in *planning viability assessments.

There are several potential approaches to development appraisal; however, at their roots sit what is essentially a *residual valuation method. Depending upon the aspect of the potential project that the developer is interested in assessing, this approach can be used to generate either an estimate of the developer's profit (i.e. the amount of profit the developer could make from the project if the cost of the site is already known or an assumption is made about it), or alternatively the amount that the developer can afford to pay for a site in order to still make a profit at a predetermined level.

The types of cost that will be included are the costs of purchasing the site which is to be developed, construction costs, planning fees and related costs, taxation costs, legal and other professional fees, marketing costs, **contingencies** (i.e. an allowance for unexpected costs), and the costs of borrowing money to carry out the development (or alternatively the related *opportunity cost i.e. the potential loss of income which could have been generated by doing something else with the money invested in the development; at its simplest, say the loss of interest had the money been left in the bank).

A figure is determined for the probable value of the completed development (known as the *gross development value), and the costs, including purchasing the site, are then taken away from this, leaving behind the figure for the predicted profit from the development:

gross development value

– costs, including purchasing the site, but excluding the developer's profit

= developer's profit

Alternatively, the developer may already know what level of profit he or she is willing to accept in order to undertake the development (bearing in mind whether there is a high level of risk attached to the project or not), in which case they may use an altered version of the same approach to assess how much they would be willing to pay for the site that they are considering developing:

gross development value

– costs, excluding buying the site, but including developer's profit

= amount that can be paid to buy the site

(In some circumstances, therefore, this approach is also used to assess the
*market value of a potential development site—*see* RESIDUAL VALUATION
METHOD—although *comparative valuation approaches are also used when
suitable comparables are available.)

At the early stages of a project, the potential profit can be predicted with
significantly less certainty than as the project continues, because the costs (and
also the final value of the development) can still be subject to changing external
factors (such as a decline in the market resulting in a reduction in value of the
final development, or an increase in *interest rates during the project which
affects the cost of financing it). Whilst it may be possible to reduce some of these
uncertainties, for example, by agreeing a fixed rate of interest, more sophisticated
development appraisal models aim to build in some of this uncertainty, by
including, at the appraisal stage, the potential impacts of some of these events on
the profitability of the project. This is referred to as conducting a **sensitivity
analysis**. In conducting such an analysis, the assessments of potential profit are
run several times with key factors altered; so say different *interest rates used.
This will tell the developer what level the relevant factor (say interest rates) would
have to change to in order for the project not to be financially attractive, and will
therefore help them to assess the risks of the project. (Although the example here
addresses the use of sensitivity analyses in development appraisal, they can also
be used in assessing the risks attached to other investments.)

At its most simple, the approach discussed above also assumes that any
borrowings are spread evenly through the life of the project and costs are
incurred right from the start of it (neither of which is likely to be the case). Hence,
development appraisals can be developed at a more sophisticated level using
*discounted cash flow approaches which try to more realistically account for the
likely cash flow patterns into and out of the project. These can then be used to
provide other measures of whether the potential project is likely to be worth
pursuing, including, for example, an **internal rate of return** (*see* PAYBACK
PERIOD for more information on internal rates of return). There are a number of
computer software packages that can be used to carry out this kind of analysis
which can also include incorporating sensitivity analyses.

Development Consent Order Form of permission granted by central
government in England and Wales, under the *infrastructure planning regime for
nationally significant infrastructure projects and incorporating *planning
permission together, potentially, with *compulsory purchase rights. They are
designed to facilitate infrastructure projects without the need for the relevant
authority to obtain multiple permissions, so may also include other such aspects
as *listed building consent or permission to divert highways. Where they
incorporate compulsory purchase rights they have the effect of a *compulsory
purchase order.

development control Term used in the *United Kingdom to describe the
process by which potential *development is controlled through the operation of

the *planning permission system. It is distinct from *building control, which describes the operation of the *building regulations system.

Development Control Advice Notes *See* PLANNING POLICIES.

development plan *See* LOCAL PLAN.

devolution Process of passing powers from central government to more localized levels. The United Kingdom Parliament has passed the rights to make laws in a number of areas to devolved administrations in Scotland (*see* SCOTTISH GOVERNMENT), Northern Ireland (*see* NORTHERN IRELAND EXECUTIVE) and Wales (*see* WELSH GOVERNMENT).

devolved *See* DEVOLUTION.

dewlap The fold of loose skin that hangs below the neck or throat of certain animals, notably cattle, dogs, rabbits, and some poultry. It appears to have little purpose.

dew pond An artificially created *pond, usually on the top of a hill, providing natural drinking water for *livestock. In construction, a layer of *straw is laid in a hollow in the ground and covered with a thick layer of *puddled clay. It is thought that the straw is critical, acting as insulation that produces the temperature difference that allows dew to form and fill the pond over time. They are common on the *downland of southern England, many of them being centuries or even thousands of years old.

Dexter A breed of *cattle, the smallest in Europe, originating in the south-west of Ireland. Standing only around 1 metre at the shoulder, it is little more than half the size of large breeds. Coloured red, black, or dun, the cow produces a rich milk, but the breed is more often kept for beef as it has the reputation for outstanding eating qualities.

diameter at breast height A measure taken at a practical height of a standing tree (typically measured at 1.3 metres from the ground) in order that the volume can be estimated. Specialist girth tapes calibrated in diameter are available. *See also* BASAL AREA; STANDING TIMBER VOLUME.

dicotyledon A flowering plant that has two *cotyledons or embryonic leaves as it emerges from the soil. This characteristic is used to categorize angiosperm flowering plants. *See also* COTYLEDON.

dieback A general description of the dying back of vegetation, starting at the plant's tips, that can be caused by disease or climatic factors. Frequently now cited in relation to Chalara dieback (*Chalara fraxinea*), which is a fatal disease affecting ash trees, causing leaf loss and crown dieback. Young ash trees are especially susceptible and can be killed quickly, whilst older trees can resist for several seasons. This disease was confirmed in the UK in 2012, but it is established and has had significant impacts on ash populations in mainland

Europe. Oak dieback or **decline** affects oaks and is characterized by dead branches and vegetation in the upper areas of the tree.

differentiation Mechanism by which an organization seeks to distinguish its brands, products, or services by creating and marketing them in such a way that they are likely to be perceived to be different from competitors' offerings and therefore attractive to target customers. Almost all elements of the *marketing mix can be adjusted to generate differentiation and create *unique selling propositions. An individual organization may seek to create differentiation in a number of ways, for example, by selling products or services under several brand names, each of which are attractive to particular parts of the market (perhaps in terms of buyers' age or income), or by creating differentiated products or brand families (overarching brand names under which an organization may offer several product ranges). *See also* SEGMENTATION.

diffuse pollution The pollution of watercourses can be divided into point source and diffuse. Point source pollution is where a pollutant enters the watercourse at a single point such as a factory or sewage works, and diffuse pollution is where the pollutants come from many sources in small quantities. The most common source of diffuse pollution is *agriculture and it is often coupled with *soil erosion, when rainfall is sufficiently heavy that it cannot percolate into the soil, and it may *runoff into a nearby ditch or stream. The water carries with it soil particles together with *nutrients and *pesticide residues. The nutrients can cause excessive growth of *weed and *algae in the *watercourse, the latter called algal bloom which may give off toxins causing further polluting effects. Pesticide residues can have a damaging impact; for example, an *insecticide may kill the aquatic insects on which certain fish depend.

digestate *See* ANAEROBIC DIGESTION.

digestibility A measure of how easily an animal feed can be digested or how much of the nutrients can be absorbed by the animal. For example, the yield of grass will rise as it matures, but the digestibility declines making the timing of cutting for hay or silage a matter of judgement.

digestion The process by which food, carbohydrate, protein, and fat, is broken down into soluble compounds that can be absorbed into the blood stream. This is achieved firstly by chewing and then by the action of enzymes in the mouth, stomach, and digestive tract.

dilapidations Repairs required during, or at the end of, a *lease or *tenancy and which are, therefore, in effect, a *breach of covenant.

diligence *See* BREACH OF COVENANT; REGISTERS OF SCOTLAND.

diploid Containing two sets of chromosomes. Most cells or nuclei contain this pair of sets, one derived from each of the parents.

dipping The immersion of sheep in a bath of water containing an insecticide to control external parasites including **blowfly**, **ticks**, **keds**, **lice**, and **scab mites**. Since the 1950s the active ingredients of sheep dips have been based on organophosphate insecticides which are highly toxic and are increasingly under review by regulators. Operators must be protected, and the spent liquid must not be allowed into watercourses.

direct drilling The sowing of a crop directly into the soil which has been untouched since the removal of the previous crop. Also known as zero tillage, it has the advantage of not disturbing the soil structure, but requires a drill specifically designed for the purpose and may not be effective when the soil is wet or heavy.

Directive, EU A *European Union legislative act that defines a particular goal to be achieved. Individual member countries are thus required to adapt or devise their own laws in order to achieve the goal. Examples include the *Habitats Directive and the *Water Framework Directive.

direct seeding A forestry term relating to the method of sowing of seed directly onto the area of ground for the proposed crop. This is normally preceded by some form of ground cultivations. It has potential benefits but it is not generally a common form of *tree establishment, and results can be less reliable and unpredictable in comparison with more conventional planting techniques. Temperate taungya is a specific description for direct seeding of trees under an arable crop.

Disadvantaged Area (DA) *See* LESS FAVOURED AREA.

disc harrow An agricultural implement designed to move the surface of the soil and to chop any plant residue. It is formed of concave metal discs, which may be scalloped, around a central spindle which rotate as the implement is pulled forward. The offset of the discs determines how vigorously the soil is moved. It may be trailed or mounted on the tractor's *three-point hydraulic linkage. *See also* HARROW; CHAIN HARROW.

discounted cash flow (DCF) An assessment of the flow of cash into and out of an enterprise, or project, which takes into account (by using *discounting) that cash received now will have a greater value than cash received in several years' time (because cash received now will have earned *interest for several years more). Discounted cash flows underpin a number of methods of appraising whether potential investments in new projects are worth pursuing, including **net present value** and **internal rate of return**. For more details, *see* PAYBACK PERIOD. *See also* DEVELOPMENT APPRAISAL.

discounting A means of accounting for the time cost of money, i.e. that where you have to wait to receive a sum of money you will have lost the *interest that would have been available had you had that money today. Using a discounting mechanism therefore allows for the effect of interest to be accounted for when determining what sum of money will need to be invested now in order to receive a defined amount of money at a future date. For example, in order to have £100 in a savings account at the end of the year, if the savings account is paying a rate of 5% interest paid annually, then you would need to invest £95.24 at the beginning of the year:

£95.24 initial balance + £4.76 interest at 5%
= £100.00 balance at the end of the year

This can be expressed using the term **present value**. In this case, the present value of £100.00 to be received at the end of one year, assuming a discount rate of 5%, would be £95.24. Depending upon the context in which it is used, a discount rate is equivalent to the *interest rate, or the *yield on an investment, or an investor's target rate of return (what rate of return an investor would accept for the risk of investing in something), or the cost of capital (effectively what rate an organization has to pay in order to borrow money).

To calculate this, a **discount factor** can be used. In this case, the discount factor would be 0.9524, so £100 x 0.9524 = £95.24 (i.e. the amount you would need to invest to get £100 at the end of one year at an interest rate of 5%). A number of different sets of discount rates are published (such as *Parry's Valuation and Investment Tables and online/computer program equivalent) in order to account for different investment scenarios. In this particular instance this discount factor comes from the **present value of £1** table which gives discount factors to calculate the amount that needs to be invested today in order to accumulate £1 at a future date at different discount rates (taking into account **compound interest**). So, if someone wished to have £200 in their savings account in four years' time, and it is in an account paying a fixed rate of 8% interest, then they would have to invest:

PV of £1 in 4 years @ 8% = 0.7350
£200 x 0.7350 = £147

So the present value of £200 in four years' time at 8% is £147. This can be broken down accordingly:

Initial balance: £147.00
End of Year 1: £147.00 balance + £11.76 interest (at 8%) = £158.76
End of Year 2: £158.76 balance + £12.70 interest (at 8%) = £171.46
End of Year 3: £171.46 balance + £13.72 interest (at 8%) = £185.18
End of Year 4: £185.18 balance + £14.81 interest (at 8%) = £199.99 (rounding error)

This assumes that the interest is compounded, i.e. that all the money (and interest) is left in the account until it is needed after four years. However, in some cases, people will want to know how much they need to invest in order to receive a regular income from their investment at a specified level of interest. A different table is produced to account for this: **years' purchase (single rate) (or the present value of £1 per annum)**, i.e. the right to receive £1 at the end of each year for 'n' years at 'i' rate of compound interest. So:

YP (single rate) 4 years @ 6% = 3.4651

gives a discount factor which will tell someone how much they have to invest now in order to earn a given amount for four years if their investment is paying a rate of 6% interest. So, if someone wishes to earn £2,000 a year for four years (and they are investing at a rate of 6%) then they would need to calculate:

£2,000 x 3.4651 = £6,930.20

So they would need to invest £6,930.20 in order to receive £2,000 a year for four years. This can be broken down as:

Initial balance: £6,930.20
End of Year 1: £6,930.20 balance + £415.81 interest (at 6%) = £7,346.01
£2,000.00 withdrawn
£5,346.01 balance
End of Year 2: £5,346.01 balance + £320.76 interest (at 6%) = £5,666.77
£2,000.00 withdrawn
£3,666.77 balance
End of Year 3: £3,666.77 balance + £220.01 interest (at 6%) = £3,886.78
£2,000.00 withdrawn
£1,886.78 balance
End of Year 4: £1,886.78 balance + £113.21 interest (at 6%) = £1,999.99
(rounding error)

It should be noted that once the last £2,000 has been withdrawn there is no money left, so a wise investor would have set up a *sinking fund in order to ensure that their original investment was not lost (and there are various tables available which can deal with this).

In some kinds of investment there will be an expectation that the income will continue to be received indefinitely, or '**in perpetuity**' (for example, the *freehold of a let office unit). In this case, a **years' purchase in perpetuity (the present value of £1 per annum in perpetuity)** table will be used.

discretionary trust *See* TRUST.

disease An illness or abnormality that may be caused by a pathogen, such as a bacterium, virus, or fungus. In the context of farming, a disease may infect animals or plants. Some are *notifiable in that once identified the incidence must be reported to Government authorities, at which point a statutory eradication programme will come into effect. These include *foot and mouth disease and avian influenza.

Display Energy Certificate (DEC) Certificate which, by law in England, Wales, and Northern Ireland, must be displayed in a building which is occupied by a public authority, is frequently visited by the general public, and which provides information about the energy efficiency of that building (in Scotland, an *Energy Performance Certificate must be displayed). The energy performance of such buildings is given in a band from A (the most efficient) to G (the least efficient). The display of such certificates is designed to raise public awareness of energy use in buildings.

Diseases of farmed animals in the UK

Cattle and Sheep		Pigs and Poultry
Anthrax	Foot rot	African/classical swine fever
Bloat	Hypocalcaemia	Aujeszky's disease
Bluetongue	Hypomagnesaemia	Avian influenza
Bovine tuberculosis	Liver fluke	Blue ear
Bovine spongiform encephalopathy (BSE, mad cow disease)	Mastitis	Coccidiosis
	Ringworm	Foot and mouth disease
	Salmonella	Newcastle disease
Bovine viral diarrhoea	Scrapie	Porcine epidemic diarrhoea
Brucellosis	Watery mouth (*E. coli*)	Rotavirus
Campylobacteriosis	Worms: lungworm, tapeworm, nematodes	Salmonella
Coccidiosis		Swine dysentery
Enzootic abortion (chlamydia)		Swine vesicular disease
Foot and mouth disease		

Diseases of arable crops in the UK

Cereals		Brassicas
Barley yellow dwarf virus	Loose smut	Alternaria leaf spot
Brown rust	Net blotch	Damping off
Bunt or stinking smut	Powdery mildew	Club root
Cephalosporium leaf stripe	Rynchosporium	Downy mildew
Covered smut	Septoria	Powdery mildew
Crown rust	Sharp eyespot	Sclerotinia
Downy mildew	Snow mould	
Ergot	Sooty moulds	
Eyespot	Take-all	
Foot rot	Yellow rust	
Fusarium		

disposition *See* CONVEYANCING.

dispute resolution *See* ALTERNATIVE DISPUTE RESOLUTION; COURTS, LAW.

distillers grains Also known as brewers' grains, these are a cereal by-product of the brewing and distillery industries that are used as animal feed. *Malting barley, or occasionally other grains such as *wheat, rye, or *maize, are used to provide the sugars in the production of beer or spirit in the process of malting, in which the grains are encouraged to germinate, converting starch into sugars. Once this process has taken place, the grains with the sugars removed are sold as animal feed.

distress Old remedy available to *landlords for non-payment of *rent in England and Wales which involved the landlord selling goods from the *tenant to pay the outstanding amount (subject to restrictions on the types of goods that could be taken, and the procedure used). It was much criticized on human rights grounds, and also that it was not consistent with the implied *covenant that landlords should allow their tenant to have quiet enjoyment. Therefore, in 2014 it was abolished and replaced by *Commercial Rent Arrears Recovery (CRAR).

district council *See* LOCAL AUTHORITY.

District Valuer Official of the *Valuation Office Agency (England,Wales, and Scotland) or *Land and Property Services (Northern Ireland) responsible for providing valuation advice to government authorities (such as Her Majesty's Revenue and Customs) in relation to the payment of property-related taxation. *See also* SCOTTISH ASSESSOR.

disturbance *See* COMPULSORY PURCHASE; AGRICULTURAL HOLDINGS ACT 1986.

diversification, farm Process where a farming business develops new enterprises using the farm's assets, with a view to creating new income streams, usually in addition to the business's traditional farming activities (such as their existing livestock or arable production enterprises). Farming businesses commonly diversify for a number of reasons, from spreading the risk of fluctuating income levels from farming (say as a consequence of a poor harvest), to a wish to use redundant assets to create additional income (such as buildings which are no longer big enough to accommodate modern farm machinery), to personal interest in a project. Common forms of diversification include the conversion of redundant farm buildings to offices to let, through to providing tourist accommodation, to starting food processing businesses (say using milk from the farm's dairy herd to make butter and cheese), to producing new agricultural products (such as fish farming or growing flowers), to farm shops, or renewable energy production.

Because the types of activities that farming businesses undertake to diversify their farm income vary so widely, data sets on diversification can include or exclude different things. For example, some farmers will undertake farm *contracting work, using their machinery to carry out specific tasks for other farms, such as baling hay or hedge cutting. Some data on diversification include contracting work as a form of diversification, whilst others exclude it, arguing that contracting work is difficult to distinguish from the normal farming activities of a farm, as they utilize the same machinery that would be used on the farm on a normal day-to-day basis. Similarly, whilst some datasets include the letting out of farm buildings to third parties for non-farm uses such as offices, others exclude this.

diversity, biological Normally refers to the variety of species within a particular *habitat, but can also describe the variety of habitats within a specific area. A simple measure of biodiversity is the number of species within a defined area, i.e. species richness. *See also* BIODIVERSITY; ECOSYSTEM.

divisible balance *See* PROFITS METHOD.

domestic microgeneration equipment Equipment located at a dwelling which is used to generate energy for that property. *See also* COMBINED HEAT AND POWER; FEED-IN TARIFF; RENEWABLE ENERGY; RENEWABLE HEAT INCENTIVE.

domestic rates Local tax paid by the owners and occupiers of domestic property in Northern Ireland. It operates on a similar basis to *Council Tax in England, Scotland, and Wales, but there are some differences. For example, unlike with Council Tax, owners of let residential properties can be liable for the domestic rates if the *capital value of the property is lower than a threshold level.

dominant Species that have an overriding influence on the *ecosystem and that have the characteristics which allow them to monopolize resources can be described as dominant species. More casually, the term can also be used to refer to species that are the largest or most abundant in a particular area. *See also* BIODIVERSITY.

dominant tenement *See* TENEMENT.

dormer window *See* APPENDIX 2 (ILLUSTRATED BUILDING TERMS).

Dorset Down A breed of *sheep native to the Dorset Downs. The fact that lambs are fast maturing and that ewes can be served by the ram between July and November make them suitable for early spring lamb production. It is recognized as a rare breed by the Rare Breeds Survival Trust.

Dorset Horn A breed of *sheep with large curly horns, although a polled strain has been bred. It is noted largely for its prolificacy, as it is able to breed all year round. It is recognized as a rare breed by the Rare Breeds Survival Trust.

double cropping The growing of two crops on the same land in a single season. The second crop may be sown immediately after the harvest of the first, or even sown into the first before harvest. In some cases the two crops are grown simultaneously. The practice is more common in vegetable and fruit production but can only be achieved in a climate where the growing season is long enough.

double lows Varieties of *oilseed rape, the seed of which have a low content of eurucic acid and glucosinolates in the oil. Oil is extracted to be used as cooking oil, whilst the residue is used as animal feed. Both of these compounds are harmful to humans and animals, so plant breeders developed varieties with low concentrations. With the exception of the limited production of oilseed rape for the industrial market, all varieties grown in the UK are now double lows.

down breeds Breeds of *sheep that traditionally grazed the *downland areas of the country. They are relatively hardy with thick fleeces but also have good meat characteristics so are often used as terminal sires. Examples are Hampshire Down or *Dorset Down.

downland A characteristic *landscape formed on the shallow lime-rich soils normally found on limestone or chalk. This is moulded by the grazing impacts of wild animals and/or *livestock, and the prevalence of calcicole *flora, i.e. plants that can thrive in lime-rich soil, and their associated *fauna. Chalk and limestone *grassland are an important farmland *habitat, but a significant proportion of the remaining downland in the UK has been lost to agricultural *intensification. *See also* CALCAREOUS; OPEN ACCESS.

downy mildew Fungal disease of a range of plants including grapevines, lettuce, onions, peas, and tobacco. It is caused by several *fungi, mainly species of peronospora, plasmopara, and bremia. Yellow or brown blotches appear on the upper side of the leaf, with whitish downy patches on the underside. The disease thrives in cool, damp conditions but can be controlled by *fungicide.

draft horse (heavy horse) A type of *horse characterized by its power and strength, together with its calm and good nature. Used in agriculture, forestry, and urban environments to pull heavy loads. Their commercial uses in developed countries have all but disappeared, and they are more commonly used today in vintage and specialist shows. The Clydesdale, the Suffolk Punch, and the Shire are amongst the more well known breeds of these horses.

drag hunting A non-competitive *cross country equestrian sport that involves the use of a specially laid trail rather than the pursuit of a wild animal. The drag is normally a piece of material which holds a scent, typically consisting of urine and aniseed. The trail is laid by a drag-man either on foot or from a horse or *quad bike. The length of the trail, or line as it is normally termed, varies, but is typically around 3 miles. Hounds, usually foxhounds, are then laid on the trail and followed by the mounted participants, or *field. Bloodhounds are another breed of hound that are used to hunt 'the clean boot', i.e. the natural scent of a human runner. Because of the similarity in that both follow artificial lines, they are frequently treated as one and the same. *See also* HUNTING; TRAIL HUNTING.

(⊕) SEE WEB LINKS
• Website of the Masters of Draghound and Bloodhounds Association: details the governance of drag hunting and bloodhound hunting.

drain A channel or pipe carrying surplus liquid or the process of that liquid. In terms of agriculture, surplus rainfall is removed from soils by means of **drainage**.

Porous pipes may be laid beneath the soil to collect the water, which then runs down the pipe into a **drainage ditch** by means of gravity. This may also by facilitated by means of a *mole plough that creates a tubular channel through the soil without the laying of a pipe. Where the land to be drained is below the level of the ditch, pumps may be used to pump the water up into the drainage ditch, which may then be known as a dyke, rhein, rhyne, or rhine.

draught 1. An animal used to pull agricultural equipment, as in draught horse. Also spelt draft, especially in American English. (*See also* OX.) **2.** A current of air or a quantity of liquid, particularly of beer from a cask or keg rather than a bottle.

drawbar A bar, usually metal, on the rear of a vehicle, such as a tractor, for towing a trailer or implement. *See also* TOWBAR.

drench A liquid medicine given to farm animals or to administer such liquid. For example, an *anthelmintic drug may be administered to sheep or cattle by mouth, using a dispensing gun, to control worm parasites.

dressage French for 'training'; a demonstration of a disciplined partnership between the rider and the horse. It has its basis in military training, and is now a competitive equestrian sport that can be ridden as a single competition or as one of the elements of *eventing. Dressage competitions are normally undertaken as a test of a predetermined set of movements which are judged by a single judge or panel of judges.

dribble bars A boom with nozzles spread along the length for the application of liquids to land. This often describes the spreading of *slurry, as the liquid comes out of the nozzle close to the ground rather than by spreading it by throwing it into the air, for example, by a spinning disc, thus limiting air pollution. Can also apply to water *irrigation.

drill 1. A farm machine for planting *seed or to plant seed in the ground. There are numerous designs of drill; some apply fertilizer to the soil at the same time as the seed, known as combine drills. The seed may be sown into a slot or channel cause by a tine or disc. The seed may move under gravity or wind may be used, in which case the implement may be called an **air seeder**. A *direct drill is one which is designed to sow seed into soil that has not been disturbed since the removal of the previous crop. **2.** A tool, usually electric, for drilling holes.

drip irrigation A form of *irrigation that delivers water to the roots of plant by pipes on the surface of the soil or just below, with nozzles to emit small quantities of water. This is more efficient than irrigating crops by spraying the water into the air, as it minimizes evaporation.

driven shooting Typically in Britain this will refer to the shooting of *game birds where a line of *beaters drive the birds towards a waiting line of shooting participants or 'guns'. A similar process can be applied to the driving of other *quarry species, including hare and wild boar. Driven shooting evolved from imported variations from continental Europe, and became established in the nineteenth century. *See also* SHOOTING; SEWELLING; WALKING UP.

driving Can be applied as a general term to normally describe the hitching of horses and ponies to a wheeled vehicle or sleigh by a harness. These can be used for a wide range of purposes that can be for working, recreational use, or competition. Carriage driving is a more familiar type of competitive driving using two- or four-wheeled carriages.

(⊕) SEE WEB LINKS
• Website of the British Driving Society: supports and assists those interested in equine driving.

drove A herd or flock of animals being walked from one place to another; for example, from farm to market. Sheep fairs used to be held in early autumn at which flock replacements were bought and sold. These fairs required sheep to be driven to and from the market. A drover is a shepherd or cowman responsible for the driving of the animals, whilst a drove road or drove way is the track along which the animals were driven. These are usually broad lanes of great antiquity, perhaps 20 metres wide, with hedges or dry stone walls on each side to prevent the stock from straying.

dry cow A *dairy cow that is not being milked, usually because it is in an advanced state of pregnancy. This is an important period when the cow can recover between lactations and gain body weight if necessary.

dry heating system Heating system which is non-water based and normally comprises heat sources such as blown hot air, or *night storage heaters (although night storage heaters are typically independent of each other, and are therefore not, strictly speaking, a central heating system). *See also* WET HEATING SYSTEM.

dry matter The constituent of an animal feed once all moisture has been removed; the solids. The nutrients of a feedstuff are contained within the dry matter so that the drier the feed, the more nutrients it may contain. **Dry matter intake** is a measure of the feed an animal can consume on a dry matter basis. However, the *digestibility of the feed may decline if it becomes too dry. The dry matter content of *silage, for example, is an important measure of nutritional quality.

dry stone wall Walls built without the use of cement or mortar are common in *farming regions across the UK, particularly where *hedges do not grow easily, and where stone and rock are found above the *soil. They are therefore more prominent in northern and western regions, where they are built as livestock barriers and to define field boundaries. They are an important and recognized feature of the farming *landscape, and the building and maintenance of walls is often grant aided using funding from *agri-environment schemes. *See also* FIELD BOUNDARIES.

duck The common name for a number of waterfowl. *See also* POULTRY.

due diligence The process of a business or organization undertaking a series of structured checks and investigations before undertaking a particular activity in order to ensure that they are fully compliant with the law and any other obligations (for example, in relation to their duty of care to their client, employees, or the general public). The structured nature of the due diligence process is designed to avoid situations in which potentially problematic issues may be accidentally overlooked, leading to allegations of breaking the law or of *negligence.

dun *See* HORSE.

dune A recognizable *habitat, the sand dune occurs where the shape of a beach allows for sand to dry out and is blown inland to create a strip, or low hills and hollows. Inhospitable to vegetation, except specialist species able to colonize and tolerate the low nutrient, instability and hydrology of sand dune systems. Arguably includes some of the only true *wilderness areas remaining in the UK, but sites are also often under severe pressure from human impact and recreational pressures.

dung The faeces of an animal. It is often taken to describe the mixture of faeces with animal bedding such as straw, when it is also known as *farmyard manure.

durum A species of *wheat, *Triticum durum*, specifically grown for pasta production. It is the hardest of all wheats, a quality in the milling process, with high protein content. Its milling qualities make it more suitable for pasta than for bread. Grown widely in Mediterranean countries, but less in the UK.

Dutch barn *See* BARN.

Dutch elm disease A devastating fungal *disease (*Ceratocystis ulmi*) of elm trees. The fungus is passed from tree to tree by elm bark beetles which infest the areas under the bark of diseased trees. The disease appeared in north-west Europe in the early part of the twentieth century and has been prevalent in Britain since the 1960s. Its major impacts, particularly on the English elm, are still very evident today, and with only a few exceptions there are no mature trees surviving.

duty of care An obligation to take reasonable care to protect the well-being, safety, or welfare of others, of property, or of the environment. The term is used widely in a number of contexts; for example, when considering the legal obligations of businesses in relation to the health and safety of their employees, customers, and the general public. Breaches of an organization's duty of care may lead to claims of *negligence being made against it. The term 'duty of care' is also used specifically in relation to the obligations placed upon farming businesses by the *agricultural waste regulations.

DV *See* DISTRICT VALUER.

D value The digestibility of a feed, used as a measure of quality of *silage. A measure of the digestible organic matter of the feed expressed as a percentage of the total *dry matter, used to determine the **metabolizable energy**—the amount of energy available to the animal. For grass silage, for example, the digestibility falls as the plant matures, especially when it starts to flower. A typical value might be 70%.

early neutral evaluation *See* ALTERNATIVE DISPUTE RESOLUTION.

easement A legal right, attaching to land, and applying in England, Wales, and Northern Ireland, which can be exercised by one party over the land of another (in Scotland the broadly equivalent right is known as a *servitude). This includes rights to either use, or restrict the use of, another's land (although creating new easements which restrict the use of another's land is now either not possible, or not viewed positively by the *courts, in the *United Kingdom). An example of an easement of the type commonly found in the UK would be one where a farmer who owns one area of land has the right to cross land owned by a neighbouring farmer in order to access the public highway. Easements are rights which normally attach to the land permanently; hence if the land is sold, the right or (where recognized) the restriction usually passes to the new owner.

In most cases, in order for there to be an easement there must be a **benefited property** (also sometimes referred to as the **dominant tenement**) and a **burdened property** (also known as the **servient tenement**). So, for example, if Farmer A, in order to access his own land from the road, has the right to cross Farmer B's land, then Farmer A's land is the benefited property, and Farmer B's land is the burdened property. Generally the two pieces of land must be close enough to have some connection between the two (though they do not necessarily need to be directly adjacent to each other).

Utility companies are a slight exception from this rule, however, in that they have *statutory powers to create easements when they install equipment on property not owned by them (or adjacent to land owned by them), as they will wish to retain some legal rights over the land through which their apparatus runs (for example, a water pipeline). A one-off payment will be made to the landowner by the utility company, and the landowner may often continue to use the land over which the pipeline runs (although *health and safety law may restrict what they can do with it to prevent damage to the pipeline). Because the easement is a permanent legal right which attaches to the land, it is not possible for the landowner to decide to terminate it.

Easements are distinct from *wayleaves (which are also sometimes used by utility companies, for example, for electricity lines). Wayleaves are a form of agreement between the utility company and the landowner for a period of time (depending on the type of wayleave, this can be for a period of decades or may run from year to year). Regular payments are typically made to the landowner for the wayleave (often on an annual basis). Whilst wayleaves can also bind future owners, there are some circumstances in which a landowner can terminate a wayleave, though whether this is possible depends on the nature and circumstances of the particular wayleave. *See also* COMPULSORY PURCHASE.

EBLEX *See* LEVY BODIES.

Ecological Focus Areas A component of the *Common Agricultural Policy *greening measures that comprise seven options where agricultural practices are required that are of benefit to the environment. These options are hedges, margins, agro-forestry, fallow land, nitrogen-fixing crops, green cover, and catch crops. Within a prescribed framework the options must add up to 5% of the *arable land. Smaller farms, those that are predominantly *permanent pasture, and *organic farms are exempt. *See also* CROSS-COMPLIANCE.

ecological networks A collection or grouping of high-quality *biodiversity sites and associated ecological linkages. These linkages provide the connectivity between the sites that enable *species to move. Crucially important is the mobility of genes within these networks. Ecological networks may be used and operate at different scales globally across and between countries and regions, or even at the level of an individual farm.

ecological zones Large global-scale areas with parameters that define approximate zones of isolation. These zones are characterized by broad groupings of *flora and *fauna, and include examples such as desert, and tropical rainforest. Where there is some equivalence in the type of flora and abiotic factors such as temperature and rainfall, the ecological zone can be described as a **biome. Agro-ecological zones** describe areas of land characterized by its physical factors and by an assessment of its agricultural suitability.

ecology The study of the relationships of living organisms to each other and their surroundings. The term is derived from the Greek *oikos*, meaning a 'house' or 'place to live', so the term 'ecology' effectively translates into 'the study of the earth's households'. It adopts a holistic approach, and although more commonly the various parts are studied separately, it is the synthesis of all the available information into an overall picture of the living organisms and their *environment that is important.

ecosystem A community of organisms or the *biotic component, and their physical environment or *abiotic component, interacting as an ecological unit. Different ecosystems are united to form the **biosphere**. *See also* ECOLOGY; NATIONAL ECOSYSTEM ASSESSMENT.

ecosystem services The benefits people derive from *ecosystems. These services can derive from a range of intermediate services and result in a direct good or benefit to humans, e.g. provision of clean water. The services are commonly categorized into **supporting services**, **provisioning services**, **regulating services**, and **cultural services**. *See also* NATIONAL ECOSYSTEM ASSESSMENT.

ectoparasite *See* PARASITE.

efflorescence Situation where a white powdery material is seen on the outside of bricks when salts within the brick leach out in the first few years after the construction of a building.

EIA *See* ENVIRONMENTAL IMPACT ASSESSMENT.

elder *See* WOODY SHRUBS.

electrical safety regulations Term commonly used to refer to *regulations concerning the safety of electrical wiring, appliances, and equipment in the *United Kingdom. These include duties on employers, and also *landlords of domestic and commercial premises, to ensure that their electrical installations and any appliances and equipment are safe. In practice this means that a testing and maintenance regime should be in place for wiring (in some circumstances the frequency of electrical wiring testing is set out in law). Portable electrical equipment and appliances should also be regularly inspected and, where necessary, tested (*see* PORTABLE APPLIANCE TESTING). Electricity safety is also addressed within the *building regulations (*building standards system in Scotland). *See also* HEALTH AND SAFETY EXECUTIVE; HEALTH AND SAFETY EXECUTIVE FOR NORTHERN IRELAND.

elephant grass *See* MISCANTHUS.

elm (English, wych) *See* TREE.

ELS/OELS *See* ENVIRONMENTAL STEWARDSHIP SCHEME.

employers' liability insurance Type of insurance which employers in the *United Kingdom are legally required to hold in order to cover the financial liabilities of an employer who is legally required to pay compensation, and legal costs, if an employee is injured, or made ill, in the course of their employment. A certificate confirming details of this insurance must be displayed at each place of work. *See also* PROFESSIONAL INDEMNITY INSURANCE.

employment law Body of *statute and *case law relating to employment. Legal disputes relating to employment law are normally heard in **employment tribunals** (*see* COURT, LAW). *See also* CONSTRUCTIVE DISMISSAL; EMPLOYERS' LIABILITY INSURANCE; HUMAN RESOURCES; MINIMUM WAGE; REDUNDANCY; STAKEHOLDER PENSION; VICARIOUS LIABILITY; WORKING TIME REGULATIONS.

employment tribunal *See* COURT, LAW.

EMS *See* ENVIRONMENTAL MANAGEMENT SYSTEM.

enabling development Provision within heritage *planning policy in the United Kingdom allowing for *planning permission to be granted for a *development which would not normally be acceptable in planning policy terms, on condition that the development is used to fund the long-term *conservation of a building, structure, or area with a heritage planning designation (for example, a *listed building). Because this provision justifies the granting of planning permission for something which would not normally be acceptable (for example, a housing development in a *Green Belt location where it would not normally be permitted, in order to fund the conservation of a listed mansion house) it is very tightly controlled. A range of conditions apply, including that the scale of the development is as small as possible, that it does not harm the heritage property, and that all other sources of grant funding have been exhausted. Enabling development will also not be permitted where the owner of the heritage property cannot afford to fund its upkeep because they have in some way acted imprudently; for example, by having paid too much to purchase it originally.

endoparasite *See* PARASITE.

endurance riding Horse races covering up to 50–150 miles, normally over a variety of terrains and held over one to three days. The health and fitness of the horse is an integral part of this type of competition, and horses are checked by veterinarians before, during, and after the ride. A successful endurance rider will complete the course within the time allowed and with a healthy and unstressed horse.

(⊕) SEE WEB LINKS
• Website of Endurance GB, the governing body for endurance riding in Great Britain: includes details of local, national, and international events.

energizer Equipment that provides an electrical current for an electric fence. Most commonly a conventional battery that can be recharged similar to that used for a vehicle, but may be more innovative, such as a small solar panel.

energy crops Crops grown for renewable energy in various forms. *Miscanthus (elephant grass) and *short-rotation coppice such as willow are grown for *biomass to be used for heat or electricity generation. *Oilseed rape may be grown for *biodiesel, and wheat, sugar beet, or potatoes for *bioethanol to be used as transport biofuels. Maize, rye, or other crops can be grown for feedstock for *anaerobic digestion plants that produce methane which can be burnt to generate electricity.

Energy Performance Certificate (EPC) Certificate required by law when a building in the *United Kingdom is either newly constructed, let, or sold, and which gives a measure of its current energy efficiency and any measures that could be taken to improve this. Whilst there are some exceptions set out in the *regulations, most types of either domestic or commercial building require an EPC when they are newly constructed, let, or sold. The purpose of the EPC is to let potential purchasers or *tenants know about how energy-efficient the property is likely to be (and therefore how much it will cost to heat and light), and the efficiency information is given as a band from A (the most efficient) to G (the least efficient). *See also* DISPLAY ENERGY CERTIFICATE.

enforcement notice *See* ENFORCEMENT, PLANNING; BUILDING REGULATIONS.

enforcement, planning Process of taking action against breaches in planning law in the United Kingdom; for example, where work has taken place without the appropriate permissions or consents, or not in compliance with planning *conditions. Enforcement action can be taken in relation to breaches of a wide range of types of permission and consents, including *planning permission, *listed building consent, breaches of *advertising controls, or failure to obtain relevant consents where required for works to trees (for example, in relation to trees in *Conservation Areas or protected by a Tree Preservation Order). Enforcement action is usually taken by the *local planning authority.

Where there has been a breach of planning law in relation to *development, the planning authority has a number of potential courses of action:

- To reach an informal agreement whereby the breach is rectified by the relevant owner or occupier (perhaps when it was as a result of a genuine mistake).
- To invite a **retrospective planning application** to be submitted (i.e. a planning application submitted after the work or change of use has been started). If permission is granted this regularizes the situation, but there is no guarantee that permission will be given.
- Serve an **enforcement notice** which sets out what the local planning authority considers to be the breach, and what activities are required to either stop that breach or remedy it. Non-compliance with an enforcement notice is an offence which is punishable by fine issued by the courts. If the responsible parties have not carried out the requirements of the enforcement notice, the local planning authority can also enter the land, carry out the works themselves, and then recover the costs from the owner of the land. There is a right of *appeal against the serving of an enforcement notice.
- If a local authority considers that it is necessary to take action to stop an activity before the deadline set out in the enforcement notice, then it may, in some circumstances, serve a **stop notice** alongside the enforcement notice which requires the activity detailed in the notice to stop more quickly.
- It is also able to serve a **temporary stop notice** independently of an enforcement notice. The effect of a temporary stop notice is to require a particular activity to stop immediately, and it is therefore used in the most urgent cases. It is usually followed by the issuing of an enforcement notice.
- If the breach relates specifically to a breach of conditions, then the local planning authority can serve a **breach of condition notice** to require compliance with the terms of a planning condition. This has a similar effect to an enforcement notice, and non-compliance is also an offence.
- In the most serious cases, the local planning authority can also apply to the courts for an **injunction**. Non-compliance with an injunction can result in a person being sent to prison.

In some cases, the local planning authority may serve a **planning contravention notice** at an early stage where it suspects there may have been a breach of planning control. Such a notice can be used to allow the local planning authority to require information to be provided about, for example, any operations on, or use of, land; or to invite the recipient of the notice to enter into communications with the planning authority about how any suspected breach of planning control may be remedied. Whilst these notices are sometimes used prior to the service of enforcement notices, this is not always the case, and sometimes an enforcement notice can be served without the previous use of a planning contravention notice.

There are several time limits which apply, beyond which enforcement action cannot normally be taken. The two most significant are:

- Four years after the substantial completion of something which amounts to operational development (broadly physical changes).
- Ten years after other breaches of planning control (most commonly changes of use).

(In Northern Ireland a standardized period of five years applies to both sets of circumstances.)

The enforcement processes for breaches of listed building consent are broadly similar to those for wider development control breaches. However, there are no time limits for the issuing of **listed building consent enforcement notices**, and listed building consent cannot be granted retrospectively. There is, however, a right of appeal against a listed building enforcement notice.

Anyone damaging or carrying out work on protected trees without the appropriate consent is guilty of an offence and may be fined. They also have a legal duty to replace the tree. Local authorities may either seek to negotiate with the owner to remedy the work done, or seek an injunction through the courts to stop any ongoing or anticipated breaches, or, in the most serious cases, carry out a prosecution.

For information about the enforcement regime for advertising, *see* ADVERTISING CONTROLS. *See also* BUILDING REGULATIONS; BUILDING STANDARDS SYSTEM.

enfranchisement, leasehold *See* LEASEHOLD REFORM.

English Heritage *See* HISTORIC ENGLAND.

English Nature *See* NATURAL ENGLAND.

English Woodland Grant Scheme (EWGS) This scheme (now closed) was part of the EU *Rural Development Programme for England (RDPE) administered by the *Forestry Commission and was directed at woodland owners and other interested parties. It included a number of separate publicly funded grants targeted to promote and support the creation and appropriate management of woodlands. (*See* COUNTRYSIDE STEWARDSHIP SCHEME for current forestry grants.) Forestry grants in Scotland are provided through the **Rural Development Contracts** administered by the Forestry Commission. Forestry grants in Wales are provided through the various *Glastir schemes and now administered through *Natural Resources Wales.

(⊕) SEE WEB LINKS

- Website of the Forestry Commission: includes full details of all grant schemes and regulations.

enhanced capital allowance *See* CAPITAL ALLOWANCE.

enterprise areas Locations in Scotland where the Scottish Government in partnership with *local authorities wishes to encourage businesses to establish themselves, or move to. At the time of writing, the Scottish Government had designated four enterprise areas situated in sixteen locations across Scotland. Similarly to *enterprise zones in England, Wales, and Northern Ireland, a number of different mechanisms exist to support businesses in these areas. These can vary between areas, but potentially include discounts on *business rates, tax relief on investments in plant and machinery, streamlined *planning, skills and training support, and improved broadband connections.

enterprise zone Area in England, Wales, or Northern Ireland designated as one where central government, in partnership with the *local authority and/or *Local Enterprise Partnership (in England), wishes to encourage businesses to establish themselves, or move to. The mechanisms used to support businesses in enterprise zones vary between them (and between England, Wales, and Northern Ireland), but can include initiatives such as wider *permitted development rights (or other mechanisms of streamlining *planning decisions), discounts on *business rates, tax relief on investments in plant and machinery, access to grant funding or loans, and improved broadband connections. A similar system of *enterprise areas also exists in Scotland.

entire A male animal that has not been castrated, that still has its testicles, and thus has relevant characteristics such as the ability to reproduce once sexually mature. *See also* CASTRATION.

entrepreneurs' relief *See* CAPITAL GAINS TAX.

Entry Level Stewardship *See* ENVIRONMENTAL STEWARDSHIP SCHEME.

environment The complete range of *biotic and *abiotic external conditions in which an organism exists. In human dimensions, the term can be more broadly used to encompass and include economic, social, and cultural considerations. For example, a person can be described as living in a *rural or a village environment as opposed to an urban or city environment. *See also* HABITAT; ECOSYSTEM.

Environment Agency A non-departmental body that is under the umbrella sponsorship of the *Department for Environment, Food and Rural Affairs. In England it is responsible for the protection and enhancement of the environment, with a focus on water, air, and soil resources. It has a particular and high profile role in promoting water quality in *watercourses, *flood prevention, river navigation, and regulating freshwater *fishing. *See also* NATURAL RESOURCES WALES; NORTHERN IRELAND ENVIRONMENT AGENCY; SCOTTISH ENVIRONMENT PROTECTION AGENCY.

environmental impact assessment (EIA) Assessment, required in some circumstances prescribed by law, of the possible environmental impacts of a potential physical change to the natural or built environment. Regulations set out the circumstances when an EIA is required, and they are typically situations where there is potentially a significant risk to the environment, either because of the environmentally sensitive nature of the location, or because of the nature or scale of the proposed change or activity.

In the United Kingdom there are broadly three types of situation in land management where an EIA may be needed: *development; forestry projects (including planting or felling in some circumstances); and proposals for some changes to agricultural/rural land, primarily intensifying the use of uncultivated or semi-natural agricultural land (for example, by ploughing or increasing *fertilizer use) or making large-scale changes to the physical structure of land (for example, removing or adding long lengths of field boundaries or moving large quantities of earth or rock).

The EIA process requires an assessment of potential *environmental impact (an **environmental statement**) to be submitted to the relevant authority prior to the change. This may result in consent not being given for the change, or being granted subject to conditions, which often involve longer-term environmental monitoring of the site.

EIAs relating to development are normally submitted to the relevant *planning authority in the United Kingdom. Those relating to forestry go to the *Forestry Commission in England and Scotland, or to *Natural Resources Wales or the Forest Service (Northern Ireland) respectively, whilst those relating to agricultural land go to *Natural England, the *Welsh Government, *Scottish Government Rural Payments and Inspections Division, or the *Department of Agriculture, Environment and Rural Affairs (Northern Ireland) respectively.

Whilst the regulations in the United Kingdom set out some circumstances where an EIA must be completed (generally those which are considered to present the highest risk of environmental damage), they also detail further situations where an EIA may be required. In order to provide clarity for potential applicants about whether an EIA is needed or not, and also the amount and type of information required, the EIA process has two stages before the full EIA is submitted. Firstly, a request for a **screening opinion** is submitted to the relevant authority; this requires the authority to confirm whether or not an EIA is required. If an EIA is required, then a **scoping opinion** can be requested; this sets out the broad nature, or scope, of information that the authority requires within the EIA. *See also* STRATEGIC ENVIRONMENTAL ASSESSMENT.

Environmentally Sensitive Area Scheme (ESA) The first main *agri-environment scheme, introduced across the UK in 1986. They were designated areas within which farmers could voluntarily participate in a ten-year management agreement to manage the land in a more environmentally sensitive way. Such operations could include restricting the use of *fertilizers and other *agrochemicals, reverting arable land back into grassland, and maintaining hedgerows and other *field boundaries. It has been superseded by new and updated schemes.

environmental management system (EMS) A system (which can be certificated against a recognized standard by an external body) for managing an organization's environmental risks and improving its environmental performance. An environmental management system typically comprises a number of elements:

- High-level management commitment to the ongoing improvement of the environmental performance of the organization.
- An initial review of the organization's environmental performance.
- The establishment of an environmental policy.
- The production of a register of relevant environmental *legislation.
- The generation of a register of **environmental aspects** and **impacts**. An environmental aspect is generally considered to be some way in which the organization, through its activities, interacts with the environment, whilst an environmental impact is any change to the environment (good or bad) as a result of that interaction. (One aspect may generate many impacts.) So, for example, an organization may wash its vehicles (which would be an aspect), which would have a number of impacts, including the use of natural resources (for example, water, and potentially fossil fuels if electricity is used), as well as the potential risk of pollution.
- The generation of objectives and targets for continual environmental improvement. The identification and selection of these targets are driven by the register of environmental aspects and impacts (i.e. those areas targeted should be those where the organization is likely to be having the biggest impact on the environment, or where the biggest risks of this happening negatively are).
- The production of a management programme and manual (incorporating aspects such as training and awareness raising) to deliver the organization's identified objectives and targets.
- Regular recording and monitoring of environmental performance against the targets and related management programme.

Whilst an organization may choose to implement an internal environmental management system, it can also choose to apply to have it recognized by an external certification/registration body (accredited by the **United Kingdom Accreditation Service** in the United Kingdom). This is done through an inspection process where the relevant inspection body visits the organization to review the processes and procedures it has in place in relation to its environmental management, and whether they meet the relevant standard (once the standard has been achieved, then follow-up inspections are made to ensure continued compliance). There are three main standards:

- **ISO14001 standard**, managed by the *International Organization for Standardization.
- **EU Eco-Management and Audit Scheme (EMAS) standard**, a *European Union standard.
- **BS8555 standard**, managed by the *British Standards Institution, and offers a number of certification stages and is therefore sometimes used by small to

medium-sized organizations in the process of implementing an environmental management system. The final stage is normally achievement of the ISO14001 standard.

See also LIFECYCLE COSTING.

environmental permit Form of permission from a government body in England or Wales (normally the *Environment Agency in England, *Natural Resources Wales, or the *local authority) to carry out some form of activity that may interact with the environment in some way (for example, potentially cause pollution, or increased flood risk). Regulations set out the situations in which an environmental permit must be applied for (which can include, for example, *waste management activities that do not meet the conditions for a *waste exemption, or some types of intensive farming). Generally, environmental permits are required for activities, or in locations (such as near to sensitive environmental sites), which are considered to pose some risk to the environment, and they may involve inspection visits from the relevant government body and are usually granted subject to conditions relating to how the activity is managed. In Scotland and Northern Ireland, similar types of activities are covered by requirements to apply for a *pollution prevention and control permit and/or *waste management licence.

Environmental Protection Act 1990 *See* CONTAMINATED LAND; STATUTORY NUISANCE.

Environmental Stewardship Scheme An *agri-environment scheme launched by the Department for Environment, Food and Rural Affairs (DEFRA) in England in 2005 that evolved and developed from the earlier and so-called classic agri-environment schemes. It is a voluntary scheme and is funded through the *Rural Development Programme for England. It is designed to encourage farmers and landowners to manage the land more sensitively, and to protect and enhance the environment. Now closed for new participants, but a number of farms are remaining in existing management agreements. It had a lower level based on a points target, the **Entry Level Stewardship (ELS)** and **Organic Entry Level Stewardship (OELS)**, and a higher level for more complex management, the **Higher Level Stewardship (HLS)**. The Scheme also incorporated a funding stream to encourage conversion to *organic farming. *See also* COUNTRYSIDE STEWARDSHIP SCHEME.

EPC *See* ENERGY PERFORMANCE CERTIFICATE.

epiphytes These include vascular plants such as ferns that are rooted on the tree itself and not rising from the ground, although many grow on rocks or even the soil. Frequently found on *pollards or other trees with wide-spreading branches that provide suitable pockets of organic matter. They use the tree for support, but are not parasitic.

equine Commonly used as an adjective to describe, for example, an activity or disease that is related to, or resembles, a horse. Respective examples include equine courses at university, and equine influenza. Equine is derived from the family *equidae*, a group of mammals that includes the domestic *horse, pony, and donkey.

equipped land Farmland with buildings (and often residential property). Care needs to be taken when looking at information relating to equipped land, as in some cases it includes residential property and sometimes does not, and where it does include residential property there may be wide variations between farms in the relative proportions of the overall farm value that are attributable to residential property. *See also* BARE LAND.

equitable lease *See* EQUITY.

equity 1. A principle existing in English law which operates on the basis of fairness and justice and that what has been done is what ought to have been done. Therefore, for example, if two parties had intended to create a *lease, but it had not been executed by *deed as the law requires, then an **equitable lease** can still exist. So whilst under *common law the lease is not enforceable, it may be so under equity. For further discussion, *see* COMMON LAW. **2.** An **ordinary share** in a *limited company. The ordinary share entitles the owner to **dividends** (effectively a share of the profit); however, if the company is wound up the ordinary shareholder is paid last after all other debts. **3.** The value of the *assets of a business entity after all liabilities have been paid off. Because, in the event of a limited company being wound up, the ordinary shareholders are the last to be paid, this also equates to the **equity capital** (effectively that part of the finance of the company received, normally from its ordinary shareholders, in exchange for shares). **4.** The part of an investment which is not funded by debt; so, a homeowner whose house is worth £200,000 and who has a mortgage with £150,000 of borrowings outstanding is said to have £50,000 of equity in the property.

erosion The process whereby material is lost or diminished by the action of wind, water, chemical, or human action. For example, the surface of a *Public Right of Way may be eroded by continual pedestrian use; buildings may be eroded by pollution and *acid rain; cliffs may be eroded by the waves of the sea; *soil may be eroded by wind or water. An example of the former is the Dust Bowl of the American and Canadian prairies in the 1930s, when millions of tonnes of topsoil were lost due to wind and dust storms. An example of the latter is when heavy rainfall cannot percolate into the soil and causes *runoff, taking soil particles into ditches and streams. Good soil structure to allow drainage and the roots of plants help to avoid soil erosion.

ESA *See* ENVIRONMENTALLY SENSITIVE AREA SCHEME.

establishment costs The cost of setting up any enterprise. In a farm context, these might include the costs of storing a crop, including the seed, tillage to prepare a seedbed, and sowing. They can also include the cost of *fertilizers or *pesticides if applied before or whilst the crop is establishing itself or emerging from the soil.

estate in land The right of possession of land (i.e. the ability to enjoy it). England, Wales, and Northern Ireland still retain elements of the system of land ownership introduced by the Normans, and all land, therefore, ultimately belongs to the Crown, with others holding estates in it. As a consequence of the **Law of Property Act 1925**, in England and Wales there are now only two estates in land: *freehold (fee simple absolute in possession) and *leasehold (*see also* COMMONHOLD, which is, in effect, a form of freehold). Northern Ireland, however, retains significant elements of the system which also existed in England and Wales before the 1920s. Therefore, alongside leasehold there are three different freehold estates in Northern Ireland: the **fee simple**, the **fee tail**, and the **life estate** (for further discussion, *see* FREEHOLD). In England, Wales, and Northern Ireland, two or more people can hold an estate in land together in one of two ways, as a *joint tenancy or as *tenancy in common. For the position in Scotland, where there are significant differences in land and property law, *see* REAL RIGHTS.

estoppel Principle existing in English law that if someone (person A) has taken some form of action as a consequence of something person B has led them to believe (say that person B will be gifting them something at a later date), and if person B does not then do what person A had been led to believe they would, and person A has suffered some detriment (harm) as a consequence of the actions they took in false expectation of what they were led to believe was going to happen, then the *courts may take the view that person A should still have the right to what was promised (i.e. the gift), or some form of monetary *compensation in lieu of the promise.

At a very simplistic level, say, person A owns a ten-year-old lawnmower. Person B has a brand-new lawnmower and says on several occasions that they will only need it for a year so they will give it to Person A next year. Person A therefore gives their ten-year-old lawnmower to a local charity. Person B then changes his mind and says that he wants to keep the lawnmower. Person A has suffered a detriment because he will now have to buy another lawnmower, as he has given away his old one in the expectation of Person B giving him the lawnmower the following year. The courts therefore may take the view that Person B should give Person A the lawnmower as promised (although in reality it is unlikely that a dispute over a lawnmower would end up in court, given the high cost of doing this relative to the costs of buying a lawnmower).

There are a number of different types of estoppel. These include **proprietary estoppel**, where an estoppel relates to some legal interest in land (say a *freehold, *leasehold, or *easement); for example, where someone acts in a certain way to their detriment in the belief that they will be given the freehold of a property at some future date and are then not given it. In an extreme case this may result in the courts awarding the freehold of the property to that person if they consider that it is an example of proprietary estoppel. In Scottish law, **personal bar** is broadly equivalent to estoppel.

estover Particular and limited rights of a tenant of land, to specific products from that land. Very commonly these can include wood for fuel or other purposes. More broadly, it can also refer to an allowance made to an individual out of an estate. *See also* COMMONER.

EU *See* EUROPEAN UNION.

EU Emissions Trading System *See* CARBON TRADING.

European Commission *See* EUROPEAN UNION.

European Federation of Associations for Hunting and Conservation Better known by its abbreviation FACE, this is a European NGO based in Brussels that represents the interests of its constituent national hunting associations that includes all twenty-eight EU member states. In this European context, hunting refers more commonly to the practice of shooting of *game species as well as hunting *quarry species using dogs. *See also* BRITISH ASSOCIATION FOR SHOOTING AND CONSERVATION; COUNTRYSIDE ALLIANCE.

((⊕)) SEE WEB LINKS
• Website of FACE: details the range of its practical and political activities.

European Union An economic union of twenty-eight member states, of particular relevance here because financial support for agriculture through the *Common Agricultural Policy makes up a very large proportion of the total budget. Its influence on farming, the environment, and wider rural development within individual member states is significant. The **European Commission** acts as the executive arm of the EU responsible for the operational activities, including proposing and implementing legislation. *See also* DIRECTIVE, EU.

European Valuation Standards (Blue Book) Set of standards, published by *The European Group of Valuers' Associations (TEGoVA) for the *valuation of real estate. The standards cover a number of aspects, including definitions used in valuations (for example, market value), how the valuation process should be approached (from accepting an instruction to value something, through the process of carrying out the valuation, to the writing of the valuation report), and the requirements demanded of qualified valuers in respect of professionalism, skills, competence, knowledge, ethics, and conduct. It also sets out requirements in relation to European Union *regulations and standards (for example, relating to financial reporting), and addresses other aspects, such as energy efficiency and property valuation. The main standards are pan-European, with additional country-specific content which reflects differences in legislation and practice in individual countries. *See also* RED BOOK; ROYAL INSTITUTION OF CHARTERED SURVEYORS.

eutrophication An increase of nutrients levels in water, causing an explosive growth of vegetation. Its subsequent die-off and decomposition by an excessive increase in *aerobic *bacteria depletes the oxygen in the water. This depletion in oxygen can lead to adverse impacts, including the death of fish and other *aquatic *fauna. Nitrogen in runoff water from agricultural land is a particular concern and cause of eutrophication.

eventing A key element of eventing is that it is characterized by a single horse and rider combination that competes in three disciplines: *dressage, *cross country, and *show jumping. It can be ridden and completed in a single day, or commonly as a **three-day event**. The sport of eventing can also be referred to as **horse trials**, in which points are accumulated and the competitor with the least penalties is the winner.

evergreen Describes *trees and *woody shrubs that retain their leaves throughout the year, as distinct from *deciduous plants that normally drop their leaves in winter. Most *conifers are evergreen.

ewe A female *sheep, especially when mature. In a lowland flock, breeding ewes are usually cross-bred to gain the attributes of both parents, such as hardiness, prolificacy, and mothering ability. Flocks in the uplands may be pure-bred which are then crossed with another breed to produce replacements for the lowlands. Such a cross-bred ewe may typically have two lambs, but this may range from one to four. *See also* SHEEP.

EWGS *See* ENGLISH WOODLAND GRANT SCHEME.

exclusive possession Term established under *case law to identify one of the characteristics of the type of occupation that a *tenant or *lessee has under a *tenancy or *lease, as opposed to someone holding a *licence.
   The principles which currently underpin, in English law, the distinction between a lease/tenancy and a licence are set out in the *court case of *Street v Mountford* (1985), and have been further refined by subsequent case law. The judgment in *Street v Mountford* suggested that in order to distinguish between a lease and a licence, a tenant/lessee had to have 'exclusive possession' of the property. This essentially means that the tenant has the right to control who enters the property and, if they so wish, to keep the *landlord out of the property (unless they wish to enter for some purpose for which they have the legal right of entry; for example, repairs and maintenance). If the tenant does not have the ability to control who enters the property, then the courts are likely to consider that they have a licence rather than a tenancy (i.e. a personal right rather than a property right).
   Exclusive possession alone, however, will not necessarily indicate the existence of a lease. Under the law a lease must be limited in time, so case law also suggests that the right of exclusive possession needs to be for a *term, i.e. a period of time (about which there is no uncertainty). In *fixed-term tenancies the term is clearly specified in the original agreement; however, for *periodic tenancies it is determined by how frequently the rent is paid, so rent paid monthly is a monthly tenancy, i.e. effectively it has a term of a month (even if this is then followed by another term of a month, and so on). In most cases a lease/tenancy will also involve the payment of *rent or some other consideration to the landlord.
   Case law has also established guidance on when a situation may not be a tenancy. This includes where there is some kind of service provided as part of the licence fee (for example, providing food, regular cleaning, linen, and so on); this

is largely because the *licensor then has the right to enter the property regularly in order to provide the service; hence the occupation is not exclusive. It has also been found that where an employer requires an employee to live in accommodation provided by the employer for the better performance of the employee's job (perhaps because someone is needed on the employer's premises at all times), then, because the employee's occupation is a condition of their job, it is not a tenancy (but is, instead, described as a *service occupancy). This may apply even if, ostensibly, the employee appears to have exclusive possession of the property in which they are living. If, however, the employer allows an employee to live in a property owned by the employer, but this is not specifically in order to allow the employee to do their job effectively, then it can be considered to be a tenancy.

In Scottish law, as in English law, exclusive possession is also normally seen as a prerequisite for a lease to exist (as opposed to a licence); however, it is not as rigidly applied by the courts as in England, and authority on exclusive possession in Scotland is found in a different body of case law from England.

executor (*fem.* **executrix**) Person named in a will who usually obtains *probate and is then responsible for **administering** (dealing with) the **estate**, including collecting the assets, paying any debts, paying any *Inheritance Tax, and distributing the estate to the **beneficiaries** (those people entitled to it).

ex-farm At the farm gate; thus the price paid for a product ex-farm does not include the cost of haulage to the purchaser's premises. The delivered price includes the cost of haulage.

exotic species *See* NON-NATIVE SPECIES.

expert witness Person (normally a professional) acting as an expert in a *court case or arbitration (*see* ALTERNATIVE DISPUTE RESOLUTION). The role of the expert witness is to provide advice to the court or arbitrator on matters within the area of expertise of the expert witness, and their duties are therefore to the court or arbitrator rather than to any of the parties within the case. Expert witnesses acting in courts in England, Wales, and Northern Ireland must comply with the *Civil Procedure Rules which set out court procedure.

express consent *See* ADVERTISING CONTROLS.

express covenant *See* COVENANT.

express obligation *See* COVENANT.

extensification The process of making production less intensive, i.e. using fewer inputs. For grazing animals, for example, it would involve lowering the *stocking rate, the number of animals per unit area, which may also mean that less *fertilizer or *pesticide is required for the growth of the *pasture. For growing *crops, it would also involve using less fertilizer and pesticide.

externality A situation where the effect of production or consumption of goods or services may have positive or negative consequences for an unrelated party. An example of a negative externality could be, for instance, where pollution from an industrial plant impacts on agricultural land and may damage plant life and affect the livelihood of farmers nearby. An example of a positive externality might be the growing of apple trees, which also provide a good source of nectar for bees to help make honey, thus benefiting a beekeeper.

e

FAAV *See* Central Association of Agricultural Valuers.

FACE 1. *See* European Federation of Associations for Hunting and Conservation. **2.** *See* Farming and Countryside Education.

factor Person managing a rural estate, most commonly in Scotland, where the role is sometimes referred to as 'estate factor'. *See also* land agent; rural surveyor.

factory farming A system of rearing farm animals with large numbers of *cattle, *sheep, *pigs, or *poultry kept within buildings using controlled conditions and economies of scale to produce meat, milk, or eggs at the lowest possible price. It can be criticized as being unnatural, with poor animal welfare, but that is not necessarily the case, depending upon factors such as stocking density and climatic conditions.

fair rent *See* Rent Act 1977; tenancy; Rent (Agriculture) Act 1976; Scottish residential tenancies; Scottish agricultural tenancies.

Fairtrade A worldwide movement that seeks to ensure that producers of foods, especially in developing countries, are treated fairly when selling their produce and not exploited by global corporations. The criteria include price and the working conditions of producers. For example, 70% of the world's coffee is grown by 25–30 million smallholder subsistence farmers in fifty of the poorest countries in the world, but is sold to consumers by a small number of multinational companies with a very large market advantage. Produce that complies to Fairtrade standards is sold with the logo or statement on its label.

fair wear and tear Principle, accepted within the law in the *United Kingdom, that *tenants of residential properties should not be expected to pay for the costs of replacing, repairing, or redecorating items or parts of a property that have been subject to fair (or reasonable) wear and tear, i.e. that degree of wear that would be expected through normal occupation of the property (say, for example, carpets gradually wearing out). Consequently, the costs of replacing, repairing, or redecorating such items, or parts of the property, should be met by the *landlord, who should not make any deduction from the tenant's *deposit at the end of the tenancy to cover them. Where damage has occurred that is in excess of fair wear and tear, then it is generally the tenant's responsibility to deal with it. Fair wear and tear is, however, a grey area as it is assessed on a case-by-case basis taking into account factors such as the condition of the property at the start of the tenancy. *See also* INVENTORY.

fairy ring Caused by some *soil-borne *fungi growing under the surface of *grassland. They can grow in arcs or in circular rings which can sometimes affect the growth of grass so as to be conspicuous from the air or a more distant viewpoint. Not a particular issue on agricultural land, but can be a concern with management of domestic and commercial lawns. There is a substantial body of folklore associated with fairy rings, and the destruction of them is thought to bring bad luck.

fallen stock Any animal that has died of natural causes or disease on a farm or that has been killed on a farm for reasons other than human consumption. It is illegal to bury fallen stock on a farm, due to the risk of spreading disease through residues in the soil, groundwater, or air pollution. Dead animals are required to be collected from the farm by a *knackerman or approved contractor. Some hunts also collect fallen stock.

fallow An area of arable land that is ploughed and tilled but not sown or cropped, and therefore not economically productive. It usually applies to a season or entire year. Land is left fallow in order to allow the soil to break down and aerate and for weed seeds to germinate, thus helping in their removal from the land with appropriate management. Traditionally, the fallow was part of the *rotation of crops but, due to problems associated with *soil erosion and the loss of income from the land, large areas of fallow land are now uncommon. A **bastard fallow** is one where a seedbed is prepared and then unsown for a relatively short period of time to allow weeds to germinate so that they can be removed before the crop is planted. *See also* SET-ASIDE; STUBBLE.

FAO *See* FOOD AND AGRICULTURE ORGANIZATION.

farm An area of land devoted to the production of food or other economically important *crops, usually for commercial purposes. There are broadly three types of farm: arable, which produces only plant based crops; livestock, which produces animal based products; and mixed, which incorporates both. Within each of these broad classifications are further sub-categories according to specialism of enterprise. Typical *arable crops include *brassicas and *cereal crops, plus other crops such as *potatoes, sugar beet, and *legumes. Livestock enterprises are classified according to the type of animal produced. *Poultry farms specialize in raising poultry either for eggs or meat, including *ducks, *turkeys, and *chickens. *Pig farms may focus solely on breeding *sows to produce *weaners that are then sold on to be reared, or they can be reared on the same farm. *Dairy farms specialize in producing milk, with the calves of the *dairy cows being used to produce *beef, either on the same farm or on another farm. An alternative beef production system is *suckler beef, where calves remain with their mothers until weaning at approximately six months old and are then reared slowly, usually at *grass. The suckler beef system often suits farms where land is unfit for arable crops, but can be improved to give good *grazing, or where large areas of usually poorer-quality land is available for relatively small numbers of animals. *Sheep farms are stratified according to topography, with those farms in mountainous areas producing the breeding stock for *hill farms. These farms produce some meat *lambs but predominantly the breeding stock for *lowland farms. It is generally the lowland sheep farmers with access to good-quality grazing and which enjoy more favourable weather conditions that produce meat lambs. Livestock farms often incorporate crops which can be used as *fodder rather than relying on other sources. These crops include *kale, *fodder beet, and *stubble turnips as well grass for conserving as *hay or *silage. Mixed farms incorporate both livestock and arable enterprises on one farm and are probably the commonest type of farm within the UK. Farms may be bare land, but most would include a farmstead with buildings around a farmyard, a farmhouse, and often workers' cottages. Most are *conventional farms but some are *organic, where the use of artificial *fertilizers and *pesticides is not permitted. Farms may be *owner-occupied, where the owner farms the land or the land is let by the landowner to a *tenant farmer. The area of a farm may range from a *croft or smallholding, medium-to-large-scale single or cooperative enterprises, through to large multinational conglomerates. Under communist regimes, land was often taken by the state and turned into very large state or **collective** farms. *See also* Catchment Sensitive Farming; Common Agricultural Policy; farm assurance schemes; farmer; horticulture; Integrated Farm Management; Land Registry.

farm assurance schemes Voluntary schemes that seek to guarantee certain standards of quality, provenance, traceability, animal welfare, and/or areas of wildlife conservation on the farm. Most UK food assurance schemes are run as product certification schemes that are accredited by the **United Kingdom Accreditation Service (UKAS)**. These schemes use regular independent inspections to check that members are meeting specific standards, and often use logos on consumer products to indicate this compliance. A farmer will join an assurance scheme and his production will then be monitored to ensure that the standards are upheld. His produce can then be described or labelled as complying with the scheme. They may cover the entire food chain from 'field to fork', thus including processors, wholesalers, and retailers, as well as primary producers. The foremost assurance label in the UK is **Red Tractor**, but others include **Freedom Foods** and **Conservation Grade**.

farm buildings The buildings on a farm currently or previously used for *agriculture, usually situated within a farmstead but occasionally in isolated locations. Traditional farm buildings were built of timber, brick, or stone, with a thatched, tile, or slate roof, many of them now redundant for farming because their size and shape are not suitable for modern mechanized agriculture. In this case, many have been converted for residential or commercial reuse. Modern farm buildings are of portal frame construction built of steel, timber, or concrete stanchions with corrugated metal, asbestos, or fibre cement roofs. Farm buildings may be used for the storage of *grain, *hay or *straw, for *livestock or for machinery and equipment. The design of the building will depend upon its intended use. *See also* BARN.

Farm Business Survey Annual surveys undertaken by the governments of England, Wales, Scotland, and Northern Ireland, providing information on the financial (and in some surveys, physical and environmental) performance of farms in order to inform the respective governments' policy making, as well as allowing farmers to carry out *benchmarking to compare their performance against other farms, and researchers to undertake research into the sector. A range of types of data is collected in each of the surveys (with some variations between each of the surveys), including farm incomes; rents; financial performance; fertilizer, energy and water use; and farmers' future intentions for their farm. Similar information is also collected by governments in the *European Union. *See also* AGRICULTURAL CENSUS.

farm business tenancy (FBT) *See* AGRICULTURAL TENANCIES ACT 1995.

farmer A person who manages or works on a farm. The farmer may own their farm, and therefore be described as an *owner-occupier, or may be a *tenant renting the farm from a landlord. *See also* FARM.

Farmers' Union of Wales A membership organization founded in 1955, dedicated to the interests of Welsh farmers and growers. Given the official right by the UK Government to represent and speak on behalf of farmers in Wales, it has done so at the highest level, at the National Assembly, in Whitehall, and in Brussels. It has a network of county branches and offices throughout Wales that provide help and advice to members. The *National Farmers' Union also has members in Wales, known as NFU Cymru.

Farming and Countryside Education A registered charity whose aim is to promote educational work for children and young people about food, farming, and the countryside. FACE was formed against increasing concerns that these groups have become disassociated from what farming involves and the production of healthy foods. It works with teachers, farmers, and a range of partner organizations. *See also* LINKING ENVIRONMENT AND FARMING.

(⊕) SEE WEB LINKS
• Details information on teaching resources, membership, partner organizations, and farms to visit.

Farming and Wildlife Advisory Group (FWAG) More popularly known by its acronym FWAG, it is a membership organization providing independent environmental advice to farmers and landowners. FWAG conservation advisers give guidance on wildlife and habitat management as well as technical advice on *agrochemicals, waste management, and pollution control. Following a period in administration in 2011, the **FWAG Association** was created, and separate FWAGs continued to operate at local and regional levels.

farm loss payment Payment made to some farmers in Scotland who are affected by *compulsory purchase, to recognize that, in some circumstances, having to move to different land may result in a short-term fall in profits. Whilst farm loss payments were previously available in England, Wales, and Northern Ireland, in these parts of the *United Kingdom they have now been replaced with **basic and occupier's loss payments** which are also available to other types of business. (For further explanation of these payments, *see* COMPULSORY PURCHASE.)

farm machinery The equipment used on a *farm for husbandry, such as a *trailer or cultivator, or mounted on the *three-point hydraulic linkage of a tractor. It may be self-propelled, such as a *tractor, or be pulled by a tractor, horse, or other motive force. It may also be static, such as a *grain drier or milking machine.

farm tenancies *See* AGRICULTURAL TENANCIES.

Farm Woodland Premium Scheme Woodland scheme introduced by the *Forestry Commission to incentivize landowners with annual and capital payments to create new woodlands on agricultural land. Now closed to new applicants. *See also* ENGLISH WOODLAND GRANT SCHEME.

farmyard The area of the farm in which the farm buildings are situated. *See also* FARM.

farmyard manure A compost or *organic *fertilizer made from animal faeces together with the bedding, usually straw. Traditionally, livestock housing would be 'mucked out' from time to time, with the soiled bedding forming a muckheap or midden. Once rotted down, this material makes a good organic fertilizer, improving the organic matter and fertility of soils. Many organic farming systems rely on farmyard manure to provide plant nutrients, as the use of artificial fertilizer is prohibited.

farrier A skilled craftsmen concerned with the specialist care and shoeing of equine hooves. In Britain, the Farriers (Registration) Act 1975 and as amended regulates the fitting of shoes to a horse only by persons registered to do so. Today, farriers use a mobile forge and travel to the horse owners' property.

farrowing rails Rails along one side of an area where *sows are to give birth, to allow a place where the newly born *piglets can avoid being lain on by the sow. The danger of loss of life is such that sow stalls with rails on both sides were used to prevent the sow from laying on her piglets, but these have now been banned in many countries, including the UK, on *animal welfare grounds. **Farrowing crates** are similar to sow stalls but allow the sow more room and can be used legally.

fatstock *Livestock that has been fattened for slaughter to produce meat; animals that have reached the required weight and level of conformation to be slaughtered for meat. As livestock fat now carries health connotations, the term 'fattening' for slaughter has largely been replaced by 'finishing'.

fauna Commonly used to refer to the wild animal life in a specified region or time period. Examples might include all the fauna of the UK or the fauna of medieval England. Very often used also in conjunction with *flora. *See also* BIODIVERSITY.

FBT *See* FARM BUSINESS TENANCY. *See also* AGRICULTURAL TENANCIES ACT 1995.

FC *See* FORESTRY COMMISSION.

feed barley *Barley that is destined to be used in animal feeds.

feeder 1. Any receptacle used to feed animals, such as a hayrack. There may be two separate parts for *forage and for *concentrate feed. **2.** A machine that distributes animal feed, some of which can mix ingredients.

Feed-in Tariff Support scheme for small-scale low-carbon electricity installations in England, Wales, and Scotland, including small-scale *solar power (photovoltaics), *anaerobic digestion, *hydro-electricity, and *wind turbine *renewable energy installations. Since the introduction of the Feed-in Tariff (FIT) in 2010, most new *domestic microgeneration equipment installations which generate electricity have been supported through it; however, it is expected that the scheme will close in spring 2019.

Under the scheme, generators are able to export any electricity they do not use for their own purposes to the electricity grid and receive a guaranteed minimum price (the **export tariff**) for it. In addition, the generators receive a **generation tariff** payment for a specified period (normally twenty years) for every unit of electricity they generate (whether they export it or not). The generation tariff payment is fixed at the point when the generator goes into the scheme and is adjusted for inflation, thereby ensuring that the generator receives a guaranteed income from the installation for the period of time they are in the FIT scheme.

Once in the scheme, generators continue to receive payments at the rate which was in existence when they entered the scheme (adjusted for inflation). However, payment rates for new entrants to the scheme are subject to **degression** (i.e. they are reduced as more examples of each technology are installed, usually because the costs of installation have fallen). Therefore, new installations may be receiving lower-generation tariff payments than equipment which was commissioned at an earlier date.

The FIT scheme operates alongside the *Contracts for Difference scheme (and its predecessor the *Renewables Obligation), which both support larger-scale low-carbon electricity generation.

The FIT scheme has not been operating in Northern Ireland, where, instead, very-small-scale generators (termed **microgenerators**) have been supported through the **Micro-Northern Ireland Renewables Obligation** scheme (for further information, *see* RENEWABLES OBLIGATION). This scheme, however, closed to new entrants in 2017.

feed wheat *Wheat that is destined to be used in animal feeds.

fee farm grant *See* FREEHOLD.

fee simple *See* FREEHOLD.

fee simple absolute in possession *See* FREEHOLD.

fee tail (entail) *See* FREEHOLD.

felling Normally used to describe the cutting down of trees which have reached the size suitable for sale as *sawlogs. *Clear felling is the simplest and least costly method of felling all trees from a site at the same time, but can adversely impact the *natural heritage. **Selective felling** involves the purposeful selection of individual trees, and is used to maintain a specific age or species mix. This felling method is usually more costly than clear felling, but with reduced negative impacts. *See also* CONTINUOUS COVER FORESTRY; FELLING LICENCE.

felling licence The *felling of growing *trees normally requires a licence from the *Forestry Commission. A licence is not required for felling as part of an approved *Woodland Grant Scheme, nor for felling or thinning of small-diameter trees. The licence allows the *local planning authorities and other organizations to be consulted, and consideration can be given to impacts on landscape and amenity aspects of the felling and restocking operations.

femtocell *See* TELECOMMUNICATIONS EQUIPMENT.

fen *See* WETLAND.

fence A barrier that encloses a *field, garden, or other area of land primarily for the containment of *livestock. Such barriers on farms were traditionally *dry stone walls or stock-proof *hedges, but now are more commonly a fence composed of wooden posts with timber rails, wire strands, or wire netting between them. For *horses, timber post and rail fences are common, avoiding wire to minimize potential damage to the animal. For *sheep or *chicken, wire netting is normally strung between the posts, the size of the netting depending upon the size of the animal to be contained. For cattle, three or more strands of barbed wire are common. Where the animals to be controlled can burrow, such as *rabbits or *badgers, the bottom of the wire netting is buried in the ground, whilst where the animals can jump, such as deer, the fence may need to be extended upwards by perhaps 2 metres. Temporary fencing may be composed of plastic or wire strands or netting through which an electric current is passed to deter the animals. Fences around gardens may be of woven or close-fitting timber known as closeboarded, or with gaps between the wooden slats, known as a paling or picket fence. *See also* ENERGIZER.

fenestration Term used to describe the arrangement of windows in a building.

FERA *See* FOOD AND ENVIRONMENT RESEARCH AGENCY.

POST AND RAIL   CLOSEBOARD   WOODEN PALISADE

POSTS WITH STOCK NETTING AND BARBED WIRE   POST AND WIRE   WOVEN WOOD

CONTINUOUS BAR (METAL)   CLEFT CHESTNUT PALING   VERTICAL BAR (METAL)

Types of fence.

fertilizer A product that contains *nutrients that plants need for survival. Fertilizer generally describes manufactured inorganic fertilizing products, as opposed to *livestock manures and other *organic *waste, which are also used as fertilizers. Fertilizer products come in four main forms: straights, where only one nutrient is available, e.g. *ammonium nitrate; blends, where individual particles of straight fertilizer are mixed together in specific proportions, e.g. 20:10:10; compounds, where a number of nutrients are mixed together in each fertilizer particle; and liquids, where nutrients are either fully or partially dissolved in water. The numbers used in blends and compounds refer to the percentage of nutrients available and are in a prescribed order. The first number represents nitrogen, the second represents phosphate, and the third represents potassium. Non-liquid fertilizers are generally applied using a fertilizer spreader, which broadcasts the fertilizer onto the *crop. Prilled fertilizer (formed into small unformed balls) gives a more accurate spread pattern than granulated fertilizer. Most nutrients are held in the *soil, and the fertilizer application is to replenish this supply. However *nitrogen is not held and is easily leached out of the soil, so needs to be applied to a growing crop. Fertilizer requirements for crops are determined by the supply of nutrients from the soil and the nutrient needs of the crop grown. As nutrients are taken up via the plant roots, placement of the fertilizer, its breakdown into the soil solution, and its availability to the *rooting system all have an effect upon its utilization and efficiency. *See also* NUTRIENT; AMMONIUM NITRATE, NITRATE; FARMYARD MANURE.

fibreboard *See* PULPWOOD.

field 1. An area of farmland used for growing *arable crops, or *pasture for *livestock *grazing, usually surrounded by a boundary such as a *hedge, *fence, or *dry stone wall. Many fields were created during **enclosure**, changing land use from the mediaeval open field strip system with *commoners' rights to the system of tenure recognized today. The main enclosures were by Acts of Parliament between 1750 and 1850, and many hedges were planted at that time to delineate land ownership and fields and to make them *stock-proof. Fields may be of any size and were originally much smaller, but over the past fifty years many have been amalgamated to allow the use of larger and more sophisticated farm machinery. **2.** The group of mounted participants following a *hunt. *See also* PADDOCK.

field boundary A functional boundary delineating field edges and commonly providing for *livestock control. They are a key and important feature of rural landscapes, but *intensification of agriculture has seen an increased loss of field boundaries as fields have been amalgamated and increased in size. Field boundaries can include *banks, *fences, *dry stone walls, *hedgerows, ditches, dykes, and *ha-has. *See also* FIELD MARGIN.

field margin The strip of land alongside the *field boundary and between this and the edge of the cropped field. Frequently a grass or wildflower mix that can provide a *buffer strip and *habitats for wildlife, including overwintering sites for beneficial invertebrates and breeding cover for *game birds. One of the commonest management options encouraged by *agri-environment schemes on *arable farmland. Can be sprayed in part or full to limit *weed and *crop pests, but this reduces habitat benefits. *See also* CONSERVATION HEADLAND; BEETLE BANKS.

field shelter A small shed or building placed in a grass *field to provide shelter for *grazing animals. They may be temporary and portable so that they can be moved as required, and are often placed in *paddocks used for horse grazing.

field sports Normally refers to activities involved in catching and/or killing a mammal, bird, or fish in which there is a level of skill and enjoyment by the participants. Common activities include *hunting, *shooting, and *angling. Hunting and shooting can also be described as blood sports, a term used more usually by opponents of these activities. Field sports can be used to describe and include a wider range of outdoor pursuits including, for example, *clay shooting and archery. *See also* GAME.

field vegetable A vegetable grown on a *field scale for human consumption. These may be root vegetables such as carrots, potatoes, or parsnips, seeds such as peas or beans, or *brassicas such as cabbage, broccoli, or cauliflower. Root vegetables require light soils without stones which might damage mechanical harvesters.

ARABLE FIELD MARGIN

Hedgerow tree

Buffer strip
Cultivated strip
Grass strip
Wildflower strip
Bird cover strip
Sterile strip

BARRIER

Hedge
Fence
Wall
Grass baulk
Windbreak
Terrace

Hedge bottom

Farm track

Hedge bank

Ditch
Drain
Stream

BOUNDARY
STRIP

CONSERVATION HEADLAND

FIELD BOUNDARY

CROP EDGE MAIN CROP

Field margin. (After Greaves and Marshall, 1987: 'Field margins: definitions and statistics', in Way, J. M. and Greig-Smith, P. J. (eds.), *Field Margins*. Monograph No. 35. British Crop Protection Council, Thornton Heath, Surrey, pp. 3–10.)

field work rate Measure of the amount of work that can be done (usually by one operator) using a specified piece of agricultural machinery per hectare (or acre) of land per unit of time (often an hour). Standard field work rates are published for different agricultural activities in order to help farmers plan their labour requirements. However, there can be significant variation depending upon the size of equipment available and the particular circumstances of the farm (for example, its soil type). When using field work rates to plan labour requirements, it also needs to be recognized that because of weather conditions the days available for fieldwork are likely to vary from month to month. *See also* STANDARD MAN DAYS.

financial institution Organization that provides financial services (for example, a bank or building society).

finish To prepare animals for slaughter for meat, to achieve the required level of age, weight, and carcass conformation prior to slaughter. At one time, to 'fatten' was a more common term than to 'finish', but the modern health connotation of animal fat has led to the latter becoming more acceptable. Animals may be *weaners or stores, which means they have been reared to the point at which they may be fattened or finished. In some cases, the animals may

be sold at this stage for a breeding unit to a finishing or fattening unit.
See also FATSTOCK.

fir (Douglas, noble) *See* TREE.

firearms Rifles and *shotguns are the normal firearms used within a variety of
*field sports activities. Their use in the UK is strictly controlled by the **Firearms
Act 1968**, which regulates the possession of all firearms and ammunition.
Shotguns are subject to a lesser degree of control than rifles. *See also* SHOOTING;
DEER STALKING.

fire break Openings in the woodland where the vegetation is kept clear and
managed to reduce the spreading of fire. Normally at least 10 metres in width,
and frequently incorporating roads and tracks.

fire safety regulations Term commonly used to refer to various *regulations
concerning fire safety within the *United Kingdom. The different parts of the
United Kingdom have some differences in their fire safety regulations, although
all follow similar principles. The scope of fire safety regulations is broad, and
includes both structures (such as buildings) and also the contents of buildings
(for example, domestic furniture which is subject to laws which set out standards
for fire resistance). Whilst the fire safety of buildings themselves is mainly
controlled by *building regulations (*building standards system in Scotland),
additional fire safety regulations may also apply depending upon the nature of
the use of the building; for example, if it is let for residential or commercial use.
For example, those people who exercise some control over commercial premises
(which can include both *landlords and their *tenants) are required to undertake
a range of fire safety measures, including completing fire safety *risk
assessments. Similarly, landlords of let residential properties are also required to
comply with fire safety requirements, for example, the provision of fire alarms. In
addition, the landlord of a *house in multiple occupation (HMO) will be subject
to further fire safety regulations over and above those applying to other let
residential property, given the higher risks attached to fire in these types of
property.

firewood Describes the market and use for lower-quality *hardwoods and
*softwoods. Frequently, firewood from *coppice woodland, lop and top, and
thinnings are sold by the tonne or by the *cord. *See also* WOODFUEL.

First-Tier Tribunal for Scotland (Housing and Property Chamber) A
specialist legal panel which has taken over the functions of the **Private Rented
Housing Panel, Scotland** and the **Rent Assessment Panel for Scotland**. It
deals with disputes involving *landlords and *tenants in Scotland (primarily
relating to *rents and repairs, including dealing with objections in relation to
*fair rents set by *rental valuation officers). *See also* FIRST-TIER TRIBUNAL
PROPERTY CHAMBER; RENT SERVICE SCOTLAND.

First-Tier Tribunal Property Chamber A specialist legal panel which is part of **HM Courts and Tribunals Service**, and which deals with applications, appeals, and references relating to disputes in relation to property and land in England and Wales. A tribunal is usually less formal and complicated than a court process. An appeal against a decision of a First-Tier Tribunal is generally heard in an Upper Tribunal. Further appeals can also sometimes be made to the Court of Appeal. The Property Chamber is one of seven chambers of the First-Tier Tribunal. It brings together work previously undertaken by **Rent Assessment Committees**, **Leasehold Valuation Tribunals**, **Residential Property Tribunals**, **Rent Tribunals**, **Agricultural Land Tribunals**, and the work of the **Adjudicator to HM Land Registry**. The Property Chamber is divided into three divisions: Residential Property, Land Registration, and Agricultural Land and Drainage.

Residential Property. The tribunal deals with settling residential property-related disputes in England. It is usually comprised of a Chair (who is a judge or chartered surveyor), one or two members who may be chartered surveyors, and other professionals or lay people. The tribunal deals with a number of types of dispute, including:

- Rent cases:
  - If a landlord or tenant of a *regulated residential tenancy has objected to a fair rent decided by the *Rent Officer, then the tribunal will decide on a fair rent.
  - If a tenant with an assured tenancy or assured shorthold tenancy (*see* HOUSING ACT 1988) applies for an open *market rent to be determined, after a landlord has served a notice informing the tenant that they intend to increase the rent.
  - If a tenant of an assured shorthold tenancy applies for a rent to be adjusted within the first six months of a tenancy, because they consider that it is too high.
  - In a situation where a landlord or tenant has served a notice proposing new terms when the *fixed term of an assured or assured shorthold tenancy has ended, then the recipient of the notice can ask the tribunal to determine suitable terms for a statutory periodic tenancy.
- Leasehold enfranchisement cases (*see* LEASEHOLD REFORM):
  - Various matters relating to leasehold enfranchisement, including determining the price to be paid, and related terms, in situations where these cannot be agreed between the parties. This covers a range of situations, including if a single leaseholder wishes to buy (enfranchise) the *freehold of their house, or extend the long *lease; or where a group of leaseholders wish to buy the leases of their flats; or where a single leaseholder wants to extend or renew the lease of their flat.
- Leasehold disputes:
  - Various matters related to disputes in relation to the management of leasehold properties (for example, relating to *service charges).

- Housing Act 2004 cases:
  - Various applications and appeals in relation to provisions within the Housing Act 2004; for example, Local Housing Authority Empty Dwelling Management Orders (the power that local authorities, have, in certain circumstances, to take over the management of privately owned houses that have been empty for long periods).

Land Registration. Tribunal which considers disputes in England and Wales, generally referred to by the *Land Registry, in relation to the registration of land. The judges in this division are specialists in *land registration and property law.

Agricultural Land and Drainage. Tribunal, comprised of a Chair and two lay members, which considers disputes in England relating to tenancies under the *Agricultural Holdings Act 1986 (AHA), and also drainage disputes between neighbours under the Land Drainage Act 1991. The tribunal cannot be used for appeals under the *Agricultural Tenancies Act 1995. Issues dealt with by the tribunal include:

- Applications for *succession to an AHA tenancy.
- Application by a landlord for a certificate of bad husbandry under an AHA tenancy.
- Applications for a direction to provide *fixed equipment under an AHA tenancy.
- Applications for consent to the operation of a *notice to quit served on an AHA tenant by a landlord.
- Applications for a direction to carry out maintenance work to ditches on neighbouring land.

See also Agricultural Land Tribunal for Wales; courts, law; First-Tier Tribunal for Scotland (Housing and Property Chamber); Rent Assessment Panel; Residential Property Tribunal Wales.

fishery A defined unit involved in raising and subsequent harvesting of wild fish or through *aquaculture. The unit can often be typically defined by the area of water or species of fish. *See also* angling.

fishing *See* angling.

FIT *See* Feed-in Tariff.

fixed costs Overhead costs that are not directly attributable to a particular enterprise. The most common form of farm accounting distinguishes between fixed costs and *variable costs which are directly attributable to a particular enterprise such as seed, *fertilizer, *pesticide, feeds, and medicines. These variable costs are deducted from the gross output to produce the *gross margin. Net margin or *profit is then determined by deducting fixed costs. Typically these would include labour and machinery costs, rent, property repairs and insurance, finance, and professional fees.

fixed equipment Term used in agricultural *landlord and *tenant law to describe buildings and structures which are affixed to the land as well as works in, on, over, or under it (for example, roads and ditches). It also includes things which are grown on the land but not normally removed for use or consumption (for example, *hedges). The exact definition of fixed equipment as it relates to each different type of *tenancy is set out in the relevant *statute.

fixed-term tenancy (or lease) A *tenancy or *lease with a duration which is fixed at the start; for example, a lease for five years. *Compare* PERIODIC TENANCY.

fixture An object which in English law is considered to be attached to the land, and therefore legally becomes a part of it. This is distinct from a **chattel** (for example, a piece of jewellery) which is not attached to the land, and therefore does not form part of it legally. The significance of whether an object is a fixture or a chattel can be seen in a number of situations. This includes when a property is being sold (the seller would normally be entitled to take a chattel away with them, but not a fixture, as the fixture would form part of the property being purchased). The principle is also significant when determining whether an object forms part of a *listed building and therefore would require *listed building consent if the owner wished to remove it (generally a fixture would).

Whether an object is considered to be a fixture or not is not always clear (for example, a set of tapestries on the walls of a room could be considered to be fixtures or chattels, depending on the particular situation). Therefore two legal principles have been established (primarily through *case law) to determine this: namely, the degree of annexation of the object, and the purpose of annexation. Firstly, the degree of annexation considers the way in which the object is fixed to the land. Whilst this may create clarity in some situations, in others it does not necessarily definitively provide an answer. Two people looking at the same method of fixing of tapestries to the wall may reach differing conclusions as to whether the tapestries are fixtures or not (depending on the degree to which they thought the tapestries could easily be removed). Consequently, the courts also consider the purpose of the annexation (i.e. the reason why the object is there). In the case of the tapestries, if they were part of the original architectural design of the house, and had always been there, then there is a significantly greater likelihood that the *courts would consider them to be fixtures, than had they been purchased and hung by the current owners and had never formed part of the original design of the property. The same broad legal principles are also found in Scottish law, although the term 'chattels' is often replaced by the term **moveable items**.

flail mower An agricultural machine that cuts grass and other foliage by means of flails or blades rotating around a central spindle. The power may come from a separate engine, but more commonly the spindle is driven by the *power take-off of a tractor to which the machine is attached, either mounted on the *three-point hydraulic linkage or trailed. Some machines, especially those in use in gardens or landscape amenities, may be self-propelled.

flighting *See* WILDFOWLING.

flock The collective noun for *sheep or *poultry.

flood Water covering land which would normally drain and be dry. Most often caused by heavy rains, melting snow, tides, storms, and so on. Low-lying agricultural land adjacent to rivers and lakes can be part of a natural **flood plain** or **flood meadow**, where excess water is essentially stored before draining out through the associated river systems. Increased agricultural *intensification has promoted drainage of these natural flooding mechanisms and has resulted in increased flooding pressures lower down the river systems and adjacent towns and cities. This in turn has precipitated increased resources given to construction of **flood defences**. *See also* CATCHMENT.

flora Commonly used to refer to the native plant life in a specified region or time period. Examples might include all the flora of the UK or the flora of prehistoric England. Very often used also in conjunction with *fauna. *See also* BIODIVERSITY.

flushing *See* BEATING.

fly fishing A lure called a fly which imitates an invertebrate such as an insect or worm, and attracts the target fish onto the hook. The fly is very light, and fly fishing uses the weight of the line to fling or cast it out into the water. This is referred to as casting and requires a skilled technique. *See also* ANGLING; GAME FISHING; SALMON.

fly-tipping Illegally dumping *waste onto land (often the side of the road) which is not licensed to take such waste.

FMD *See* FOOT AND MOUTH DISEASE.

foal 1. A young *horse or other *equine that is less than a year old. **Mare-in-foal** describes a pregnant mare. **2.** As a verb in the context of giving birth to a foal.

fodder Animal feed, especially conserved *forage fed to ruminant animals. It is usually applied to *hay, *silage, *haylage, *straw, or other bulk feed rather than concentrated foodstuffs. *See also* FORAGE; FODDER BEET.

fodder beet A large coarse yellow-to-red beet grown as animal feed. Also known as **mangel wurzel**, **mangold**, and **mangold wurzel**. Developed in the eighteenth century as a *fodder crop, either grazed in the field or lifted and carried to housed cattle, it is much less common today than other beet such as swedes and turnips.

followers Farm animals that make up the herd or flock but are not in production at the time of use. Usually applied to offspring such as lambs or calves that will become productive in future. Thus the followers in a dairy herd are those *heifers that have yet to produce a calf and thus milk.

Food and Agriculture Organization (FAO) An agency of the *United Nations with a key objective to improve nutritional standards of human populations by increasing sustainable agricultural productivity and achieving food security. It acts as a coordinating body to help individual governments to develop agriculture, fisheries, forestry, and natural resources. It also undertakes and funds research, and provides technical assistance. Based in Rome and with a network of offices in 130 countries throughout the world.

(((•))) SEE WEB LINKS

• FAO website: includes a breakdown of activity in each individual country.

Food and Environment Research Agency (FERA) A major centre for interdisciplinary research covering sustainable agriculture, food, plant health, and the environment. It was formerly an Executive Agency of the *Department for Environment, Food and Rural Affairs (DEFRA), and although now independent of government it continues to work closely with it. Reformed as **Fera Science Limited (Fera)**. *See also* ANIMAL AND PLANT HEALTH AGENCY.

food chain 1. The processes that food passes through from primary production on the farm to final consumption. Thus the first link in the chain is the farmer who produces the raw material such as *milk, meat, or *grain. This is sold to one or more processors, such as the *dairy that prepares milk for liquid consumption or processes it further into butter, cheese, yoghurt, or other dairy products. Meat passes through an abattoir and **cutting plant** before moving on to a butcher. Grain is sold to an animal feed compounder, flour miller, or **maltster** who processes it further. The processed product then may pass through a wholesaler to a retailer who sells it to the consumer. The longer the food chain, the more people handling the goods and looking to make a profit, so the smaller proportion of the final sale price that the farmer receives. There have been concerted efforts to shorten the food chain by vertical integration or by farmers taking on the processing and marketing, for example, through farm shops. **2.** The transfer of energy from producers through consumers to the apex predator. Usually linear, they can in their simplest form consist of just two or three links. Producers are nearly always plant species that harness the sun's energy through photosynthesis; these are then eaten by the primary consumer (a herbivore or omnivore). This is predated upon by a secondary consumer (carnivore or omnivore) which in turn may also be predated upon by tertiary consumers until the chain ends with the apex predator. Food chains are often depicted as a triangle or pyramid, as the number of individuals at each level decreases with progression through the food chain. Understanding food chains is of critical importance when considering the *ecology of a *habitat, as manipulation of one *species is likely to have a significant impact on the species which depend upon it as a food source or on the populations of species which it feeds upon.

Food Standards Agency A non-ministerial department, its key responsibility is for food safety and food hygiene across the UK. It operates in UK meat plants and works with *local authorities to advise and enforce food safety regulations. The FSA also has responsibility for food labelling policies in the devolved UK governments, but in England this is undertaken by the *Department for Environment, Food and Rural Affairs (DEFRA).

(((⊕))) SEE WEB LINKS

• Food Standards Agency website: details varying responsibilities within each of the UK countries.

foot and mouth disease A highly contagious viral disease of cloven-hoofed animals, notably cattle, sheep, pigs, goats, and deer. It is caused by a virus of the *Picornaviridae* family, of which there are seven strains. Immunity to one does not give immunity to any of the others, so individual vaccines are necessary to protect against disease. Symptoms include blisters in the mouth or on the feet, fever, loss of weight, loss of production such as milk, loss of appetite, frothing at the mouth, and lameness. The disease is rarely fatal in adults, but the young and weak may die. Spread of the disease is by direct contact with infected animals, by contamination on boots or wheels, or airborne. From infection to the first symptoms is usually 3–6 days. It is endemic in many parts of the world, including Asia and Africa, but in many countries, especially in Europe and North America, there are very strict quarantine and slaughter policies. It is a *notifiable disease, which means that it must be reported to the appropriate authorities, which may then slaughter all susceptible animals that may have come into contact with it, and impose movement restrictions over a wider area. The carcasses are incinerated, rendered, or buried, and the premises thoroughly disinfected. Vaccine may be used, especially in a cordon around an outbreak, but is not accepted in those countries free of the disease because of the difficulty of distinguishing between those animals vaccinated and those infected. Export restrictions are normally imposed on infected countries.

footpath A right of way on foot, permissible under a range of different regulations or conditions. Most frequently used to describe public footpaths, which are the commonest type of *Public Right of Way. These include the more publicized UK long-distance footpaths as well as the fifteen National Trails in England and Wales and equivalent in Scotland.

forage Animal feed, particularly pasture or other crops grazed by ruminant animals. To an extent there is a distinction between fresh forage and conserved *fodder. *See also* FODDER.

forage harvester An agricultural machine that cuts and chops grass, maize, and other crops for conservation as *silage. Smaller versions may be trailed behind a tractor and driven by the *power take-off, whilst larger ones are self-propelled.

foreshore Essentially that part or strip of the seashore that lies between the average low-tide mark and the average high-tide mark. In Britain the foreshore is owned mainly by the **Crown**. Public rights over it are for navigation and fishing when covered by the tide. Indirectly this guarantees *public access when not covered by the tide, and in aggregate the foreshore provides a significant and important amenity resource.

forest The distinction between a *wood or *woodland and a forest is essentially one of scale and is not strictly defined, and the terms are commonly interchangeable. Forests cover a relatively larger area, and usually with a denser collection of *trees. Forest is an historic term and formerly described a royal *hunting ground on which deer were maintained for the enjoyment of hunting by the crown, and hence the larger size. The New Forest and Sherwood Forest were two examples. *See also* ANCIENT WOODLAND; ANCIENT SEMI-NATURAL WOODLAND; WILDWOOD; COPPICE.

Forest Enterprise The government agency set up in 1996 to manage Britain's forest estate. The responsibilities have since devolved to the respective countries of England, Scotland, and Wales, and these are now represented respectively by **Forest Enterprise England**, **Forest Enterprise Scotland**, and *Natural Resources Wales. They are tasked with looking after the public forest estates, and are the public point of contact and information on timber and related production, recreational activities, wildlife control, and commercial developments. *See also* FORESTRY COMMISSION.

forester *See* INSTITUTE OF CHARTERED FORESTERS.

Forest Research A government agency that is part of the *Forestry Commission and provides forestry research relevant to the UK and international interests.

- Forest Research website: includes separate themes of research, and relevant news and information.

forestry Commonly summarized as the science and practice of managing *forests and *woodlands. Timber production is a preeminent objective, but increasingly forestry includes multipurpose management to deliver *biodiversity benefits, *landscape enhancement, and other *ecosystem services such as *carbon sequestration. *See also* FORESTRY COMMISSION; INSTITUTE OF CHARTERED FORESTERS.

Forestry Commission Founded by statute in 1919 in the aftermath of the First World War, it was set up by the need for the expansion of a strategic reserve of timber for use in any future conflicts. Large areas of *conifer *plantations were established on low-value agricultural land which included *heath and *moorland, land types whose importance for environmental benefits was not recognized as it is has been in more recent times. The Forestry Commission has been through a series of structural changes and is now a non-ministerial government department responsible for forestry in England and Scotland. In Wales, the responsibility for forestry lies with *Natural Resources Wales.

Although it retains a key priority for timber production it promotes multifunctional woodland both on its own land managed by *Forest Enterprise and on private land. It acts as a regulator and promotes its activities through a range of grants and incentives. *See also* FORESTRY GRANTS.

(⊕) SEE WEB LINKS
• Website of the Forestry Commission: details information on silvicultural systems and matters of woodland and forestry-related regulation and legislation. It also outlines the respective departments operating separately, covering England, Scotland, and Wales.

forestry grants Grants to landowners to encourage *tree planting in the UK were started by the *Forestry Commission in 1919. They have evolved and diversified from relatively simple grants to encourage new planting, through to management for multiple-purpose *woodland promoting *environmental and access benefits as well as timber production. The *Forestry Commission is the government body most involved with administering government grants in England, for example, now through the new *Countryside Stewardship Scheme. *See also* ENGLISH WOODLAND GRANT SCHEME; NATURAL RESOURCES WALES.

forfeiture Right of a *landlord in England, Wales, and Northern Ireland to end a *lease early if the *tenant has breached (broken) any of the *covenants in the lease. In Scotland, *irritancy is broadly equivalent to forfeiture. *See also* BREACH OF COVENANT.

forwarder A large purpose-built articulated machine designed for heavy-duty forestry operations. It is used to load and transport logs from the floor of the forest to a suitable site for stacking and subsequent onward transport by road truck.

forward selling The selling of a product in advance of final production, often on a forward contract. Thus an arable farmer might sell his grain before harvest, to be delivered at some specified point in the future to achieve what he perceives to be a good price, higher or at least equal to that which he could achieve at the time of delivery. *See also* GRAIN MARKETING.

fossil fuels The group of carbon and hydrogen-based *non-renewable fuels, including coal, oil, and natural gas, which formed from the remains of dead organisms that lived millions of years ago. *See also* CARBON CYCLE.

foul water Dirty water that is contaminated with waste and normally comes from the internal drainage system of a building (for example, from bathrooms), and which requires cleaning up before it can be released into the environment. This is distinct from rainwater (which is also known as *surface water) which might, for example, be collected through the gutters on the outside of a building, and which is generally considered to be cleaner.

Some types of foul water are sometimes described as **grey water**. Grey water (for example, that from baths or sinks) is foul water which is considered to have a lower risk of containing dangerous pathogens than other types of foul water (for example, that from toilets, which is sometimes known as **soil water**). **Grey**

water harvesting systems harvest grey water, treat it on-site (through the use of chemical, biological, ultraviolet, and/or filtering processes), and then pump it back into the building for uses such as flushing toilets. Greywater recycling systems are distinct from *rainwater harvesting systems, as rainwater harvesting systems rely on the collection and reuse on site of rainwater, rather than grey water.

foundations The part of a building which supports it on, or in, the ground and spreads the weight of the building over the ground on which it sits. There are several different types of foundation, and the choice of which is used depends on factors such as the weight of the building and the type of soil and its typical levels of moisture:

- **Strip foundations** consist of strips of *concrete (sometimes reinforced with high-tensile steel rods) running under the length of the load-bearing walls (on sloping sites, strip foundations can be stepped).
- **Raft slab foundations** consist of a large concrete slab lying under the whole area of the building.
- **Pad foundations** consist of a series of concrete pads, and the building structure comprises a series of columns which are constructed on top of the pads
- **Pile foundations** are used when the ground close to the surface is much more unstable than that further down. The process of installing pile foundations involves the drilling of a cylindrical shaft deep underground which is then filled with concrete and forms a concrete column down to the sub-soil. Reinforced concrete beams (or a reinforced concrete slab) are then used to link the piles together and support the load-bearing walls.

See also SUBSIDENCE.

4 Ps *See* MARKETING MIX.

fox The red fox (*Vulpes vulpes*) is a member of the family **Canidae**; it is widespread throughout the world and is native to the British Isles. Its relatively large size and abundance and its ability to exploit a diverse array of prey species can bring it into conflict with people where their diet is of economic or conservation interest. This may include *livestock, more especially *lambs and *poultry, but in particular *game birds. *See also* FOX HUNTING.

fox hunting An activity that has been taking place in different forms and across many different countries for centuries. It relies on the chase of the fox, using dogs or hounds that hunt by smell. It is perhaps best known and at its most controversial in the UK. Fox hunting is organized into different geographical areas or '**countries**', each with its own pack of hounds. Hunting is organized to cover a particular area of the 'country', and starts at a prearranged venue or **meet**. Autumn hunting, also less commonly known as cub hunting or cubbing, takes place at the start of the *season in early to mid-autumn. This period gives the opportunity for young hounds to be trained. Each 'hunt' essentially operates like a club with subscribing members and elected officials or *Master of Foxhounds. The larger packs or 'hunts' employ staff to hunt the hounds and to care for them. Since 2004, fox hunting has been illegal in England and Wales, and also in Scotland under different legislation (though it remains legal in Northern Ireland). However, the structure of hunting and the individual packs of hounds remain and continue to hunt within the legal requirements, such as *trail hunting. *See also* STAG HUNTING; DRAG HUNTING; HARE HUNTING; BEAGLING.

(((●))) SEE WEB LINKS

• Website of the administrative hub for the Council of Hunting Associations, with dedicated site of the Masters of Foxhounds Association, and full details of the rules and codes of conduct of fox hunting.

fracking Physical process (also known as **hydraulic fracturing**) of extracting oil or gas from shale rocks deep underground by injecting liquids under high pressure in order to fracture the rock, thereby releasing the gas. The gas extracted is sometime referred to as **shale gas**.

freehold Legal *estate in land in England, Wales, and Northern Ireland giving the holder the most complete right to possession of that land (i.e. the ability to occupy and enjoy it). In England and Wales the term freehold is usually used when owners have a **fee simple absolute in possession** giving the owner (the *freeholder) absolute rights over the land during their lifetime, and entitling them to leave the property to someone else at their death. There are no limits or restrictions on the rights of a freeholder holding a fee simple absolute in possession in connection with their land (apart from those imposed by the law, such as in relation to *planning), and the freeholder also has immediate enjoyment of the property. Fee simple absolute in possession is one of only two legal estates in land recognized in England and Wales (the other being *leasehold).

In Northern Ireland, several types of freehold are currently recognized (although the Northern Ireland *Law Commission has recommended revisions to this): namely, the fee simple, the fee tail, and the *life estate. (This system mirrors that of England and Wales before the **Law of Property Act 1925**.) Fee simple has a similar meaning as in England and Wales, though is not necessarily always 'in possession' (for example, if the holder of the fee simple grants a life estate to someone else; see below). Fee tail is considered to be a lesser freehold interest and provides for the land to be held by one person, and at their death for the **entailed estate** to pass to only another member of the family (usually the eldest son); however, it is unlikely that new fee tails would be created today, not least as they would offer very little flexibility in planning for *capital taxation. Under a life estate, the *life tenant is able to hold the estate for their life; however, at their death it passes to the owner of the fee simple. Some property in Northern Ireland is also held as a *fee farm grant. This is similar to a fee simple absolute (in that the holder of a fee farm grant holds the property without any time limit on their ownership), except that an annual *ground rent has to be paid to the freeholder. Holders of fee farm grants are, however, now able to buy out, or **redeem**, their ground rents through the payment of a capital sum, so the ground rent no longer needs to be paid, and it is also no longer possible, under law in Northern Ireland, to create a new fee farm grant.

For the position in Scotland, where there are significant differences in land and property law, *see* REAL RIGHTS.

See also COMMONHOLD; TERMS USED WHEN PROPERTY IS SUBJECT TO LEASE OR TENANCY; TITLE.

freeholder The owner of a *freehold. *See also* LANDLORD; LESSOR; TERMS USED WHEN PROPERTY IS SUBJECT TO LEASE OR TENANCY.

free-range eggs Eggs produced from *poultry that have access during daytime to the open air, usually a field. Under European Union regulation, to be described as free range there must be a minimum of one hectare of outdoor range for every 2,500 hens (equivalent to 4 square metres per hen), and at least 2.5 square metres per hen must be available at any one time if rotation of the outdoor range is practised. There must be continuous access during the day to this open-air range, which must be 'mainly covered with vegetation', and there must be several pop holes extending along the entire length of the building, providing at least 2 metres of opening for every 1,000 hens.

freeze marking The branding of an animal using a tool cooled by liquid nitrogen or dry ice and alcohol to create a permanent mark on the skin for identification. The most common animals to be branded are *horses to prevent theft or misidentification, and *cows so that they can easily be identified in the milking parlour. Such branding was traditionally carried out by means of an extremely hot branding iron, but freeze marking is now more popular because it causes far less distress to the animal.

FRI Abbreviation of **full repairing and insuring**; a description of the nature of *lease and what the *lessee is responsible for. In a lease described as an FRI lease, the lessee is responsible for all repairs and for insuring the property. *See also* IRI; IRT.

FRICS *See* CHARTERED SURVEYOR.

front loader A type of *tractor, tracked or usually wheeled, that has a front-mounted loading implement. This is normally a square or rectangle wide bucket that is used to load material from the ground, such as *soil or *manure. It can be mounted with *tines for easier handling of loose material such as *silage. *See also* FORWARDER.

frustration of contract A situation where an unforeseen event means that it becomes impossible, or illegal, to perform a *contract for some reason, and the contract is then normally treated as being discharged (effectively ended) unless the contract specifies otherwise. The law provides for how any sums of money already paid, or due, at the point when the contract is frustrated, are to be dealt with.

fuelwood Wood that is used as a fuel. More commonly refers to modified timber products such as charcoal, *wood chip, and *pellets. *Firewood is also a fuelwood. *See also* BIOMASS.

full planning permission A type of *planning permission granted in England, Wales, and Northern Ireland when full details of the application have been provided at the point of application to the *local planning authority. Full planning permission establishes the principle that the development for which permission has been granted can be undertaken without the need to apply for further planning permission, providing that it is carried out in exactly the same way as proposed in the original application. Full planning permission is normally, however, granted subject to planning *conditions, and these may require the developer to obtain the further approval of the local planning authority in relation to specific matters (for example, submitting samples of particular building materials such as roofing tiles, for approval prior to construction). However, these approvals relate only to very specific aspects of the development, as the granting of full planning permission establishes that the overall development is acceptable.

Full planning permission is distinct from both *outline planning permission and *planning permission in principle, both of which only establish the overarching principles of what would be permitted in planning terms at a site (for example, that it would be acceptable to the local planning authority to change the use of an area of land from agriculture to housing). These types of permission, therefore, do not allow the developer to proceed with the development until they have made further application to the local planning authority which includes the outstanding details which were not comprised within the outline planning permission, or planning permission in principle. Providing these additional details (such as housing layout or appearance) are approved by the planning authority, the development may proceed (though again, potentially subject to conditions). *See also* DETAILED PLANNING PERMISSION.

fungi A kingdom of organisms distinct from both plants and animals, ranging from single cells such as yeast and moulds to complex multicellular organisms such as mushrooms. It is estimated that there may be anywhere between 700,000 and 5 million species of fungi, of which only around 100,000 have so far been identified. They are of great importance both as a source of food, medicines, and enzymes, and as a cause of disease of significance to agriculture. *See also* FUNGICIDE.

fungicide A chemical compound used to control *fungi. *See also* PESTICIDE.

furlong Old unit of measurement, used in the United Kingdom, equal to ten *chains; eight furlongs equal 1 mile.

furrow A long, narrow depression or trench in the soil caused by the action of a *plough. The soil is inverted by the *mouldboard of the plough, leaving the cut edge as a furrow. *See also* RIDGE AND FURROW.

FWAG *See* FARMING AND WILDLIFE ADVISORY GROUP.

FYM *See* FARMYARD MANURE.

G

GAEC *See* GOOD AGRICULTURAL AND ENVIRONMENTAL CONDITION.

game Normally describes mammals and birds that are killed for sport and usually also for food, such as *deer and *game birds. **Game shooting** is the practice or sport of shooting game. *See also* GAME MANAGEMENT; GAMEKEEPER; FIELD SPORTS; OPEN SEASON.

game bird Wild birds that are classified as *game and are legally allowed to be shot during specified periods of the year or *seasons. In the UK they are defined under legislation and include *pheasant, *partridge, red and black *grouse, ptarmigan, capercaillie, common snipe, and woodcock.

Game Conservancy and Wildlife Trust (GCWT) A membership-funded organization with more than 20,000 members predominantly concerned with all aspects of *game shooting. It has a strong emphasis on scientific research and works with government departments and other organizations to promote the interests of game and the wider environment and its wildlife.

game cover A variety of seed-bearing *crops such as *maize or *kale that are sown in strategic patches and locations on farmland to enhance the carrying capacity and sporting potential of *game birds. They provide shelter or 'cover' for *pheasants and *partridges and hence the description, but should not be confused with *cover crops, which have a primary agricultural purpose. *See also* GAME; GAME MANAGEMENT; COVERT.

game fishing The term used for fishing or angling for game fish which includes trout, sea trout, *salmon, and char. This is distinct from coarse fish, which refers to all other freshwater fish which are not normally eaten, although increasingly game fishing practises catch and release to help conserve fish stocks. The commonest type of game fishing is *fly fishing. Game fish can also be caught by spinning, i.e. using flashing metal or spinners.

gamekeeper Commonly employed on estates that actively promote commercial *game shooting. They may also be employed by a group or *shooting syndicate who lease land for game shooting. The gamekeeper is involved in all aspects of *game management with a primary objective to ensure that there is enough game for the owner or clients to shoot. A gamekeeper on a large estate might have an assistant or under-keeper working alongside. *See also* NATIONAL GAMEKEEPERS' ORGANISATION.

game management Description of activities related to providing the best or optimum conditions to promote numbers and availability of *game. Management might include habitat creation and/or maintenance, supplementary feeding, restocking of game, predator control, or anti-poaching measures. *See also* GAMEKEEPER.

Gas Safe Register Official list of gas engineers in the *United Kingdom, Isle of Man, and Guernsey. All gas engineers must, by law, be on the Gas Safe Register. The Gas Safe Register issues licences to businesses to allow them to carry out gas work. It replaced **CORGI** registration in 2009–10.

gas safety regulations Term commonly used to refer to *regulations concerning the safety of gas appliances and fittings in the *United Kingdom. The scope of the regulations includes a range of duties applying to *landlords of domestic and some commercial premises to ensure that gas appliances, fittings, and flues are safe. These include requirements for regular servicing and annual gas safety checks, and that all checks and work are undertaken by a *Gas Safe Register registered engineer. One of the objectives of the gas safety regulations is to reduce the risk of dangerous *carbon monoxide leaks as a consequence of problems with equipment or flues.

gate fees Fees charged, usually on a per tonne basis, by the operators of waste-processing facilities to those who are taking waste to the site. *See also* LANDFILL TAX.

GATT *See* GENERAL AGREEMENT ON TARIFFS AND TRADE.

gazumping *See* CONVEYANCING.

gazundering *See* CONVEYANCING.

GCWT *See* GAME CONSERVANCY AND WILDLIFE TRUST.

GDV *See* GROSS DEVELOPMENT VALUE.

gelding A castrated animal but more commonly referring to a stallion. The surgical procedure to remove the testicles of the male horse is usually carried out between six and twenty-four months of age. The procedure makes the horse more temperamentally suitable for a wider range of uses, and also helps to ensure that only the intended and best individual horses are used for breeding purposes.

General Agreement on Tariffs and Trade (GATT) A trade agreement initiated in 1948 and now with over 160 countries signed up. The main purpose of the treaty is to promote international trade by requiring member countries to reduce or remove tariffs and other trade barriers. It has since been, in effect, superseded by the establishment of the *World Trade Organization in 1995.

General Permitted Development Order (GPDO) An order which, in effect, grants blanket planning permission for certain types of *permitted development, subject to their meeting certain criteria (for example, relating to size or layout). England, Wales, Scotland, and Northern Ireland each have their own GPDO. In Wales there are currently proposals to consolidate their current *Use Classes Order and General Permitted Development Order.

General Register of Sasines *See* LAND REGISTER OF SCOTLAND.

General Vesting Declaration *See* COMPULSORY PURCHASE.

Genetically Modified Organism (GMO) A plant, animal, or other organism in which the genetic code has been altered to gain a desirable attribute. This may be done by **transgenic** modification, where a gene from another organism is transplanted into the target species, or by **cysgenic** modification, where the genes that are modified are part of the organism's own genetic code. For example, a gene may be transplanted from a wheat plant to a barley plant which would be transgenic. Or the genetic code of a wheat plant may be modified to increase resistance to disease by adjusting the relationship between existing genes, which would be cysgenic. The process of genetic modification is very limited by regulation in many countries, particularly in Europe.

geodiversity The natural variety of geological elements, most commonly including soils, rocks, and the range of natural features. Essentially includes the *abiotic component of the natural world, and is closely associated with *biodiversity.

Geographic Information System (GIS) A system designed to store, retrieve, manage, display, and analyse all types of geographic and spatial data. Usually a computer programme or web-based system that allows information to be presented or gives a better understanding of information provided by maps or satellite imagery.

geopark A designated area of land principally focused on promoting awareness of *geodiversity, and sustainable **geotourism**. There are a small number within the UK, which do not in themselves provide any level of statutory protection but are frequently located within other protected areas such as *National Parks. They are part of the **Global Network of National Geoparks**, a voluntary network of more than a hundred geoparks.

(⊕) SEE WEB LINKS
• Part of the UNESCO website, detailing the geoparks in the UK.

geothermal energy Form of *renewable energy utilizing the heat of the Earth. Geothermal energy can be harnessed at a variety of scales, including large plants (providing energy to multiple users) where wells are drilled deep into underground reservoirs or hot rocks in suitable locations in order to utilize their heat in the form of water or steam, through to ground and water source *heat pumps serving individual properties which use the warmth of surface soil and water bodies.

GHG Common abbreviation for *greenhouse gas.

gift hold-over relief *See* Capital Gains Tax.

gilt 1. A young female *pig that has yet to give birth. **2.** A fixed-interest loan, usually issued in units of £100, to the British government, which may be either short-term (less than five years) or long-term (thirty years plus). Gilts are generally thought to be amongst the safest forms of investment, given that the British government has never failed to pay the interest or return the amount invested when due, and hence they are also known as **gilt-edged security**.

girth A band of material that encircles the horse or other animal's body and attaches either side to the saddle. It is the main means of securing the saddle, and it is therefore vital that it is good condition. It is usually made of leather, webbing, or a man-made fibre. *See also* TACK.

girthing tape A specialist tape that measures the *diameter at breast height (DBH). The numbering on the tape converts the measured circumference into the diameter.

GIS *See* Geographic Information System.

glade A small opening in *woodland that may perhaps be caused by a fallen tree or failed establishment of new trees. These are of a temporary nature, but openings of a more permanent nature are often kept open by specific management such as cutting or *grazing. Glades provide important *habitat diversity to woodland that creates their own microclimate. For example, they can provide sunlit areas and protection from the wind that supports some invertebrate groups such as butterflies. Many species thrive on the woodland edge, and glades also increase the internal lengths available within a woodland.

Glastir The *agri-environment scheme operating in Wales and funded through the Welsh Government and the European Union. It has developed from the earlier schemes, Tir Cymen and Tir Gofal, into a broader scheme targeting farmers and land managers to encourage more sustainable land management practices. This includes combating climate change, improving water quality, and improving biodiversity. The scheme operates through **Glastir Entry**, which is a simpler whole farm scheme; and **Glastir Advanced**, which is a more targeted improvement scheme. The related **Glastir Woodland Creation Scheme** provides capital and management grants to promote environmental goals in woodlands. *See also* Natural Resources Wales.

glebe land A piece of land owned to provide additional income to support the parish priest. The word 'glebe' originates from Latin, meaning 'clod', 'land', or 'soil'. It could include complete farms, individual fields, or residential or commercial property. The land or property would be passed on from vicar to vicar, but from 1978 onwards the ownership of all Church of England glebe lands passed from the individual parish to the respective diocese, and income was used and shared by all parishes in the diocese.

Global Positioning System (GPS) A system that uses signals from satellites to pinpoint a precise geographic position. The receiver may be handheld or mounted into a vehicle such as a *tractor or *combine harvester. By linking this information with a *Geographic Information System, maps can be presented giving data such as *crop yield maps, mineral deficiencies, or the presence of *pests or *diseases. This facilitates *precision farming whereby applications of *fertilizers, *pesticides or other inputs can be precisely tailored to the requirements of the crop.

glyphosate An active ingredient that is widely used in *herbicides. It is a non-selective, broad-spectrum compound that is effective in controlling a wide range of plant species. Glyphosate is generally used as a foliar application, where it is absorbed into the plant and translocated (moved internally around the plant) to various sites within the plant. It kills the plant by inhibiting photosynthesis, respiration, and protein synthesis. In chemical terms, glyphosate is an organophosphate compound but is not toxic in the same way that organophosphate insecticides are. The European Food Standards Agency states that its use poses no threat to human health. Nevertheless, there have been concerns about its safety since a report published by the World Health Organization in 2015 claimed that the chemical is 'probably carcinogenic'. Its licence by the *European Commission has been renewed. *See also* PESTICIDES.

GMO *See* GENETICALLY MODIFIED ORGANISM.

goat (*Capra aegagrus hircus*) A hardy domesticated *ruminant mammal kept for its milk and meat. An *ungulate, it is closely related to domestic *sheep. Goat's milk is valued by some dieticians for its health benefits, but much of it is processed into cheese. Originally from Southwest Asia and Eastern Europe, the goat eats a wide range of vegetation and thus can be kept on poor *grazing and can also be used to control *scrub growth.

Good Agricultural and Environmental Condition A range of conditions that together with *Statutory Management Requirements make up the *cross-compliance rules with which a recipient of *Common Agricultural Policy agricultural or environmental payments must comply. These apply across all EU member states, though there is variation with the individual implementation detail for each country. The conditions are predominantly concerned with providing a level of environmental and resource protection above the level of basic legislation, but below that provided through funded *agri-environment schemes.

good agricultural practice *See* CODE OF GOOD AGRICULTURAL PRACTICE.

good husbandry 1. General term directed at the activity of growing crops or raising animals for food. Thus good husbandry is farming in a sustainable way with concern for the health of soils and of animals. Maintaining the structure and fertility of soils with high organic matter content by a sound rotation of crops is good husbandry, as is the rearing of animals with regard to their health and welfare; for example, appropriate stocking densities. The *Department for Environment, Food and Rural Affairs (DEFRA) has published a *Code of Good Agricultural Practice providing detailed guidance for farmers, and there are also a number of *farm assurance schemes which guarantee that food is produced by good husbandry. **2.** Specific term within the *legislation concerning agricultural tenancies, particularly *Agricultural Holdings Act 1986 tenancies, and hinging on the tenant maintaining a reasonable standard of efficient production and keeping the holding in a condition to allow for this standard to be maintained in future. In some circumstances, the *landlord can regain *vacant possession of an Agricultural Holdings Act 1986 *tenancy if it is considered to be in the interests of good husbandry; however, such events are rare and are subject to significant *statutory controls. A similar provision also exists within *Scottish agricultural tenancies.

good leasehold *See* LEASEHOLD; TITLE.

good neighbour agreement Agreement in Scotland between a landowner or developer and a community body in relation to the use or *development of an area of land and making commitments in relation to that land or development process, either permanently or temporarily (for example, that information is provided to the community body about the progress of the development). They are distinct from *section 75 planning obligations in that they are between the landowner and/or developer and the community (rather than *local planning authority), although they may address similar areas (but are not permitted to require the payment of any monies).

goose A large group of *waterfowl, including many species both wild and domesticated. *See also* POULTRY.

gorse (*Ulex sp.*) An evergreen coarse and prickly shrub distributed throughout the UK, in which there are found three species. It has very limited agricultural value, but does provides valuable habitat for many species of birds and insects that benefit from the shelter and food that it can provide. Also referred to as furze, hence not an unusual historical name given to *fields, *barns, or *farms. *See also* SCRUB.

GPDO *See* GENERAL PERMITTED DEVELOPMENT ORDER.

GPS *See* GLOBAL POSITIONING SYSTEM.

grading The sorting by quality of commodity products such as timber and *grain to aid valuation. In addition to size measurements, other quality factors for timber need to be considered, such as **Knot Volume Ratio**, staining, and wood-grain characteristics. There are numerous grading rules and guidelines specific to types and species. Technical methods including X-rays and ultrasound are now also available to aid the grading process. In addition to grading of timber products, tree seedlings from nurseries can be graded for quality. Grain is graded to measure factors such as moisture content and non-grain material for consideration in the final valuation by weight.

grain The edible seed of grasses, particularly *cereals such as *wheat and *oats. The term is also used for the seed of other plants such as quinoa. Part of the staple diet in most parts of the world, grain is used as a source of starch in a wide variety of foods from bread and *rice to beer and spirits. They are also the basis of *concentrate feed.

grain marketing The selling of grain produced on a farm. This may be done by selling to a *grain merchant for more or less immediate collection of the grain at the price applicable at the time, known as the spot price, or it may be for collection as a future point in time, known as *forward selling. Grain, or indeed any other product, may be grown on contract whereby the price, or a range of prices dependent on factors such as quality, is set out in a contract before the crop is planted.

graminicide A chemical compound used to control *grasses. Some are non-selective, killing a broad range of species, whilst others are selective, killing only certain species. This latter is important when controlling grass weeds such as *blackgrass or *wild oats in a *cereal crop. *See also* PESTICIDES.

grass (grassland) A group of plants of the *Gramineae* family, or the land on which they are growing. Areas of farmland may not be suitable for *arable cropping and are thus in grass on a non-rotational basis, known as *permanent pasture or *rough grazing. Other areas are sown to grass species, sometimes with other species such as *clover or *herbs, for food for *grazing animals, or for making *hay or *silage. Grass species are also used for recreational purposes such as lawns or sports grounds.

Types of grasses in the UK

Annual meadow grass	Meadow foxtail
Barren or sterile brome	Onion couch or false oat grass
Blackgrass	Perennial ryegrass
Cocksfoot	Red fescue
Common bent	Rough-stalked meadow grass
Common couch	Sheep's fescue
Creeping bent	Soft brome
Crested dog's-tail	Tall fescue
Italian ryegrass	Timothy
Giant fescue	Yorkshire fog
Meadow brome	Wild oat
Meadow fescue	

grassum *See* PREMIUM.

grazier The *licensee who takes a *grazing licence.

grazing The consumption of growing *pasture or other plant material by herbivores. In a farming context, this is the eating of *grassland by domestic *ruminant animals such as *cattle and *sheep. The density of animals on the pasture is known as the *stocking rate. Grass varieties bred for yield need to be grazed or cut regularly for optimum use, and the nutritional value will decline if this is not done. When the stocking rate is such that the grass is degraded and cannot grow quickly enough to sustain the grazing animals, this is known as **over-grazing**. When a grassland becomes rank and over-mature because the stocking rate is too low to consume the growth, this can be described as **undergrazing**, which may result in the encroachment of unwanted species such as bracken.

grazing/mowing licence Type of *licence used for agricultural land which gives the *licensee (also known as a **grazier**) the right to enter the *licensor's land in order to allow the licensee's livestock to graze it, and/or for the licensee to mow it in order to produce *hay or *silage. The licensor maintains responsibility for actively growing the grass crop (for example, *fertilizing, *weed control, rolling, etc.), as well as activities such as maintaining the fences and water supply.

Great Britain Collective name for England, Wales, and Scotland and their associated islands (but not including Northern Ireland). It is distinct from the term *United Kingdom, which describes England, Wales, Scotland, and Northern Ireland.

Green Belt *Planning designation applying to specified areas at the edges of some towns and cities in England, Scotland, and Wales which acts to preserve the openness of the countryside by restricting the type and scale of development which can be undertaken in rural locations around the main settlement. The designation was introduced after the Second World War amidst concerns that large tracts of countryside were being lost to urban sprawl as towns and cities spread outwards. National *planning policy sets out the types of *development which are allowed in Green Belt areas, and *local planning authorities must follow these policies when preparing their *local plans and making decisions on applications for *planning permission. In Green Belts the construction of new buildings is heavily restricted, although some which are considered appropriate in rural areas (for example, agricultural buildings) may be allowed. Whilst the Green Belt designation was also used in Northern Ireland, large-scale development in rural areas of Northern Ireland is now controlled through other planning designations, primarily the *Special Countryside Area designation.

green box 1. A common *World Trade Organization term referring to permitted *subsidies because they have limited impact on trade. Typically they include environmental programmes such as *agri-environment schemes. **2.** A management tool created by *Linking Environment and Farming for farmers to help measure positive impacts on the environment of their management techniques.

greenfield land Land which has never had any form of *development. *See also* BROWNFIELD LAND.

greenhouse effect Natural phenomenon whereby *greenhouse gases in the earth's atmosphere trap infrared radiation generated from sunlight hitting the earth's surface, preventing it from escaping back towards space and therefore maintaining the earth's temperature. The greenhouse effect is needed in order to keep the earth's temperature at a stable level, and changes in the levels of greenhouse gases in the earth's atmosphere are thought to result in alterations in the amount of infrared radiation which is held around the earth, thereby contributing to *climate change.

greenhouse gas One of the atmospheric gases which together contribute to the *greenhouse effect. There are a number of greenhouse gases including *carbon dioxide (CO_2), methane, nitrous oxide, ozone, and fluorinated gases, all of which vary in their potency in terms of their contribution to the greenhouse effect. The amount of a particular greenhouse gas is often expressed as a 'CO_2 equivalent' figure, i.e. a number which takes into account both the physical quantity of a gas in the atmosphere (for example, in grams or tonnes), and also its global warming potential relative to carbon dioxide. As a consequence, greenhouse gas management is sometimes colloquially referred to as 'carbon management'. Water vapour is also a greenhouse gas; however, humans have very little known influence on its levels in the atmosphere. *See also* CLIMATE CHANGE; KYOTO PROTOCOL; UNITED NATIONS FRAMEWORK CONVENTION ON CLIMATE CHANGE.

greening measures Under the reformed *Common Agricultural Policy, and in order to be able to claim all or part of the *Basic Payment Scheme, farmers in the EU are required to comply with certain greening requirements. These are commonly referred to as greening measures and involve farmers in three main activities: retaining permanent *grassland, creating *Ecological Focus Areas, and growing at least three crops. There are a number of exemptions to the need for compliance with greening measures, including size and organic status. *See also* CROSS-COMPLIANCE.

green lane A general description of an unsurfaced track commonly running between walls, *hedges, or ditches, and where infrequent use permits vegetation to grow, hence 'green'. A green lane itself is not a legal definition, though it may also often be a type of *Public Right of Way giving legal access to walk, ride, or use a motorized vehicle.

Green Paper Paper issued by government in the United Kingdom setting out initial proposals for a possible change in the law and inviting comments from the general public or interested parties, such as professional bodies or interest groups. A Green Paper is often followed by a *White Paper, which sets out more detailed proposals for a change in the law, and then subsequently by the issuing of a *Bill for consideration by the relevant elected representatives of either the

*Houses of Parliament, the *National Assembly for Wales, the *Scottish Parliament, or the *Northern Ireland Assembly. Once the Bill has been agreed by vote and is given *Royal Assent (i.e. formal approval by the monarch) it becomes an *Act of Parliament or *Act of Assembly (i.e. law).

green top milk Milk in the UK that has not been *pasteurized, and so called because the bottle or carton has a green top to distinguish it from other milk. It can sometimes be referred to as 'raw' milk. The practice is strictly regulated to prevent pathogens that would otherwise be killed in the pasteurization process entering the human food chain. It is legal in England, Wales, and Northern Ireland, but distribution is illegal in Scotland. There are about 200 producers that sell green top milk direct to consumers, either at the farm, at a farmers' market, or through a delivery service.

grey water See FOUL WATER.

grey water harvesting See FOUL WATER; RAINWATER HARVESTING.

gross development value (GDV) See DEVELOPMENT; RESIDUAL VALUATION METHOD.

gross margin (GM) The difference between gross output and *variable costs. In this system of farm accounting, widely adopted throughout agriculture, costs are divided between variable costs that are specifically attributable to the individual enterprise, such as feed, *fertilizer, seed, and *pesticides, and *fixed costs that are spread across the entire farm business, such as labour, machinery, and administration. Thus the gross margin of a wheat enterprise, for example, usually expressed per hectare or acre, is the value of the sale of the produce, wheat and perhaps straw, less the specific variable costs of seed, fertilizer, and pesticides. The gross margin of each enterprise can be compared to determine relative profitability. The gross margin of each enterprise is aggregated and fixed costs deducted to determine the profitability of the farm business as a whole.

ground rent See RENT.

grounds for possession Circumstances, recognized by the *courts, as ones where the *landlord of a property which is subject to a *lease or *tenancy should be able to regain *vacant possession when the tenant has *security of tenure (or in some cases, during the *fixed term). The grounds vary, depending upon the type of tenancy and the particular *statute under which it is operating. However, they commonly include factors such as persistent non-payment of *rent. The landlord will, however, generally have to obtain a *court order in order to rely on them.

ground source heat pump See HEAT PUMPS.

ground vegetation Term frequently used to describe the ground level *flora in relation to *woodland sites. The natural or semi-natural vegetation can provide good indications of the site in terms of its climate, soils, and land form, and can provide a general indication of potential for *tree growth. *See also* NATIONAL VEGETATION CLASSIFICATION.

groundwater *Water which is found underground in the spaces between *soil particles and within rocks. It is distinct from *surface water, which is water found on the surface of the ground (including that in rivers, streams, lakes, ponds, and springs). *See also* ABSTRACTION LICENCE; IMPOUNDMENT LICENCE.

grouse The red grouse (*Lagopus lagopus scotica*) is a medium to large-sized *game bird restricted predominantly to *heather moorlands. It is a subspecies of the willow grouse and is unique to the British Isles. Other members of the family found in smaller numbers include the black grouse, ptarmigan, and capercaillie. Red grouse feed on shoots, buds, and seeds of small shrubs, and a particular characteristic is that its diet is especially associated with heather. The 'glorious twelfth', i.e. 12 August, is the start of the grouse shooting *season. In some upland areas it can be an important part of the rural economy, mainly involving birds driven over the waiting shooters placed in hides which are referred to as butts. *See also* GAME; GAME MANAGEMENT; SEASON; SHOOTING.

grove Usually means a small wooded area, deriving from the Anglo-Saxon *graf*, but can also refer to the type of product derived from the *woodland. *See also* COPSE; COPPICE; SPINNEY.

growth regulator A chemical applied to a growing crop to strengthen the stems of the plants to prevent *lodging. *See also* HARVEST INDEX.

guano Accumulated excrement and remains of seabirds, cave-dwelling bats, and seals, used as *fertilizer. Bird guano comes mainly from islands off the coasts of Peru, California, and Africa heavily populated by cormorants, pelicans, and gannets; bat guano is found in caves throughout the world; and seal guano has accumulated to great depths on islands off the coast of north-west Peru. Bat and seal guano are lower in nutrients than bird guano, which may contains 11–16% nitrogen, 8–12% phosphate, and 2–3% potash. As an entirely natural product, guano is an acceptable fertilizer in *organic farming systems.

guelder rose *See* WOODY SHRUBS.

guidance note Document providing detailed information, and often indicating good procedural practice in relation to a specific professional, governmental, or legal matter. Guidance notes are produced by a variety of different types of organization, including government authorities and professional bodies, and their significance or status varies depending upon the publishing organization. In the context of the *Royal Institution of Chartered Surveyors (RICS), the term 'guidance note' has a specific meaning and is one of the five categories of professional guidance produced by the RICS: namely, *international standards, *professional statements, *codes of practice, guidance notes, and *information papers. Under the bye-laws and regulations of the RICS, its members must comply with RICS international standards and professional statements. Codes of practice and guidance notes provide advice and 'best practice' guidance for practitioners (often in relation to procedures), with some codes of practice being mandatory. Information papers provide information and explanations to RICS members on various matters relating to surveying (but do not recommend or advise on professional procedures). Although the different types of guidance have slightly different statuses, all practitioners are expected to familiarize themselves with new or updated professional guidance, and the RICS reiterates that if it is alleged that a surveyor has been professionally negligent, it is probable that the courts will take into account the professional guidance produced by the RICS in deciding whether or not the surveyor has acted with reasonable competence.

gymkhana A horse show where riders compete in *show jumping, races, and related events, predominantly aimed at children and ponies, with a focus on fun as well as competition. The term 'gymkhana', meaning 'games on horseback', has its roots in India. *See also* EVENTING.

habitat At the simplest level it refers to the place where an organism lives; for example, the habitat of the *badger (*Meles meles*). More commonly, the term has broadened to cover identifiable types of *landscape or features such as *grassland habitat, arable habitat, or *hedgerow habitat. With no or limited human interference, the nature of the dominant vegetation depends predominantly on soil and climate, which in turn is a major influence on the vegetation and associated wildlife that can live there. *See also* HEATH; WETLAND; WOODLAND.

Habitat Action Plan *See* BIODIVERSITY ACTION PLAN.

Habitats Directive A European Council Directive (92/43/EEC) to promote *biodiversity by means of the conservation of *habitats and of wild *fauna and *flora, the means by which the European Union meets its obligations under the **Bern Convention**. It also provides the mechanism for designation of *Special Areas of Conservation (SAC). The Habitats Directive is implemented in each of the EU member states through domestic legislation; for example, in the UK by means of the Conservation of Habitats and Species Regulations 2010. *See also* BIRDS DIRECTIVE; SITES OF SPECIAL SCIENTIFIC INTEREST.

hack (hacking) At its simplest, a hack is to ride out a horse mainly for recreation and/or for fitness, and which usually involves some roadwork and bridleways. Hacking is not a formal riding type. A hack is also an old-fashioned term to describe a horse rented out for riding, or a type of lightweight horse.

ha-ha A sunken and vertical stone wall that, together with an adjacent shaped ditch, provides a *stock-proof *field boundary which allows and preserves an uninterrupted view across the *landscape. These ditches are most commonly found delineating and separating the formal gardens of country estates and the adjoining grounds or parkland.

halter *See* HEAD COLLAR.

hardcore model (layer model) *See* INVESTMENT METHOD.

hardwood General description of wood products derived from *broadleaf timber or of broadleaves themselves, in direct contrast to *softwoods derived from *conifers. The description reflects the relative hardness of this type of wood, generally linked with its higher density and hence its greater strength and durability.

hare hunting Can be done on foot or on horseback, using packs of hounds that follow the brown hare's scent. Mounted packs of hounds are called harriers (some have switched to hunt foxes), while the foot packs are smaller and slower beagles or bassets. Harriers and beagles share the same governing body, the Association of Masters of Harriers and Beagles. The governing body for basset hound packs in the UK is the Masters of Basset Hounds Association. The pursuit of hares using dogs that follow by sight and not scent is described as hare coursing. It has developed into a competitive sport in which dogs are tested on their abilities. This and the other forms of hare hunting have been illegal in the UK since 2005, but a form of coursing in which large sums of money are gambled continues to be practised in some areas. *See also* HUNTING.

harrow A farm implement for breaking down the soil for a seedbed or for levelling the surface. It may be trailed behind a *tractor or mounted on the *three-point hydraulic linkage. *See also* CHAIN HARROW; DISC HARROW; FARM MACHINERY.

harvest The process of gathering ripe produce. The most common form is the cutting and threshing of arable crops in the field, usually by means of a *combine harvester, but the term also includes the picking of fruit, the picking or collection of vegetables, and the felling of mature timber. These processes may be carried out by hand, particularly in the case of fruit, or by machine, as with arable field crops.

harvest index The weight of *grain or harvested product expressed as a proportion of the total weight of the plant. One way in which plant breeders have increased the yield of crops is to improve the harvest index.

harvesting of woodland *See* HARVEST.

hawthorn *See* WOODY SHRUB.

hay *Grass preserved by drying so that it can be used during the winter months when there is limited *forage available. A traditional method where grass is allowed to grow until flower heads are produced and is then cut before it begins to set *seed. This is the optimum point of compromise between *yield and digestibility, and therefore ensures the best nutritive value for *livestock. The *swath of grass is allowed to dry from 80% moisture content to approximately 20% before being baled and stored. The quality of the hay is determined by the ratio of plant leaf to stem (leaf is more nutritious than stem). Drying is usually done in the field and is dependent on the weather, but generally takes two or three days. During this time the drying process can be speeded up through turning the swaths and ensuring that the maximum surface area is exposed to the sun and wind. Alternatively, grass can be barn dried, where swaths are dried on the field to approximately 30% and then baled and placed over a grid or ducts through which air is blown. However, this is expensive and is best limited to producing a small quantity of high-quality hay. *See also* HAY MEADOW; HAYLAGE; SILAGE; BALER.

haylage Conserved grass or other *herbage that is a cross between *hay and *silage. Thus the *forage is cut and allowed to dry but is collected usually by *baling whilst still at a higher moisture content than hay. It is then wrapped in polythene, and some fermentation will take place within the wrapped bale. It is stored and used as animal feed at times when *grazing is limited. *See also* HAY; SILAGE.

hay meadow A field, usually of permanent *grass, that is cut for *hay each summer. It may also be *grazed at other times of year and often has a rich *sward that includes wild flowers and herbs.

hazardous waste *Waste that is considered to present a more significant risk to the environment and human health than other forms of non-hazardous waste. As with other forms of waste, the definition of what constitutes hazardous waste is set out in law in the *United Kingdom, and the management of such waste is controlled through the *hazardous waste regulations (which generally impose stricter controls on hazardous waste management than those that exist for non-hazardous waste). In Scotland, hazardous waste is known as **special waste**.

hazardous waste regulations *Regulations controlling how businesses and organizations in the United Kingdom must manage any *hazardous waste they produce or handle. Regulations set out what is considered to be hazardous waste (known as **special waste** in Scotland). These regulations operate alongside the regulations for non-hazardous waste (*see* AGRICULTURAL WASTE REGULATIONS for more details) and follow similar principles, but with some variations in detail. These include more stringent rules about the storage of hazardous waste (as opposed to non-hazardous waste), and the requirement, if passing on hazardous waste to a party authorized by the government to handle it, for a *consignment note to be exchanged and a copy retained (as opposed to a *waste transfer note for non-hazardous waste). In Northern Ireland and Scotland, some movements of hazardous waste must also be prenotified to either the *Northern Ireland Environment Agency or the *Scottish Environment Protection Agency.

hazel *See* WOODY SHRUB.

head collar The common method of control of the *horse or pony, usually made of leather or nylon and attached around the head using a buckle on one or both sides. They are normally used with a rope that attaches to the head collar. The halter is very similar but is commonly made of rope and has the rope as an integral part of the halter. These are also usually employed to control sheep and cattle and at livestock shows.

headland The area round the edge of an arable field on which tractors and other machines turn when carrying out field operations. With the standard size of equipment in the United Kingdom, particularly *sprayer booms, now 24 metres, this is the most common width of the headland, allowing for one width of the machine. There are various options in *agri-environment schemes for *conservation headlands where *fertilizer and the use of *pesticides is strictly limited, or adjacent to headlands such as *field margins. These options would normally be for widths less than the full headland, such as 6 metres. *See also* FIELD MARGIN.

heads of terms The fundamental points of a *contract (including a *lease) for negotiation between the parties prior to the signing of a formal legal agreement between them. Typical heads of terms in a lease will include information such as the property address, the *landlord's and *tenant's details, the *rent (including any provisions for *rent reviews), the length of the lease, any *break clauses, any limitations on the use of the property, and responsibilities for repairs and insurance.

Health and Safety at Work Act 1974 *Act of Parliament concerning health and safety within workplaces in *Great Britain. It imposes duties on employers to ensure that the workplace is safe (including providing *personal protective equipment where necessary) and to provide appropriate information, training, and supervision of employees. It also places obligations upon employees to take reasonable care of the health and safety of themselves and others and to cooperate with their employer on health and safety matters. *See also* HEALTH AND SAFETY AT WORK (NORTHERN IRELAND) ORDER 1978; RISK ASSESSMENT.

Health and Safety at Work (Northern Ireland) Order 1978 Main piece of *legislation concerning health and safety in workplaces within Northern Ireland. Its provisions echo those of the *Health and Safety at Work Act 1974 which covers *Great Britain. *See also* HEALTH AND SAFETY EXECUTIVE FOR NORTHERN IRELAND; RISK ASSESSMENT.

Health and Safety Executive (HSE) Public body of the Department for Work and Pensions in *Great Britain which exists to reduce work-related injury and ill health. They provide advice and guidance on health and safety matters as well as reviewing, advising on, and enforcing health and safety *regulations in Great Britain. In Northern Ireland, the equivalent roles are carried out by the *Health and Safety Executive for Northern Ireland, which is a separate body but which liaises with The Health and Safety Executive.

Health and Safety Executive for Northern Ireland (HSENI) Public body sponsored by the Department for the Economy in Northern Ireland which exists to ensure that health and safety risks from work activities are controlled effectively in Northern Ireland. They provide advice and guidance on health and safety matters as well as reviewing, advising on, and enforcing health and safety *regulations. In *Great Britain, the equivalent roles are carried out by the *Health and Safety Executive, which is a separate body but which liaises with The Health and Safety Executive for Northern Ireland.

HEAR *See* HIGH EURUCIC ACID RAPESEED.

hearing *See* APPEAL, PLANNING.

heath (heathland) These usually occur on acid *soils in areas of low *grazing intensity, and are characterized by *scrub. Heathland occurs on drier and usually sandy soils at lower altitudes, whereas the more typically wetter and *upland heath is generally mostly covered with *heather or *bracken and is referred to as *moorland, particularly in the UK. They are an important *habitat with a diverse *fauna and *flora. Management by controlled burning regimes can be practised to maintain heathlands. *See also* MOOR.

heather (*Calluna vulgaris*) There are numerous ornamental varieties of heather, but the common type of heather, sometimes known as ling, is dominant on *heathlands in the UK and across Europe. It is an evergreen *shrub with woody stems, and the flowers are mauve, which gives the distinctive colour to these *landscapes in late summer. Heather can live up to twenty years, and the stems become tough with reduced leaves and flowers. Younger heather can provide food for *sheep and *grouse, and hence older heather can often be subjected to controlled burning regimes. The rotational burning of patches of heather can produce a characteristic patchwork quilt of heathland.

heat pumps Group of pump-based *renewable energy technologies that transfer heat between an outside material (most commonly the outside air, a water body, or under the ground) and the inside of a building, in order to provide heating (and, in some cases, cooling) for it. There are several different types of heat pump. **Ground source heat pumps** consist of circular loop of pipework which is buried underground and has fluid circulating around it. The system utilizes the warmth underground in order to warm the fluid and provide low-level heat to the building. In the summer, when temperatures inside the building are higher than those underground, some systems can be reversed to provide cooling. **Water source heat pumps** operate on the same principles; however, the outside pipework runs through a water body (for example, a river) instead of being located underground, and the system extracts heat from the water. Whilst both ground and water source heat pumps rely on having sufficient space available externally in order to locate the pipework, **air source heat pumps** operate on similar principles to domestic refrigerators and can therefore be used in situations where there is less space available (with the relatively compact units sitting on the outside of buildings). They operate by extracting heat from the outside air, even at low temperatures. As heat pumps tend to provide low-level continuous heat, buildings need to be well insulated in order for the systems to be effective, as they use energy themselves in order to operate. *See also* RENEWABLE HEAT INCENTIVE.

heaves A severe form of bronchitis in *horses, also referred to as recurrent airway obstruction. The condition can cause severe breathing difficulties, and is more common in horses kept in poorly ventilated and dusty environments.

hectare The metric unit measurement of area increasingly more commonly used as a replacement for the imperial *acre on UK *farms and *forestry. It represents 10,000 square metres and equates to approximately 2.47 acres. *See also* APPENDIX 1 (METRIC–IMPERIAL CONVERSION)

herb

hedge and ditch rule Rule, established in nineteenth century English *case law, that where the boundary between two properties consists of a hedge and an artificial ditch then the boundary is along the edge of the ditch furthest from the hedge or bank. The principle behind this is that when the ditch was dug in the first place, the person who dug it would have dug it on the very edge of their land and put the soil on their side of the boundary, forming a mound on which the hedge may have later been planted; hence the boundary lies along the ditch on the far side of the hedge. However, this presumption will not apply if there is evidence that the ditch is natural, or if the land on both sides of the ditch was in the same ownership at the point when it was dug. *See also* AD MEDIUM FILUM RULE.

hedge (hedgerow) Usually a line of *shrubs that define *field boundaries and enclose *livestock. Often comprise several different species that typically might include hawthorn, blackthorn, or hazel, but might also vary according to region and when planted. A small number of hedges that have survived for hundreds of years may be remnants of original woodland retained to mark a boundary. Increased mechanization and the decline of mixed farming has seen the removal of hedges and the loss of function. Their value as a characteristic feature of the *landscape and as an important *habitat has encouraged their continuation and management.

hedge management Hedges need to be trimmed on a regular two- or three-year cycle. Overgrown hedges may also need to be laid; that is, the stem cut almost right through and laid at about 45 degrees. Hedges may also be rejuvenated by *coppicing. Management of hedges to promote their *landscape and wildlife benefits is a commonly supported practice through *agri-environment funding.

Hedgerow Regulations 1997 Gives protection to important *hedgerows in the countryside through a system of notification to *local planning authorities in England and Wales of any proposed removals.

(((●))) SEE WEB LINKS
• A partnership of land and environment organizations detailing key interests and knowledge of hedgerows and their management.

heel in The practice of temporary planting of bundles of whips in a narrow trench or 'sheugh' in the soil to protect the roots from drying out or other damage. The bundles are placed in the trench and the soil replaced to cover the roots or 'heeled-in' with the heel of a boot. *See also* PLANTING.

heifer A young female of *cattle and related species; specifically, one that has yet to have a *calf.

herb A plant, the green leaves or foliage of which is valued for flavouring in cooking, for medicinal purposes, or as a source of trace elements or other *nutrients in animal nutrition. Examples include fennel, mint, and yarrow.

herbage Vegetative plant growth, particularly the leaves, that is used for *grazing by herbivorous animals such as cattle and sheep. Usually *grass, but may be other plant material.

herbicides Chemicals that are used to kill plant species, especially *weeds. These may act by contact, where only the part of the plant covered by the chemical is affected, or may be systemic, where the active ingredient is translocated throughout the plant. Some are short-lived, breaking down quickly once applied. Others are residual and may be applied to the soil where the chemical remains active for weeks or even months. Selective herbicides are those that only affect the target species and are not active against others. Non-selective herbicides are those that are active against a wide range of species. *Graminicides are herbicides that are active against grass weeds, whilst others are active against broad-leaved weeds. *See also* PESTICIDES.

herd The collective noun for *cattle or other related species. It can also mean to live or collect together rather than singly or in small family groups.

herd basis *See* STOCKTAKING VALUATION.

hereditament 1. Under English law, *real property which can be inherited. **2.** Under the law relating to local taxation in the *United Kingdom, a separate unit of property which is liable to be taxed; for example, in one building there may be several different *leaseholders, each paying *business rates separately, hence there are several hereditaments. *See also* CORPOREAL HEREDITAMENT; INCORPOREAL HEREDITAMENT.

heritable property Term used in Scottish law to describe land and the rights connected to land; also described as **immoveable property. Moveable property** is all property other than heritable property, so, for example, physical objects other than land (say a painting) and contractual rights (other than those concerning most rights in land). These terms are often used in combination with the terms **corporeal** (meaning physical items of property) and **incorporeal** (everything else). So, for example:

- A **corporeal moveable** item would be a painting.
- A **corporeal heritable** item would be a parcel of land.
- An **incorporeal moveable** item would be some right established by a *contract (other than one concerning land).
- An **incorporeal heritable** item would be a non-physical right in land; for example, the right to fish for freshwater fish.

heritage asset Collective term, used in the *National Planning Policy Framework in England and also national heritage planning guidance in Wales, for a range of heritage *planning designations which are covered by heritage *planning policy. These include national statutory designations (including *listed buildings, *World Heritage Sites, *registered parks and gardens, *registered

battlefields, *scheduled monuments, and *Conservation Areas), and also local
heritage assets which can be designated by *local planning authorities.

Heritage Coast Coastal areas in England and Wales defined as being worthy of
being conserved because they are considered to be the best stretches of
undeveloped coast. The designation is not statutory, hence there is not a formal
designation process set out in law (as with National Parks or Areas of
Outstanding Natural Beauty). Instead they have been developed by agreement
between local authorities and *Natural England/*Natural Resources Wales (or
their predecessors). There are a number of objectives behind defining an area as
Heritage Coast, including conserving and enhancing the natural beauty,
biodiversity, and heritage of the area, supporting the public to enjoy it,
maintaining and improving the health of the inshore waters and beaches, and
taking into account other needs (such as those of agriculture, forestry, fishing,
and local communities). The existence of a Heritage Coast designation is a
potentially significant consideration in planning applications, and is addressed in
planning policy.

heritage impact statement Supporting statement required, in Wales, with
applications for *listed building consent, or *Conservation Area Consent. The
requirement for such a statement was introduced in 2017, and replaced the need,
in Wales, for *design and access statements in support of applications for listed
building consent. The precise aspects that must be covered differ slightly
depending on the type of building, and are set out in *legislation; however, they
normally include, for listed buildings, information about the proposals, including
why they are needed and what design principles have underpinned them, a
schedule of works, how the proposals may benefit or harm the special
architectural or historic interest of the building, plus any other options that were
considered, and how any access issues have been addressed. For applications for
Conservation Area Consent, the requirements are similar, and focus on the
impacts that the proposed demolition will have on the character and appearance
of the Conservation Area, plus why they are proposed, and any alternative
options considered.

Heritage Partnership Agreement *See* LISTED BUILDING CONSENT.

Her Majesty's Revenue and Customs (HMRC) *United Kingdom
government body responsible for collecting national taxation from individuals,
businesses, and organizations. This includes *capital taxes, *Income Tax,
*Value Added Tax, and environmental taxes such as *Landfill Tax, as well as
*National Insurance. Scotland and Wales also have their own tax collection
bodies (*Revenue Scotland and the *Welsh Revenue Authority respectively).
These organizations specifically collect the taxes that the Scottish and Welsh
governments are entitled to charge. However, other than these specific taxes,
the tax paid by individuals, businesses, and organizations in Scotland and Wales
continues to be collected by HMRC.

Heston bale A *bale of *hay or *straw that has come from a *baler made by the manufacturer Heston.

HGCA (Home Grown Cereals Authority) *See* LEVY BODIES.

hide A camouflaged construction of natural or net material in which to remain hidden from view from *quarry species. Frequently used for *pigeon shooting and for driven *grouse shooting where shooters are concealed behind structures called butts which are made of stone, wood, or turf. Hides of varying size and material may also be purposively built and sited to assist bird and other wildlife watching, for example, overlooking wetlands.

High Court Civil Court in England, Wales, and Northern Ireland. *Judicial reviews are considered by this court. Appeals against decisions of the High Court are considered by the *Court of Appeal or **UK Supreme Court**. *See also* COUNTY COURT; COURT OF SESSION; SHERIFF COURTS.

Higher Level Stewardship *See* ENVIRONMENTAL STEWARDSHIP SCHEME.

high eurucic acid rapeseed (HEAR) *Oilseed rape with a high eurucic acid content used for industrial purposes.

high forest Woodland derived from seed as opposed to *coppice. This may be a single species or even age and present a uniform appearance, or a mixture of species and ages to provide what can be described as 'irregular high forest'. *See also* SILVICULTURAL SYSTEMS.

Highland pony *See* HORSE.

high oleic low linolenic Varieties of *oilseed such as *oilseed rape or *soya bean that have a high concentration of oleic acid but a low concentration of linolenic acid, thought to provide a healthier oil for cooking.

high seat A platform erected at a height of 2–5 metres or more in a tree, or commonly a purpose-built wooden or metal construction. It is sited to allow a hunter an elevated shooting position that commands a view over a piece of open ground where deer are known to pass or gather. Shooting from a high seat also provides a safer backstop, as the hunter will be shooting downwards into the ground. *See also* DEER STALKING.

highways authority National or local body responsible for the management and maintenance of highways. **Highways England** has responsibility for motorways and major roads in England (in Scotland it is **Transport Scotland** and in Wales the *Welsh Government), whilst responsibility for smaller roads rests with local highway authorities (who are normally based within county councils or unitary authorities). In Northern Ireland the **Department for Infrastructure Roads** is responsible for all public roads. As well as being responsible for the management and maintenance of the highways, highways authorities are also *statutory consultees in relation to planning applications (particularly in relation to the safety of proposed access and egress routes onto and off the public highway).

hill farmer A person who farms in the hills and uplands. *See also* FARMER.

hill farming Farming that takes place in the uplands or hills. The land is usually unsuitable for arable cropping, and is predominantly livestock farming with grazing of cattle and, particularly, sheep on *permanent pasture or *moorland. Traditionally, sheep flocks in the uplands provide the cross-bred ewes for replacements for lowland flocks often sold at particular fairs or markets, leading to significant animal movements in the autumn. In the European Union, much of this land comes within the designation of *Less Favoured Area, which qualifies it for extra financial support. *See also* HILL FARMER; COMMON AGRICULTURAL POLICY.

Historic Buildings Council *Statutory body operating in Northern Ireland which advises the Northern Irish government on matters relating to heritage management, and in particular, *listed buildings and *Conservation Areas.

Historic England Public body responsible for England's historic environment. Its official name is the **Historic Buildings and Monuments Commission for England**, and it is the expert adviser to the government in relation to the historic environment. It manages, on behalf of the government, the heritage designation system (including the process of *listing buildings), and publishes guidance on the management of *heritage assets, in addition to providing educational resources and training in relation to heritage. Staff provide advice in relation to applications for *listed building consent and planning permission involving heritage assets, and the organization also operates a grant system for the protection of heritage at risk. In addition, it manages the *National Heritage List for England, which lists around 400,000 heritage items (ranging from buildings to battlefields) which have legal protection. Formally known as **English Heritage**, it changed its name to Historic England in 2015, whereupon a new charity called the **English Heritage Trust** was established to manage the **National Heritage Collection** (comprising state-owned historic sites and monuments in England), under licence from Historic England.

Historic Environment Records Records (sometimes known as **Sites and Monuments Records**) held for each local area of the United Kingdom, of local archaeological sites, and other aspects of the historic environment, such as information relating to historic buildings, landscape character, and Ancient Woodland.

SEE WEB LINKS
• Heritage Gateway website: in England, the majority of local historic environment records, plus the National Heritage List for England, as well as other databases relating to the historic environment which can be searched.

Historic Environment Scotland Public body responsible for Scotland's historic environment. It was established in 2015, bringing together the responsibilities of the **Royal Commission on Ancient and Historic Monuments of Scotland** and **Historic Scotland**. In addition to managing a collection of properties on behalf of the *Scottish Government, it carries out a range of activities (equivalent to those undertaken by *Historic England) relating to the legal protection of heritage in Scotland (for example, managing the process of *listing buildings).

((∰)) SEE WEB LINKS
• Canmore website, managed by Historic Environment Scotland, comprising an online searchable catalogue of information about the historic environment in Scotland.

Historic Environment Scotland Portal Online database, managed by *Historic Environment Scotland, of *listed buildings, *scheduled monuments, the **inventory of gardens and designed landscapes**, the **inventory of battlefields**, and historic marine protected areas in Scotland.

((∰)) SEE WEB LINKS
• Historic Environment Scotland Portal website.

Historic Scotland *See* Historic Environment Scotland.

hitch A mechanism on the back of a *tractor or other towing vehicle for attaching the equipment to be towed. It can be a simple tow bar, but may also be a ball and socket form where the trailed vehicle has a socket that fits on a ball hitch protruding from the back of the towing vehicle. Also used as a verb meaning to attach a towing vehicle to a trailed vehicle.

HLS *See* Environmental Stewardship Scheme.

HMO *See* house in multiple occupation.

HMRC *See* Her Majesty's Revenue and Customs.

hobby farmer A person who farms without the expectation of making a living from their activities. Traditionally, such farms were smallholdings where the farmer produced small quantities of produce, often by hand. In more recent times there has been a large increase in the purchase of farmland by those whose primary income comes from elsewhere and whose main reason for owning the land is amenity rather than farming profit. These can be more commonly referred to as lifestyle farmers. *See also* farmer.

hog A *pig, especially in North America. A hog roast is the barbecue of a whole pig, usually on a spit.

hogget Strictly, a *sheep between one and two years old, but often taken to mean a *lamb that is fully grown even if not quite a year old. For example, a lamb born in April might be considered a hogget if slaughtered for meat in late winter or early spring.

hogging Complete removal of the mane of a *horse or pony by clipping. It is favoured as a labour-saving operation, making it easier to keep the animal clean and looking smarter. It is a common method used on polo ponies and *cobs.

holding Refers to land and buildings held as *freehold or *leasehold and commonly used for agricultural purposes. The holding will have a designated number and is particularly relevant to describe the whole farmed unit for the purpose of receiving *Common Agricultural Policy support payments such as *BPS or participation in *agri-environment schemes. *See also* FARM.

holly *See* TREE.

Home Grown Cereals Authority *See* LEVY BODIES.

home loss payment *See* COMPULSORY PURCHASE.

Homes England Public body (formerly known as the **Homes and Communities Agency**), sponsored by the *Ministry of Housing, Communities and Local Government, that has responsibility, in England, for direct government activities in relation to home building, and the provision of new business premises. Its core objectives are to increase the number of homes and business premises available (with the major focus being on homes), and it undertakes a number of different activities in support of these objectives, including schemes to invest public money in the building of housing and business floorspace, and managing the sale of redundant publicly owned land to build housing on. It currently has a regulation directorate (known as the **Regulator of Social Housing**) that regulates social housing providers (mainly **housing associations**) (*see* AFFORDABLE HOUSING). The functions of the Regulator include ensuring that social housing *tenants are protected, and also that the social housing system is financially secure. It is anticipated that, at a future date, the work of Homes England and the Regulator of Social Housing will be formally split into two separate bodies.

Equivalent activities are carried out by the *Northern Ireland Executive, the *Welsh Government, and the *Scottish Government (with social housing regulation in Scotland being undertaken by the **Scottish Housing Regulator**).

homogenized milk A mechanical process that breaks the fat globules in the milk into smaller droplets so that they stay suspended rather than separating out and floating to the top in the form of cream. In the dairy process the milk is usually first *pasteurized to kill some of the bacteria present, and then homogenized to produce a consistent texture.

hope value The additional element of value of land or buildings, over and above the value of them for their existing use, which is attributable to the hope that their use will be changed or that they will be developed at some future date into something which is more valuable. *See also* ACCOMMODATION LAND.

Hoppus measure A measure used in the timber trade and in particular in relation to *hardwood. Named after Edward Hoppus, who published the calculation in a book published in 1736, it estimates the volume of a log in relation to the volume of sawn timber that it can produce. The log volume (abbreviated to 'Hoppus') is calculated as: Hoppus volume = (mid quarter girth)^2 × length. Measurements for the mid quarter girth and length are in imperial inches and feet respectively. A special girth tape is normally used, calibrated in quarter girths.

hornbeam *See* TREE.

horsebox A lorry or van designed to transport horses; as distinct from a horse trailer, which is towed by a separate motorized vehicle.

horse (*Equus ferus caballus*) The domestic horse is a hoofed ungulate mammal. There are many different types of horse, with a variety of different task-related attributes; for example, the *cob, Irish Draught, *shire, *hunter, and pony. Previously a mainstay of farming operations and long since replaced by *tractors and other machinery. In industrialized countries it is predominantly used in recreational pursuits, including racing, *show jumping, *hacking, and *hunting. A number of common terms are used to define age and sex differentials. For example, a **stallion** is an uncastrated male of four years or more, a **colt** is an uncastrated male of three years or less, and a **gelding** is a general term for a castrated male. A **mare** is a female of four years or more, a **filly** is less than four years old, and a **broodmare** is more commonly used to describe a female used for producing *foals, but can refer to any *equine. *See also* GYMKHANA; HORSEBOX; HORSE PASSPORT.

horse passport A legal requirement throughout the EU for all *horses and other equids. The purpose of these passports is to ensure that particular drugs used to treat horses do not find their way into the human food chain. It is also intended to prevent the sale of a stolen animal, as the passport is a proof of its identity. The passport should be kept with the horse if it is travelling to a show, for example, and it is illegal to buy or sell a horse without a valid passport. *See also* MICROCHIPPING.

Types of horses and ponies in the UK

Horse	Pony
Clydesdale	Connemara
Cleveland Bay	Dales
English Thoroughbred	Dartmoor
Hackney	Exmoor
Irish Cob	Highland
Suffolk Punch	New Forest
Shire	Shetland
Welsh Cob	Welsh Mountain

Main colours and descriptions of horses and ponies

Colour	Description
Bay	Brown coat with black mane and tail
Black	Black
Chestnut	Ginger coat and similar mane and tail
Dun	Cream golden coat and black mane and tail
Grey	Vary from almost white to dark grey
Palomino	Golden coat with white mane and tail
Piebald	General term of black with white pattern
Skewbald	Any colour other than black with white pattern
Tobiano	White coat pattern combined with any colour
Roan	Coat pattern combined with any colour

Horticultural Development Company *See* LEVY BODIES.

horticulture The growing of crops, fruit, and vegetables for food, or ornamental trees, shrubs, and flowers for show as in gardens. Top fruit are grown on trees, such as apples or oranges, and soft fruit are small fruit grown on bushes, such as strawberries or blackcurrants. Root vegetables grow underground, such as carrots or onions usually grown in light soils. Fruit trees and bushes, ornamental trees and shrubs, and flowers seeds may be bought at garden centres. An orchard is a field of fruit trees such as apples or pears, and a vineyard is a field where vines are grown for wine production.

house in multiple occupation (HMO) Residential property occupied by several people who are not in some way connected to each other (so are not, for example, all in the same family). Private houses occupied by students will, for example, often fall into the definition of an HMO. The *landlords of HMOs throughout the *United Kingdom must hold a *licence (this is over and above any general requirement for *landlord registration for private residential landlords, and unlike landlord registration, applies throughout England). The landlords of HMOs must also comply with additional legal requirements over and above other residential let properties (for example, in relation to fire safety). The definitions and licensing requirements of HMOs are set out in law (though there are some variations in the licensing requirements in different parts of the UK).

Houses of Parliament The United Kingdom Parliament comprising the House of Commons and House of Lords. The Houses of Parliament pass laws which can apply across the whole of the United Kingdom; however, under devolution, law-making powers in some specific areas have also been given to the elected representatives of the *National Assembly for Wales, the *Scottish Parliament, and the *Northern Ireland Assembly.

Housing Act 1988 *Legislation governing residential occupation and *tenancies which were entered into after 15 January 1989 in England and Wales (for the position before that, *see* RENT ACT 1977 and *RENT (AGRICULTURE) ACT 1976). There are three types of occupation covered by the Housing Act 1988: **assured tenancies, assured shorthold tenancies (ASTs**; a form of assured tenancy), and **assured agricultural occupancies**.

Much of the framework for assured tenancies is the same as for ASTs (as they are, in effect, a form of assured tenancy). In order to be an assured tenancy a house, or part of a house, must be let as a separate dwelling to an individual or individuals who occupy the dwelling as their only or principal home (so not a second home). Some tenancies are specifically excluded from being assured tenancies or ASTs. This includes any tenancy created before 15 January 1989, and a number of other criteria which are very similar to those under the Rent Act 1977.

Prior to 28 February 1997, any tenancy satisfying the conditions of the Housing Act 1988 would automatically have been an assured tenancy unless a notice was served before the start of the tenancy stating that it was to be an AST. Since 28 February 1997, however, the default position is that any tenancy meeting the criteria will be an AST unless a notice is served to the effect that it is an assured tenancy (the exception to this is an agricultural worker, when a notice must still be served as otherwise the occupier will be an assured agricultural occupant; see below).

Although sharing the same roots, assured tenancies and ASTs, in practice, operate slightly differently:

- **Assured tenancies**: when the original *fixed term of an assured tenancy expires, then the landlord and tenant may agree to renew the tenancy for a further fixed term, or the tenant may remain in occupation under a **statutory periodic tenancy** (on the same terms as the fixed term tenancy). The rent under an assured tenancy is not regulated; however, disagreements about *rent reviews of periodic assured tenancies can be referred to the **Rent Assessment Committee** (*see* FIRST-TIER TRIBUNAL PROPERTY CHAMBER). The tenant of an assured tenancy has *security of tenure because the landlord cannot gain *vacant possession without a *court order, and to obtain such an order they will need to demonstrate one of the *grounds for possession included in the Act. As with the Rent Act 1977, the grounds are divided into **mandatory** and **discretionary** ones (though there are some differences in the grounds themselves). There is also, potentially, the opportunity for one *succession under an assured tenancy.

- **ASTs**: as noted above, an AST is effectively a form of assured tenancy and therefore generally shares the same qualifying criteria. As with an assured tenancy, at the end of an initial fixed term, unless the parties agree to renew for a further fixed term, the tenant may remain in occupation under a statutory periodic tenancy (again on the same terms as the fixed term tenancy).

There is also no regulation of rent for an AST, although tenants can appeal to the Rent Assessment Committee within the first six months of their tenancy if they consider that the rent is unreasonably high. There is no imposed security of tenure under an AST, because the landlord has the right to serve two months' notice to terminate the tenancy at the end of the fixed term. The rules for obtaining possession of a property let under a periodic tenancy from the beginning are slightly different, however; again the landlord does not have to demonstrate any grounds for possession in order to obtain vacant possession of the property (although the courts will not grant an order within the first six months of a tenancy, thereby effectively meaning that ASTs must be for a minimum of six months). If, however, a landlord wishes to regain possession during the fixed term they must apply to the court for a court order and must demonstrate one of a number of grounds for possession included within the Housing Act 1988. Whilst, theoretically, succession rights apply to ASTs in the same way as assured tenancies, in practice, because the landlord has the right to terminate the tenancy the successor has no security of tenure.

- **Assured agricultural occupancy**: the successor to the Rent (Agriculture) Act 1976. It entitles a qualifying agricultural worker of a property in qualifying ownership (essentially using the same definitions as the 1976 Act) to an assured tenancy if they would otherwise be entitled to one were it not for their paying a low rent or no rent; or the dwelling house being part of an agricultural holding or farm business tenancy (thereby not qualifying as an assured tenancy); or if they have a licence with exclusive occupation; and they are not occupying as an AST tenant (i.e. no notice was served before the start of the tenancy confirming that they are occupying under an AST). As with an assured tenancy, the rent under an assured agricultural occupancy is subject to only limited control, and again, as with an assured tenancy on the termination of a contractual assured agricultural occupancy, a statutory periodic tenancy will arise. One succession is also available, and the assured agricultural occupant has security of tenure for their lifetime, with the landlord only being able to obtain a court order for possession if they are able to satisfy one of the grounds for possession. These are effectively those of an assured tenancy with the exception that the landlord is not entitled to use the ground potentially allowing a landlord to obtain possession of a dwelling house let to a tenant as a consequence of their employment when that employment has ceased.

See also RENTING HOMES (WALES) ACT 2016.

housing health and safety rating system (HHSRS) System used by *local authorities to assess health and safety risks within residential properties in England and Wales. The system operates on *risk assessment-based principles and takes into account the nature of any hazard, and the potential for, and severity of, any possible harm. The system also reflects that the likelihood of any harm happening may vary, depending on the occupiers, or potential occupiers, of a particular property. For example, an uneven path might present a particularly high risk to an elderly occupier, as the elderly are more vulnerable to falls. The system covers a wide range of potential hazards, including, for example, excess heat and cold, crowding, water supply, and noise. If a local authority considers that a let residential property has deficiencies which are causing unacceptable levels of risk to the occupiers, or potential occupiers, then the local authority has powers to ensure that the *landlord takes steps to remedy the situation (or in some situations the local authority may itself take action and then recover the costs from the landlord). In severe situations the local authority may prevent the property from being occupied, or even potentially issue an order that it be demolished.

HSE *See* HEALTH AND SAFETY EXECUTIVE.

HSENI *See* HEALTH AND SAFETY EXECUTIVE FOR NORTHERN IRELAND.

human capital All of the skills, attributes, abilities, training, knowledge, and experience possessed by individuals or a population which can be utilized by organizations and wider society.

human resources Term used to describe the people employed within an organization, and also sometimes the department (also sometimes known as **personnel**) within that organization that has responsibility for advising senior managers about best practice, and legal obligations, with regard to staff employment.

hunter 1. A common and general description of a type of horse. The term means a horse or pony with all the expected traits that make them suitable for the *hunting field, including hardiness, temperament, jumping ability, fitness, and stamina. A more modern relevance is to the rising popularity of classes for show hunters. *See also* FOX HUNTING; STAG HUNTING; DRAG HUNTING; HARE HUNTING. **2.** A general term to describe someone who is *hunting using hounds or dogs, or more normally and outside of the UK to refer to someone who is shooting live *quarry species.

hunting In the UK this specifically refers to the chasing of live or simulated quarry with hounds or dogs. This may be conducted on horseback or on foot. These activities are termed *field sports, and by opponents they are termed blood sports. Hunting has been, and continues to be, a highly controversial activity, and after significant pressures the hunting of live quarry with hounds and dogs was banned in the UK; in Scotland under the Protection of Wild Mammals (Scotland) Act 2002, and in England and Wales under the Hunting Act 2004. However, hunting in all its forms continues, using the exemptions within the legislation. It is noteworthy that in many other countries the term 'hunting' is interchangeable, or refers specifically, to *shooting. *See also* FOX HUNTING; STAG HUNTING; DRAG HUNTING; HARE HUNTING; BEAGLING; MINK.

husk The protective outer coating or shell of a seed. Usually applied to the outer casing of *grain, especially *cereals, that is separated by *threshing or winnowing.

hybrid The offspring of parents of different species or varieties. Hybrids are produced in plant and animal breeding to bring new genetic material from which desirable characteristics can be selected. Some crops such as winter barley are grown as hybrids to gain extra yield from hybrid vigour.

hybridization The *crossbreeding of two different varieties or strains of plant or animal in an attempt to bring or improve certain desirable characteristics in the offspring.

hybrid vigour The extra vigorous growth that often results from the *crossbreeding of two parents of different varieties or strains. This vigour declines after one generation, so hybrid varieties of crops such as cereals have to be bred from the original parents for each generation.

Hybu Cig Cymru Meat Promotion Wales *See* LEVY BODIES.

hydraulic fracturing *See* FRACKING.

hydro-electricity Electricity generated through hydro-electric plants and generally considered to be a form of *renewable energy. Examples of systems which utilize the energy in flowing water in order to turn a turbine to generate electricity extend back into the nineteenth century, and rural estates were some of the earliest adopters of the technology. Today, hydro-electric projects are found on a wide variety of scales, from the very large hydro-electric dams that in some countries meet a significant proportion of national demand for electricity, through to smaller-scale projects where a turbine might generate electricity for an individual property. *See also* FEED-IN TARIFF; RENEWABLES OBLIGATION.

hydroponics A method of growing plants without soil. The roots of the plant are in a solution of the appropriate nutrients to provide all the nutrition and water the plant needs. The plant may be supported by the root being in an inert medium such as gravel or simply in the solution. The solution is usually kept in the dark to prevent the growth of *algae, and may be static or flowing through the roots. The practice is normally carried out in a controlled environment such as a greenhouse to further encourage growth.

hypsometer An instrument used to measure the height of a tree, reliant on the principles of trigonometry. It can be a very simple device, although modern versions utilize laser technology. *See also* CLINOMETER.

IAAS *See* INSTITUTE OF AUCTIONEERS AND APPRAISERS IN SCOTLAND.

IACS *See* INTEGRATED ADMINISTRATION AND CONTROL SYSTEM.

ICF *See* INSTITUTE OF CHARTERED FORESTERS.

ICM *See* INTEGRATED CROP MANAGEMENT.

IDB *See* INTERNAL DRAINAGE BOARD.

IFM *See* INTEGRATED FARM MANAGEMENT.

immoveable property Property which cannot be moved, i.e. land and things attached to land. *See also* HERITABLE PROPERTY.

impact, environmental *See* ENVIRONMENTAL MANAGEMENT SYSTEM.

implied covenant *See* COVENANT.

implied obligation *See* COVENANT.

impoundment licence Form of permission required in the *United Kingdom to **impound** (store) water using structures within inland waters which permanently or temporarily change the water flow or level (for example, dams, weirs, lock gates, or reservoir embankments). Licences are issued by the relevant government environmental agency (the *Environment Agency in England, *Natural Resources Wales, the *Northern Ireland Environment Agency and the *Scottish Environment Protection Agency). *See also* ABSTRACTION LICENCE.

improvements *See* LANDLORD'S IMPROVEMENT; TENANT'S IMPROVEMENT.

in-bye Enclosed land bounded by a hedge or fence normally close to a farmstead, and most commonly *permanent pasture. It usually applies to land in the *uplands below the open *moor and thus more sheltered. *Cattle or, more often, *sheep may be grazed on the open upland moor in summer, but brought down to the in-bye land in winter or during inclement weather. *See also* FARM.

income foregone A frequently used term within the payment calculations of *agri-environment and other land management schemes. Within this specific context it is usually the agricultural income that has been foregone in adapting to required management prescriptions to comply with the scheme conditions.

Payments might also include costs of additional management operations and incentivization.

income statement *See* PROFIT AND LOSS ACCOUNT.

Income Tax Tax paid by individuals in the *United Kingdom on the basis of their income. The Government sets three rates for Income Tax in England and Northern Ireland (for Scotland and Wales, see below), with the lowest earners paying a smaller percentage of their income as Income Tax than those with the highest earnings. All tax payers (apart from the very highest earners) also have a **personal allowance**. This is an amount of income that can be earnt without having to pay Income Tax, and for some of the lowest earners this can mean they pay no Income Tax at all.

The majority of people pay their Income Tax through the **PAYE** (pay as you earn) system where the tax is deducted by their employers (alongside their *National Insurance payments) before their salary or pension is paid. *Her Majesty's Revenue and Customs (HMRC) issues taxpayers with a **tax code** which tells employers how much to deduct.

People who are self-employed (and the highest earners) must submit a **self-assessment tax return**, detailing their earnings, to HMRC, and must pay any tax and National Insurance owed directly to HMRC.

In 2016 the *Scottish Parliament was given powers to set different rates of Income Tax for Scotland (apart from the personal allowance, which it does not have the power to change). However, **Scottish Income Tax** continues to be collected by HMRC before being paid to the *Scottish Government. From April 2019, the *Welsh Government will also be able to set different rates of Income Tax for Wales (also collected by HMRC).

Whilst companies pay Corporation Tax on their profits in the UK, many farms operate with a *partnership structure, which means that instead of paying Corporation Tax, each of the partners in the farming business pay Income Tax on their share of the profits of the business. There are some specific rules relating to farmers in this situation in terms of Income Tax. Farmers (and some creative artists) have the opportunity to use **profit averaging**. This allows them to elect to pay Income Tax on the basis of the average of their profit over consecutive years (as set out in *legislation), rather than on the basis of how much profit they have had in any particular year. This is designed to allow for the possibility that profits may fluctuate significantly in a farming business if, for example, harvests were very poor one year.

incorporeal hereditament Under English law, a right in property which does not physically exist; for example, an *easement. *See also* CORPOREAL HEREDITAMENT; HEREDITAMENT.

incorporeal property Term used in Scottish law to describe all property which is not *corporeal property. *See also* INCORPOREAL HEREDITAMENT.

independent expert determination *See* ALTERNATIVE DISPUTE RESOLUTION.

indicator species An organism whose presence or absence may indicate the quality of a specific range of environmental conditions. Examples include specific invertebrates that indicate the levels of oxygen in watercourses and hence a level of its pollution status, and specific bird species that are monitored annually on UK farmland to indicate the quality of habitats for these and other species. Frequently, trends in relative abundance over time are measured, so an important and practical attribute of an indicator species is that it can be monitored easily and accurately.

inflation An increase in prices over a period of time. *See also* CONSUMER PRICES INDEX; CONSUMER PRICES INDEX INCLUDING HOUSING COSTS; RETAIL PRICES INDEX.

information paper Document providing detailed, sometimes discursive, information and explanations about a particular professional, legal, or procedural topic. Information papers are produced by a variety of different types of organization, including government authorities and professional bodies, and their significance or status varies depending upon the publishing organization. In the context of the *Royal Institution of Chartered Surveyors (RICS), the term information paper has a specific meaning and is one of the five categories of professional guidance produced by the RICS: namely, *international standards, *professional statements, *codes of practice, *guidance notes, and information papers. Under the bye-laws and regulations of the RICS, its members must comply with RICS international standards and professional statements. Codes of practice and guidance notes provide advice and 'best practice' guidance to practitioners (often in relation to procedures), with some codes of practice being mandatory. Information papers provide information and explanations to RICS members on various matters relating to surveying (but do not recommend or advise on professional procedures). Although the different types of guidance have slightly different statuses, all practitioners are expected to familiarize themselves with new or updated professional guidance, and the RICS reiterates that if it is alleged that a surveyor has been professionally negligent, it is likely that the courts will take into account the professional guidance produced by the RICS in deciding whether or not the surveyor has acted with reasonable competence.

infrastructure planning Planning system for regulating the control of large-scale infrastructure projects, usually considered to be of either national or regional significance. These include large-scale energy, water, waste, transport, and harbour projects. Whilst the majority of planning applications are dealt with by *local planning authorities, most nationally significant infrastructure projects in England are managed under a system operated by the Planning Inspectorate on behalf of central government. In Wales, the Planning Inspectorate examines energy and harbour development applications, with other types of application being dealt with directly by Welsh Government Ministers. In Northern Ireland, regionally significant developments are dealt with centrally by the **Northern Ireland Department for Infrastructure**. In Scotland, whilst national developments are dealt with by local planning authorities, *Scottish Government Ministers have powers to call in applications to determine them themselves.

Inheritance Tax Tax paid, in the *United Kingdom, on the estate (i.e. the assets including property, money, and possessions) of someone who has died, or, in some circumstances, on the transfer of assets into or out of *trusts, or if assets have been held in a trust for a lengthy period of time.

Inheritance Tax (IHT) is paid on the value of an estate by the *executors of that estate, and before the estate is distributed to the beneficiaries (i.e. those entitled to it). It is a complex area of tax with a number of **reliefs** which may reduce the amount of tax paid. These include:

- The **threshold** value of an estate, below which Inheritance Tax is not paid (this threshold is increased for most estates if an individual gives their home to their children or grandchildren, and is known as the **residence nil rate band**). These thresholds are also normally transferable between spouses, which has the effect of increasing the total threshold for a couple (i.e. the total value of a couple's estate which is free of Inheritance Tax).

- **Potentially exempt transfers (PET)**: when the value of an estate is determined in order to calculate the IHT payable most gifts made by the deceased within the last seven years are included in the value of the estate (and hence IHT is theoretically payable on them). This is in order to avoid people giving away their estate shortly before death in order to avoid paying IHT. However, for gifts made to individuals (and into some *trusts) within the previous seven years, then the PET rule applies. This essentially means that the amount of IHT payable drops in a sliding scale after year three, and after seven years the gift becomes completely exempt from IHT.

- **Agricultural property relief (APR)**: this relief applies to agricultural land, buildings, farm woodland, and farmhouses that are 'character appropriate' to the farm and occupied for the purposes of agriculture (although there are strict rules and extensive *case law concerning the eligibility of farmhouses for APR). Relief is available, either at 100% or 50%, depending on whether the farm is farmed directly or is subject to a tenancy (and if so, when the tenancy commenced).

- **Business property relief (BPR)**: this is available on land, buildings, stock, plant, and machinery used in a business (plus any shares) and is available at 100% or 50% depending on whether the business itself owns the assets or they are owned by an individual but used in a business. BPR, however, does not apply to investment property (in most cases this means that it will not apply, for example, to let property). For farms, both APR and BPR may be relevant, although APR usually comes into play before BPR. So, for a field with development value, APR may apply to the agricultural value of that field (so what it would be sold for were it purely to be useful for farming), with BPR applying to the additional element of value that could be attributed to its potential for development (i.e. the additional amount that someone might pay for it on the basis that it may in future be sold for development).

- **Woodlands relief**: this applies to woodlands (other than woodlands which are ancillary to a farm which are covered by APR, or large commercial woodlands which may qualify for BPR). It allows the value of the timber, but not the land, to be excluded from IHT, so the tax on the timber will have to be paid at a later date if the timber is sold.
- **Relief for heritage assets (conditional exemption)**: some land, buildings, works of art, and other objects can be exempt from IHT (or *Capital Gains Tax) if they are passed to a new owner as a consequence of death or a gift. These are generally assets considered to be of nationally significant heritage interest. The owner must agree to maintain them, and allow public access to them.

IHT can also be payable on certain activities involving trusts. The key principle behind these charges is to avoid taxpayers effectively transferring assets on to the next generation via a trust, without paying any IHT. Consequently, for most types of trust, IHT is charged when assets are transferred into, or out of, a trust, and most trusts are also subject to a charge to IHT every ten years (although reliefs can apply in these circumstances so that in some cases no tax is actually paid).

Inheritance Tax valuation Valuation prepared, in the *United Kingdom, for submission to Her Majesty's Revenue and Customs (HMRC) for the assessment of any *Inheritance Tax due on a deceased's person's estate. *Valuations for Inheritance Tax must use the definition of *market value which is set out in *legislation (and also take into account relevant *case law), rather than using the definition of market value based on the International Valuation Standards and set out in the *Red Book. Inheritance Tax valuations are also sometimes referred to as **probate valuations** because they are sometimes submitted to HMRC alongside applications for *probate; however, their more correct name is Inheritance Tax valuation.

injunction Order, under English law, issued by a *court, either forbidding a person from doing something (for example, causing a disturbance to other flat dwellers by playing loud music in the middle of the night) or requiring them to undertake a positive action to address an issue. The equivalent under Scottish law is an **interdict**.

injurious affection *See* COMPULSORY PURCHASE.

inorganic Not consisting of or derived from living matter, and hence in chemistry, inorganic compounds are broadly those that contain no carbon atoms. In farming it is the opposite to *organic. So, for example, an inorganic *fertilizer is one that comes from raw materials such as **rock phosphate** or *potash, or is manufactured, such as *ammonium nitrate, as opposed to organic fertilizer that is derived from *manure or plant residues.

inquiry, planning appeal *See* APPEAL, PLANNING.

insecticide A chemical compound that is used to control insect pests. *See also* PESTICIDES.

Institute of Auctioneers and Appraisers in Scotland (IAAS)

Professional body representing the Scottish livestock auctioneering sector. The Institute, which was founded in 1926, has corporate and individual members and represents the sector at national and European level, sitting on a number of stakeholder groups and developing guidance on best practice in livestock auctioneering. Members of the IAAS, in addition to marketing livestock through the sale ring and on an agency basis, also undertake other work including *valuations, land sales, and compensation claims, and some are also involved in sales of antiques and collectables.

(()) SEE WEB LINKS
• IAAS website: includes information on the work done by members of the IAAS, a digest of news and government announcements relating to livestock and rural matters, a directory of Scottish livestock markets, and market data on livestock sales in Scotland.

Institute of Chartered Foresters (ICF) The professional body representing foresters and **arboriculturists** in the UK, and that can award chartered status. It has key objectives that include maintaining and improving standards of practice of forestry, the promotion of professional status, and the protection of the public interest. Qualified members can use the nomenclature MICFor after their name.

(()) SEE WEB LINKS
• ICF website: includes eligibility details and organized events.

Institute of Revenues Rating and Valuation Professional body with members working within the public and private sector in the fields of revenues, benefits, and *valuation. The Institute has several grades of membership and offers a range of vocational and examination-based qualifications. Whilst the roots of the organization extend back to nineteenth-century London rate collectors, and much of its work relates to the UK, it also has members based outside the UK. A key aim of the Institute is to support its members' professional and personal development and share best practice, and it also seeks to have an influential voice in dialogue relating to professional matters and the development of legislation, in part, by commissioning, conducting, and publishing research in relevant areas. IRRV is a member of *The European Group of Valuers' Associations and has the right to award Recognized European Valuer status to its members. *See also* RED BOOK.

(()) SEE WEB LINKS
• IRRV website: includes information on the work done by members of the IRRV, and a digest of news and government announcements relating to national and local taxation, benefits, and valuation.

insurance valuation *See* REINSTATEMENT COST ASSESSMENT.

Integrated Administration and Control System (IACS) Essentially a database of farms and agricultural land parcels, the system was adopted by the EU in 1992 to organize agricultural support payments to farmers under the

*Common Agricultural Policy. The systems have been subsequently updated and are now managed using the Basic Payment Scheme.

Integrated Crop Management The management of crops using the techniques of *Integrated Farm Management, combining profitability with the production of safe and healthy food with the best environmental practice. Improving and maintaining the health and structure of soils by sound husbandry techniques such as *crop rotation and *minimum tillage is a major component, as is the most efficient use of resources such as fertilizer and pesticides. Care for the environment is encouraged by the adoption of green manure cover crops, field margins, beetle banks, and other *agri-environment options.

Integrated Farm Management A philosophy of farming promoted by *Linking Environment and Farming (LEAF), which describes it as a whole farm approach that combines the best of traditional methods with beneficial modern technologies, to achieve high productivity with a low environmental impact. Attention to detail is essential with the efficient use of resources, good soil management, crop rotations, and high animal welfare standards. The aim is to be as sustainable as possible, optimizing the use of non-renewable resources using traditional farming husbandry integrated with the results of the latest technological research. It is an approach endorsed by a wide range of conservation organizations, some of which operate related *farm assurance schemes.

Integrated Pest Management A holistic approach to the control of pests and diseases by the appropriate use of *pesticides and medication together with

Components of an integrated farm. (James Hutton Institute, Integrated Farm Management.)

preventive techniques such as *crop rotation, *tillage, sound *biosecurity, and *biological controls to avoid the build-up of weeds and pathogens. Resistance to chemical controls is a constant threat, as weeds and pathogens mutate and are thus no longer controlled. An example of this is a significant increase in *blackgrass in *cereal crops for which the use of *herbicides is becoming increasingly ineffective. A longer rotation of crops, including those sown in the spring, together with cover crops and tillage, offers the best approach to reducing the burden.

intensification The increase in production per unit of input, typically area or labour. *Arable cropping may be intensified by increasing the yield per unit of land by growing crops, often winter sown, and by using higher application rates of *fertilizers and *pesticides. This may also involve the removal of *hedges or other *field boundaries to increase field size and by cropping all available area. For *livestock, *flocks, and *herds it may be expanded with little or no increase in the labour involved. *Stocking rates may be increased by the use of higher-yielding forage using more fertilizer. Increase in size and sophistication of machinery and equipment may allow greater yield per unit of labour; for example, the use of computers and robots, as in **robotic milking**. The intensification of British agriculture in the second half of the twentieth century has been blamed for declines in farmland *biodiversity.

intercrop (intercropping) The growing of two or more crops in close proximity to each other, most commonly in alternate rows. The aim is to increase productivity per area of land, but it is important that the two crops do not over-compete with each other for water, nutrients, or sunlight. Vegetables may be the crops most frequently grown in this way. *Agroforestry is where rows of trees are grown within a crop. A variation of this concept is where one crop deters pest or disease infestation of the other.

interdict The equivalent, under Scottish law, of an *injunction.

interest 1. In land law, a right in, or over, land. This may include a right of ownership (for example, a *freehold or *leasehold; *see also* ESTATE IN LAND), or another form of right such as a legal *easement or legal *mortgage. It may also include additional rights in *equity (for example, those held by the *beneficiary of a *trust of land). In Scotland, the term is used slightly more loosely than in the rest of the *United Kingdom as a way of describing a *real right. **2.** Money paid for a loan. The **borrower** normally pays interest to the **lender** in return for the loan; whilst the **saver** or **investor** is paid money by the individual or organization they have deposited money with or invested in. *See* INTEREST RATE.

interest in possession trust *See* TRUST.

interest rate The amount of **interest** (money) that will be paid by one party (the borrower) to another party (the lender) so the borrower has the right to use the lender's money for the term of the loan. Interest rates set by the central banks (for example, the **Bank of England**) will usually be tracked by the various *financial institutions that lend money (for example, consumer banks and building societies). People or organizations saving money with the financial institution will receive interest on their savings, whilst those borrowing money (for example, a mortgage) will pay interest.

Interest rates are usually expressed as a percentage of the amount borrowed and represent what will be paid in a year in interest. For example, someone borrowing £100 from a lender at an interest rate of 8% will pay £8 interest per year. Periodically, the central bank will adjust interest rates in response to what is going on in the economy; for example, if they wish to encourage consumers to spend money they may reduce the interest rate, which will then mean that consumers have to spend less of their income on interest payments (and also have less incentive to save money as the interest they are paid on any savings will be lower), so they will tend to spend more money. The key interest rate set by the Bank of England is the **bank rate** (sometimes also called the **Bank of England base rate**). As this is the rate the Bank of England usually pays to commercial banks who have money on reserve with them, these commercial banks usually pass on the change in the bank rate to their customers.

A saver who earns interest in an account and then leaves that interest in the same account will start to receive interest both on the original amount they put in the account and also on the interest that they have earned. This interest is known as **compound interest**. For example, in a situation where a saver deposits £1 in a savings account at a rate of 10% interest paid annually:

End of Year 1: £1 initial savings + £0.10 interest = balance of £1.10
End of Year 2: £1.10 balance + £0.11 interest = balance of £1.21
End of Year 3: £1.21 balance + £0.121 interest = balance of £1.331 and so on

If the saver had removed the interest from the account at the end of year 1, then they would continue to earn only £0.10 interest per year; but because they have retained the interest in the account, then **compounding** has happened and so the amount of interest earned every year increases. The same situation will apply to borrowings, but in the opposite direction, with the debt effectively increasing every year if the sum borrowed is not being progressively paid back.

Internal Drainage Board (IDB) Some agricultural areas that require a special need for drainage have local independent public bodies that manage water levels. IDBs operate in England and Wales, and equivalent bodies are found in many other countries. The areas in which IDBs operate are known as internal drainage districts (IDD), and works include maintenance of river channels, pumping stations, and other associated infrastructure. Funding is raised from levies charged on agricultural landowners and *local authorities, together with some *Environment Agency grant funding. *See also* SUSTAINABLE DRAINAGE SYSTEMS.

(())) SEE WEB LINKS
• Website of the Association of Drainage Authorities, including representation of IDBs.

internal rate of return *See* PAYBACK PERIOD.

International Organization for Standardization (ISO) International
non-governmental organization, based in Switzerland, that develops standards
(identifiable by having an ISO number) relating to products, services, and
systems. These standards are documents which set out specifications and
requirements relating to products, materials, processes, or services with the
objective of ensuring that they are fit for purpose. Examples of standards include
ISO140001 *Environmental Management Systems, and ISO9000 Quality
Management. Businesses and organizations can apply for certification against
some standards to demonstrate that they meet them. This is done via a separate
certification body that will carry out a range of checks and inspections to ensure
that the organization complies with the standard for which they wish to gain
certification. *See also* BRITISH STANDARDS INSTITUTION.

international standard (RICS) Document detailing a set of high-level
principles which are mandatory for RICS members to follow, agreed
internationally between the *Royal Institution of Chartered Surveyors (RICS) and
other relevant bodies. International standards are one of the five categories of
professional guidance produced by the RICS: namely, international standards,
*professional statements, *codes of practice, *guidance notes, and *information
papers.

intestacy Situation when a person has died without leaving a **will** to indicate
how they would like their estate (for example, their money, possessions, and *real
property) distributed after their death. In this situation in many countries the law
sets out the principles by which the estate of someone who has died **intestate** is
distributed, with immediate family members usually being entitled to the estate
(after the payment of taxes and any debts). However, in the absence of any family
the estate usually goes to the state. The rules by which estates are distributed in
situations of intestacy vary between different countries, and there is some
variation even between different parts of the *United Kingdom. *See also* PROBATE,
GRANT OF; INHERITANCE TAX.

invasive species *See* NON-NATIVE SPECIES.

inventory 1. A list of *assets or property. **2.** Document listing anything owned
by the *landlord on or within a property at the start of a *tenancy or other
occupation agreement. An inventory is often combined with a **schedule (or
record) of condition**, which details the condition of a property at the start of a
tenancy. These documents are agreed between the landlord and *tenant in
order to try to avoid subsequent disputes between them in relation to any
damage to the property, as they provide a record as to whether or not that
damage was present before the tenant took occupation. *See also* DEPOSIT; FAIR
WEAR AND TEAR.

investment appraisal *See* PAYBACK PERIOD.

investment method Method of assessing the *capital value of the *freehold or *leasehold interests in properties that are let, or normally let. The investment method is commonly used for offices, shops, retail premises, and let farms where the value of the property is essentially a function of the right to receive an income (*rent) from that property.

Underpinning the investment method is the concept of *capitalization, i.e. that it is possible to convert an annual *market rent into a capital value by multiplication by a factor (Years Purchase) which takes account of the acceptable *yield for the property and also reflects that money (rent) received now is worth more than money (rent) received in the future because the money held now is earning interest immediately. (For a more detailed explanation of how capitalization operates *see* CAPITALIZATION.)

Investment valuations can also be structured to reflect that the rent currently being received may not be a market rent, but at a future date the rent will usually be reviewed and changed to market rent. There is a number of ways in which this can be addressed within an investment valuation of the freehold interest in a property which is subject to a lease, with two of the most commonly cited being **term and reversion** and **hardcore (or layer) model**:

- **Term and reversion**: using this approach, the value of the period before the rent is reviewed to market rent is treated as a separate 'term' before the rent goes to market rent (the reversion). So, referring to the diagram:

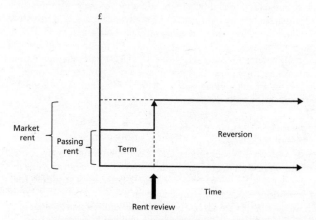

- ○ The term: the right to receive a rent at the current level (i.e. the passing rent) for the number of years left until the rent is reviewed, is capitalized.
- ○ The reversion: the right to receive the rent at the market rent (because it is after the rent review) in perpetuity is also capitalized, taking into account

that this rent will not be received for a delayed period (i.e. until the end of
the term) and therefore would not be exactly equivalent in monetary terms
to the same level of rent today (given the loss of potential interest on the
money in the intervening period).

○ The capitalized rents for the term and reversion are then added together to
   give the capital value of the freehold.

• **Hardcore (or layer) model**: in this approach it is assumed that the current
  (or hardcore) rent will continue in perpetuity. At the point of the rent review it
  is then considered that a 'reversionary increase' will happen (as the rent
  increases to market rent). This reversionary increase is then also considered to
  continue in perpetuity. So, referring to the diagram:

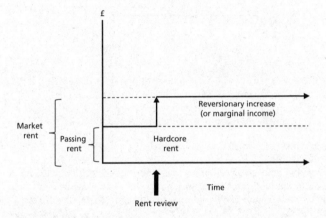

○ The hardcore rent is capitalized in perpetuity.
○ The reversionary increase is also capitalized in perpetuity taking into
   account that this rent will not be received for a delayed period (i.e. until the
   rent has been reviewed).
○ The capitalized hardcore rent and reversionary increase rents are added
   together to give the capital value.

As noted above, it is also possible, using the investment method, to derive a
capital value for a leasehold interest. A leasehold will only have a value if the rent
that the lessee is paying to their landlord is lower than market rent. If this is the
case then the leaseholder can sub-let the property to a sub-lessee at the market
rent. The difference between the rent the leaseholder is paying and the market
rent is known as a **profit rent** (as shown in the Figure below). The profit rent is
unlikely to last indefinitely, as the leaseholder's landlord is likely to review the
rent to market rent at some point, therefore eliminating the profit rent (unless
there is a provision within the leaseholder's lease to either fix the rent or review
it on a basis which will be lower than market rent, whilst at the same time the
sub-lessee's rent is regularly reviewed to market rent). However, in the meantime

the profit rent can be capitalized (taking into account how long it will be before the leaseholder's rent is reviewed to market rent).

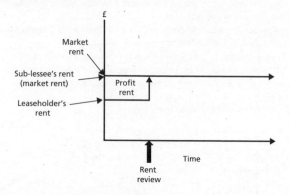

In reality there does not actually need to be a sub-lessee in existence for there to be a profit rent. The leaseholder may themselves be using the property, but there is still, in effect, a capital value to their lease because they are benefiting from a reduced rate of rent in comparison to the market rent.

IPM *See* INTEGRATED PEST MANAGEMENT.

IRI Abbreviation of **internal repairing and insuring**; description of the nature of *lease and what the *lessee is responsible for. In a lease described as an IRI lease, the lessee is responsible for internal repairs and insuring the property. *See also* FRI; IRT.

Irish Draught *See* HORSE.

irremediable breach *See* BREACH OF COVENANT.

irrigation The application of water to a growing crop to ensure optimum growth. There is a wide range of methods from using irrigation channels, to flooding the land, to aerial spraying by means of a boom or gun, and trickle or *drip irrigation through a pipe with nozzles. In global terms, in the region of 80% of all water consumed is for agricultural irrigation, though it is much lower in Europe.

irritancy Right of a *landlord in Scotland to end a *lease early if the *tenant has breached (broken) *covenants in the *lease. Irritancy is broadly equivalent to *forfeiture in England, Wales, and Northern Ireland. *See also* BREACH OF COVENANT.

IRRV *See* INSTITUTE OF REVENUES RATING AND VALUATION.

IRT Abbreviation of **internal repairing terms**; description of the nature of a *lease and what the *lessee is responsible for. In a lease described as an IRT lease, the lessee is responsible for internal repairs only. *See also* FRI; IRI.

ish The end-date of a *lease in Scotland.

ISO *See* INTERNATIONAL ORGANIZATION FOR STANDARDIZATION.

ivermectin An **anthelmintic** medicine of the **avermectin** family used to control *endoparasites, such as intestinal worms in farm animals, and *ectoparasites, such as head lice in humans. The drug may be taken orally, injected, or poured onto the skin. Avermectins appear naturally, but concerns have been raised about increased levels excreted by animals after treatment. This may delay the decomposition of dung, but studies show little other adverse environmental impact. *See also* WORMING.

Japanese knotweed (*Fallopia japonica*) A large *perennial plant native to eastern Asia. It has been introduced and become established in many areas of the world, including Europe and North America. In the UK it was introduced in the nineteenth century and found favour with gardeners, but has subsequently created difficulties and adverse impacts on the environment and on properties. It is very difficult and expensive to control. Its invasive root system can damage buildings, leading to particular problems surrounding the sale of properties. *See also* NON-NATIVE SPECIES; WEEDS ACT.

JNCC *See* JOINT NATURE CONSERVATION COMMITTEE.

Joint Nature Conservation Committee The statutory organization that advises the government on UK-wide and international issues relating to *biodiversity and related nature conservation. It has a particularly important role in complementing the individual national statutory advice bodies by providing advice across the four devolved governments of the UK. The advice and evidence is taken forward at European and global levels. It also has a key role in maritime nature conservation. *See also* NATURAL ENGLAND; SCOTTISH NATURAL HERITAGE; NATURAL RESOURCES WALES; NORTHERN IRELAND ENVIRONMENT AGENCY.

(()) SEE WEB LINKS
• Website of the Joint Nature Conservation Committee, providing a comprehensive breakdown of its roles and responsibilities.

joint ownership *See* JOINT TENANCY.

joint tenancy Under English property law, ownership of an *estate in land by multiple owners as a whole, such that in the event of the death of one of the joint tenants the interest of that joint tenant automatically **vests** (goes into the possession of) the other joint tenants. This contrasts with a *tenancy in common where each of the tenants in common can leave their share of the property in their will to whomever they chose. A joint tenancy can be **severed** (split) to create a tenancy in common. Under Scottish law, **joint ownership** of **joint property** operates in a similar way to a joint tenancy under English law (with title held in **pro indivisio** shares in Scottish law also being broadly similar to the English law tenancy in common). *See also* TRUST.

joint venture Business arrangement when two or more businesses or individuals work together in order to undertake a specified project or activity over a defined period of time, with each party contributing some element to the project (be it cash, land, buildings, labour, or machinery). Each of the parties involved often continues to operate as a separate business entity in their own right, undertaking other activities, whilst taking returns from the joint venture in proportion to their level of input to it (with each party's responsibilities and rights to take returns normally set out in an agreement between them). In a joint venture, both parties will take some element of business risk, so if the joint venture is not successful then they are unlikely to be able to take their full entitlement to returns from it, and may incur losses. *See also* CONTRACT FARMING; SHARE FARMING.

judicial review The process whereby an interested party can challenge, through the *High Court in England, Wales, or Northern Ireland (or *Court of Session in Scotland), a decision made by a public body on the grounds that that body did not have the legal power to make such a decision, or had made an error in law. Judicial reviews are used in a range of situations, including, for example, in relation to the refusal of *planning permission where an applicant may have lost a *planning appeal but considers that the appeal body has made a legally incorrect decision. *See also* ULTRA VIRES.

kale A plant of the *brassica family that is grown as a vegetable for human consumption and as a crop for livestock, particularly cattle, to graze, or as a component of a cover crop grown as habitat for farmland or *game birds. It is planted in late spring or early summer and can be grazed in autumn or winter. As a habitat for birds, it can last for two years or even more, producing a mass of seeds in the second year which can be a significant feed source. As a brassica, it is susceptible to flea beetle attack which can make it difficult to establish. This can be controlled by *insecticide, but, since a ban on neonicotinoid seed dressings there is a shortage of effective chemicals. If kale is grown on the same ground for several years, there is the danger of problems arising from *clubroot, a fungal disease for which there is no chemical control, so *rotation of crops is the best protection.

kicking boards Traditionally, timber is used on the lower half of stable walls to a height of 1.2 metres, particularly if the walls are constructed of brick or equivalent material. This is in order to minimize any injury that the *horse might cause itself, especially if it were to get stuck or *cast in the corner.

kid The young of a *goat or related species. Sometimes used to describe the meat or leather from a young goat, as in kid gloves.

killing out percentage The weight of a carcass expressed as a proportion of the *liveweight of the animal before slaughter. Typically, killing out percentages vary between 45% and 55% for ruminant species such as cattle and sheep. That of pigs is higher, around 75%, due to differences in the preparation of the carcass.

knackerman Someone who collects dead or dying farm animals, including horses. The animals, unfit for human consumption, are taken to the knacker's yard, where they are cut up into individual parts for further use; for example, the hide might be used for leather. There is a distinction between a knacker's yard and an abattoir or slaughterhouse, where animals are killed for human consumption. *See also* FALLEN STOCK.

Kyoto Protocol A *United Nations protocol (international agreement) adopted in Kyoto, Japan, in 1997, which legally binds countries to greenhouse gas emissions reduction targets, in accordance with the *United Nations Framework Convention on *Climate Change. The first commitment period ran from 2008 to 2012, and, following the **Doha Amendment to the Kyoto Protocol**, the second commitment period runs from 2013 to 2020 (though at the time of writing an insufficient number of countries have accepted the Doha Amendment for it to come into force). The Protocol places a heavier obligation on developed nations to deliver emissions reductions on the basis that they are mainly responsible for current levels of greenhouse gases in the atmosphere, as a consequence of more than 150 years of industrial activity.

k

Lady Day *See* QUARTER DAYS.

lake An expanse of water surrounded by land and not connected to the sea. There is no accepted standard of size to differentiate a lake from a *pond. A lake is usually larger, with some sources maintaining that they are at least two *acres in area, but a range of varied factors will influence its definition between lake or pond. A common equivalent term in Scotland is a **loch**, which also is applied to sea inlets, as is the equivalent Irish term **lough**.

lamb 1. The young of *sheep until a year old when they become *hoggets.
2. The meat from the young sheep.

Lammas *See* QUARTER DAYS.

land agent Person who manages a rural estate on behalf of its owner or owners, or provides professional advice on matters relating to the management of land and property in rural areas to a number of clients. The term land agent is commonly used interchangeably with the term *rural surveyor. *See also* FACTOR.

Land and Buildings Transaction Tax (Scotland) *See* STAMP DUTY LAND TAX.

Land and Property Services Organization, forming part of the Northern Ireland government's Department of Finance, with responsibilities for mapping, *land registration, *valuation, and rating in Northern Ireland. They are responsible for maintaining the *Land Registry, *Registry of Deeds, and Statutory *Charges Register in Northern Ireland, and also produce a range of paper and digital maps for the Ordnance Survey of Northern Ireland. In addition they provide and maintain a valuation list for all properties in Northern Ireland (domestic and non-domestic) in relation to rates, and are responsible for the billing and collection of these rates. They also carry out valuations for public sector bodies. *See also* DISTRICT VALUER; LANDS TRIBUNAL; SCOTTISH ASSESSOR; VALUATION TRIBUNAL.

land charge In land law in England and Wales, a third-party right relating to **unregistered land** (i.e. land where the *title is not registered at the *Land Registry). Land charges include a range of rights in addition to financial charges, so can include items such as *covenants affecting *freehold land. In order to be binding on third parties, the law sets out certain classes of land charge that must now be registered on the registers of the **Land Charges Department** of the Land

Registry (these registers are separate from the main registers of *title held by the Land Registry as they do not record ownership of unregistered land). In Northern Ireland a similar system in relation to unregistered land operates through the **Registry of Deeds** (*see* LAND REGISTRY). *See also* LOCAL LAND CHARGE.

land classification *See* AGRICULTURAL LAND CLASSIFICATION.

land cover Description of the physical land type covering the surface of the earth. It could include for example *grass, *woodland, *bare land, or *wetland. Data are most commonly presented in mapped documents using satellite and aerial imagery. Not to be confused with *land use, which describes how the land is used.

land degradation A process of the deterioration in the quality of land and the degradation of soil, vegetation, and water resources. The loss of fertile *soil through erosion can be a particular concern, as well as *compaction, loss of soil structure, *nutrient loss, and increases in salinity. The reduction in value of the **biophysical** environment also results in a reduction in the ability of land to provide *ecosystem services.

Landfill Disposals Tax (Wales) *See* LANDFILL TAX.

Landfill Tax Tax paid for each tonne of *waste going to landfill in the *United Kingdom. Since its introduction in 1990s, the rate payable has been progressively increasing. This has been done with the objective of making it more expensive to send waste to landfill, thereby incentivizing waste producers to reduce the amount of waste they produce and/or to increase the percentage of waste they manage using mechanisms which are at the more desirable end of the *waste management hierarchy (such as recycling or reusing items). Landfill Tax in Scotland is a devolved matter and is known as **Scottish Landfill Tax**. In Wales in 2018, Landfill Tax was replaced by the **Landfill Disposals Tax**, which operates on similar principles to the landfill tax.

landlord Someone allowing another party to occupy their property under the terms of a *lease, *tenancy, or *licence. The word *lessor can be used as an alternative to landlord when there is a *lease in existence. Often the landlord is the *freeholder or owner, but they can also themselves be a lessee if they have a sub-tenant. *See also* TERMS USED WHEN PROPERTY IS SUBJECT TO LEASE OR TENANCY.

Landlord and Tenant Act 1954 *Legislation governing the operation of *business tenancies in England and Wales (although some types of commercial tenancy are excluded from it; see below). Whilst business tenancies in England and Wales are subject to the same general *common law principles as other tenancies, in some circumstances, business tenants have additional protections under the 1954 Landlord and Tenant Act (commonly referred to as the **54 Act**). The 54 Act sets out the conditions which much exist in order for a business *tenant to be eligible for protection under the Act. In essence, the tenancy must be one where the property which is the subject of the tenancy is, or includes, premises occupied by the tenant for the purposes of carrying on a business (premises can include land as well as buildings, or parts of buildings, see below concerning agricultural tenancies). The **holding**, which is the part of a property covered by the 54 Act, is the property in the tenancy which is occupied by the tenant, or by an employee of the tenant who is employed in the business for which the property is occupied. The 54 Act includes a broad definition of 'business': it can include a trade, profession, or employment, plus any activity carried on by a body of persons whether corporate or unicorporate. In practice, this means that the 54 Act could apply, in certain circumstances, to a situation, for example, where the *trustees of a sports club take a tenancy for their clubhouse and sports facilities. Both tenancies and subtenancies (*see* TERMS USED WHEN PROPERTY IS SUBJECT TO LEASE OR TENANCY) are potentially eligible for protection under the 54 Act. There are, however, some types of tenancy which are expressly excluded from it. These include *Agricultural Holdings Act 1986 tenancies and *Agricultural Tenancies Act 1995 tenancies (farm business tenancies), *service tenancies where the tenancy agreement specifies the purpose for which the tenancy exists, and tenancies with a term of less than six months in some circumstances.

A 54 Act tenant will potentially benefit from *security of tenure. The basis of this is that the tenancy will not end at the end of the contractual term but will automatically continue as a periodic tenancy on the same terms. The only way that the landlord of a 54 Act tenancy can bring it to an end is by following the procedures for termination permitted in the Act. In essence, these divide into the *common law methods (i.e. the tenant serving a *notice to quit, or the tenancy ending through *forfeiture or by agreed *surrender), and those *statutory methods specified in the 54 Act. There are four statutory methods of terminating a 54 Act business tenancy:

- The landlord and tenant entering voluntarily into an agreement for a new tenancy (the relevant landlord is defined in the legislation in order to address a situation where there may be sub-tenancies).
- The tenant giving notice to terminate the tenancy (this only applies to fixed-term tenancies, periodic ones being covered by the common law notice to quit provisions).
- The tenant making an application for a new tenancy under section 26 of the 54 Act.
- The landlord serving a section 25 notice stating whether or not they oppose the granting of a new tenancy (and either setting out the proposed *terms of a new tenancy, or if they do oppose it, then specifying the ground, or grounds, on which they are opposing it). If the tenant does not wish to remain in occupation of the property, then no further action is needed after the landlord has served a section 25 notice, as the tenancy will terminate automatically in accordance with the notice.

After the serving of a section 25 or section 26 notice, the landlord and tenant are free to agree the grant of a new tenancy; however, in the absence of an agreement an application must be made to the *courts (if the landlord wishes to oppose the grant of a new tenancy they must also apply to the courts for a *court order to terminate the tenancy). If the landlord wishes to oppose the grant of a new tenancy then they must rely on one of the seven **grounds of opposition** in section 30 of the 54 Act. These include persistent delays in paying the rent in the past or other breaches of the tenant's obligations, including repairing obligations (the severity of these breaches must be sufficient to warrant the court not granting a new tenancy), or that the landlord is able to provide suitable alternative accommodation, or that they intend to demolish or reconstruct the premises (the landlord needs to demonstrate that this goes beyond a vague intention), or they intend to occupy the holding themselves (although a recent acquirer of the landlord's *interest would not be able to use this ground). Where the landlord is successful in demonstrating one of the grounds which do not involve any fault of the tenant, and the tenant effectively, therefore, sustains a loss to which they have not contributed (because they are not granted a new tenancy), then the tenant is entitled to *compensation from the landlord. If the court grants a new tenancy then the parties are free to negotiate the terms; however, in the absence of an agreement between them then the court can resolve any unagreed terms, including the length of the new tenancy and the *rent (though there are some limitations on the scope of the court to decide these matters).

It is possible, however, for the *landlord and tenant, by agreement, to *contract out of the security of tenure provisions of the 54 Act. This is done through the *lease and a process before the start of the tenancy whereby the landlord serves a notice to this effect, and the tenant signs a declaration agreeing (these are in a prescribed form, and the law sets out a specific procedure). The effect of this is that the tenancy will not continue beyond the end of the contractual term, and the tenant will have no right to apply for a new tenancy.

Under the **Landlord and Tenant Act 1927**, tenants are also entitled to claim *compensation on quitting the holding for some *tenant's improvements to the premises (although there are some slight differences in eligibility for compensation under the 1927 Act in comparison to that for general protection under the 54 Act, and the tenant must have followed a defined procedure when making the improvement for it to be eligible). Modern leases often avoid the provisions of the 1927 Act, for example, by including an express *covenant requiring the tenant to reinstate the premises at the end of the term. *See also* BREACH OF COVENANT; CODE FOR LEASING BUSINESS PREMISES IN ENGLAND AND WALES.

landlord registration Legal requirement for private *landlords of most let residential property in Wales, Scotland, and Northern Ireland, to provide their details, and those of properties they rent out, plus who is responsible for managing them, to the respective government public register which operates in the area in which they rent out property. There are separate registers operating in Wales, Scotland, and Northern Ireland. Although there is not, currently, a nationwide landlord registration system operating in England, private landlords in some *local authority areas of England are required to register on a local authority register for their area. *See also* HOUSE IN MULTIPLE OCCUPATION.

landlord's hypothec Right, in Scotland, for the *landlord of some types of commercial (but not agricultural) property to hold the *tenant's *moveable property in the leased premises as security in respect of *rent. In this instance, the hypothec is effectively a form of *lien.

landlord's improvement An improvement to a property which is subject to a *lease or *tenancy and which is to the *tenant's benefit but is made at the *landlord's expense (and is over and above the *landlord's *obligations for repair and maintenance). Generally, the law provides for the landlord to reflect an improvement in an increased level of *rent, but this is normally subject to *statutory controls (including, usually, that the improvement cannot have been imposed upon the tenant against their wishes, unless required by the law). *See also* TENANT'S IMPROVEMENT.

landowner A person who owns land *freehold. Usually used in reference to a large area of land; for example, a *farm. Landowners may manage and farm the land themselves, or let it out to a third party. *See also* COUNTRY LAND AND BUSINESS ASSOCIATION.

Land Register of Scotland Map-based register of land and property in Scotland which records and guarantees the *title (and hence ownership) of land and property. It is one of the registers held by *Registers of Scotland, which is a government department that also holds the **General Register of Sasines** as well as other public registers. As land is transferred it must now be registered in the Land Register of Scotland (and hence the proportion of land registered in the Land Register of Scotland is increasing over time). The Land Register of Scotland is designed to replace the older General Register of Sasines, which was established in the seventeenth century as a record of title *deeds. Over time it is anticipated that all property on the General Register of Sasines will be registered on the Land Register of Scotland. *See also* LAND REGISTRY.

land registration *See* LAND REGISTRY; LAND REGISTER OF SCOTLAND.

Land Registry 1. England and Wales. Department of government that registers and guarantees land *title (and hence ownership). Any land in England and Wales that is sold, or *mortgaged, or where a *lease is granted out of it (other than for a short period), must be registered at the Land Registry. Therefore, whilst there is still some land that is not yet registered (because it has not been the subject of one of the events requiring it to be, or the owner has not voluntarily done so), the Land Registry now holds records of title (and hence ownership) to the majority of land in England and Wales. The entry relating to an individual property is broken down into three parts: the **property register**, the **proprietorship register**, and the **charges register**. The property register details the property (usually by reference to a plan) and the respective *estate in land; say, for example, that it is *freehold. It also shows any rights that the land benefits from; say a right to access it by walking over a neighbour's land. The proprietorship register shows the name of the owner and the class of title that the Land Registry is willing to guarantee, for example, **absolute freehold**. It also shows any restrictions in relation to how the property can be dealt with by the owner; for example, if a mortgage lender's permission is needed to transfer it. The charges register details any *charges (*mortgages) and other third-party rights affecting the property (for example, any *easements, or *covenants). *See also* LAND CHARGE; LOCAL LAND CHARGE. **2. Northern Ireland**. Registry in Northern Ireland which is part of *Land and Property Services and which records and guarantees the *title (and hence ownership) of registered land and property plus details of *mortgages or other rights which may affect it. When a property in Northern Ireland is sold it must now be registered with the Land Registry, hence the percentage of land registered is increasing over time. The Land Registry in Northern Ireland exists alongside two other registers held by Land and Property Services; the **Registry of Deeds**, and the **Statutory Charges Register**. The Registry of Deeds holds copies of title *deeds to unregistered property but does not guarantee legal title. The Statutory Charges Register holds records of restrictions (normally those created by *local authorities or government) which may affect a property; for example, the existence of a *Tree Preservation Order or *listed building *enforcement notice. *See also* LAND REGISTER OF SCOTLAND.

landscape Includes all the natural features within a defined area, and might include the mountains, hills, rivers, lakes, and native vegetation. It also includes the human-created or influenced features including buildings, and the wide variety of land uses and different vegetation. Hence an agricultural landscape such as a *lowland farm is likely to include natural features and *fields, *crops, *livestock, *hedges, and associated buildings.

landscape character assessment (LCA) The process of applying a standard approach to the identification and description of the distinctive features which are characteristic of a specific *landscape. An LCA will incorporate documents and mapping to explain the landscape.

Lands Tribunal, The (Northern Ireland) Tribunal within the Northern Ireland Courts and Tribunals Service considering appeals, applications, and disputes in Northern Ireland in relation to property matters. They deal with a range of issues including applications for modifying and extinguishing *restrictive covenants; appeals against ratings valuations made by *Land and Property Services (*see* VALUATION TRIBUNAL), and matters relating to the relationship between *landlords and business *tenants. They also assess compensation in relation to land that has been subject to *compulsory purchase, and can act as an *arbitrator in relation to other matters affecting the value, use, or development of land. Appeals against decisions made by the Lands Tribunal are usually only made to the *Court of Appeal. *See also* FIRST-TIER TRIBUNAL PROPERTY CHAMBER; LANDS TRIBUNAL FOR SCOTLAND; VALUATION TRIBUNAL.

Lands Tribunal for Scotland Panel with *statutory powers to deal with a range of disputes relating to land and property in Scotland. The main areas in which the Tribunal works are in considering appeals against *valuations for *business rates; appeals in relation to the valuation of land where community bodies may have pre-emptive rights of purchase, and where farms may be sold to sitting tenants under the **Agricultural Holdings (Scotland) Act 2003**; appeals in relation to *land registration issues; disputes in relation to *compulsory purchase compensation; disputes relating to the rights of public sector tenants to buy their homes; disputes relating to *title conditions; and other related disputes where both sides voluntarily agree that the Tribunal may act as an arbiter to come to a decision. The Lands Tribunal for Scotland does not deal with all disputes in relation to land in Scotland. The *Scottish Land Court has responsibility for dealing with many matters in relation to agricultural *landlord and *tenant issues, including some matters relating to *crofts. Matters relating to ownership of land are dealt with by the *Sheriff Court and *Court of Session. *See also* CROFTING COMMISSION (SCOTLAND).

Land Transaction Tax (Wales) *See* STAMP DUTY LAND TAX.

land use Description of how land is used involving degrees of management of the natural environment, e.g. agricultural, transport, residential, commercial, or recreational land use. In the context of farmland, it can be further categorized, for example, as arable, permanent pasture, or rough grazing. Land use codes are used in describing land in the Basic Payment Scheme or *agri-environment scheme applications; for example, agricultural and non-agricultural land uses. Land use zones refer to types of activity or development permitted. Land use should not to be confused with *land cover, which describes the physical land type. *See also* AGRICULTURAL LAND CLASSIFICATION; PLANNING.

larch (European, Japanese, hybrid) *See* TREE.

Law Commission *Statutory public bodies, sponsored by the United Kingdom's governments, which review areas of law that are out of date, overly complicated, or unfair, and make recommendations to the respective government for reforms. The Law Commission reviews law in England and Wales, whilst the Scottish Law Commission reviews those in Scotland. At the time of writing, the Northern Ireland Law Commission is non-operational.

law courts *See* COURT, LAW.

Lawful Development Certificates *See* CERTIFICATE OF LAWFULNESS OF EXISTING USE OR DEVELOPMENT; and CERTIFICATE OF LAWFULNESS OF PROPOSED USE OR DEVELOPMENT.

layer model (hardcore model) *See* INVESTMENT METHOD.

leaching Loss of material washed out by a liquid. In *farming this relates to the loss of soluble *nutrients such as *nitrate, as it is washed through the soil profile by the drainage of rain or irrigation water. Nitrate is highly soluble, so free nitrate in the soil can readily be leached by rainfall into *drains or *watercourses. To minimize leaching, soluble nutrients should only be applied when the plant is able to absorb them. Where it is planned to sow a spring crop, the use of a green *manure or *cover crop may reduce leaching by taking up the nutrients over winter which then become available to the following crop as the residue breaks down.

LEADER Approach, advocated by the European Union, of forming local groups comprising community, public, and private sector organizations in order to identify how to spend small amounts of public money designed to support projects for the benefit of rural businesses and communities. Part of the *European Union's *Common Agricultural Policy funding must be distributed through LEADER groups (which are also known as Local Action Groups in the *United Kingdom). LEADER is a French acronym (Liaison Entre Actions de Développement de l'Économie Rurale) meaning 'Liaison among Actors in Rural Economic Development'. Whilst the European Union and individual nation states set out general priorities for spending under the LEADER stream, it is up to each LEADER group to draw up specific plans for spending within their particular area (in general compliance with their government's overall objectives for the scheme). Business and other organizations then usually bid for funding, demonstrating how their projects meet the priorities for expenditure by their local LEADER group.

lead shot The traditional and commonest type of pellet used in shotguns for *game shooting and *clay pigeon shooting. There is an increasing awareness and concern with the build-up of lead in sensitive areas such as *wetlands and areas subject to concentrated shooting activity. Lead is toxic, and the major problems surround birds, particularly *waterfowl that swallow and ingest lead pellets, resulting in death. There are regulations governing the use of lead shot in certain areas and where non-toxic alternatives such as bismuth must be used. *See also* WILDFOWLING.

LEAF *See also* LINKING ENVIRONMENT AND FARMING.

lease Contract between a *lessor and a *lessee which grants the lessee *exclusive possession of a property for a *term (i.e. a defined period), usually for *rent or some other consideration (under Scottish law there must be a rent). Under English law, exclusive possession is an essential prerequisite for a lease to exist; however, in Scotland, whilst this is normally the case there are some exceptions such as in the case of leases of fishing or *mineral rights. The term 'lease' describes both the right and the *contract document itself. Most leases (other than the very shortest) must be made by a *deed, and in the United Kingdom, leases for longer periods must be registered at the respective *Land Registry (although the lengths of leases which trigger a requirement for registration also differ between the different parts of the United Kingdom). The terms lease and *tenancy are used interchangeably, and in practical terms there is little difference between the two rights, as both grant the right to exclusive possession of a property for a term usually for a consideration; however, only a lease must normally be made by deed. *See also* LEASEHOLD; TERMS USED WHEN PROPERTY IS SUBJECT TO LEASE OR TENANCY.

leasehold Legal *estate in land in England, Wales, and Northern Ireland, which, unlike a *freehold, is for a **term of years absolute** (i.e. it does not grant rights which are unlimited by time). The owner of a leasehold is a **leaseholder**. In England and Wales the only other legal estate in land recognized is **freehold (fee simple absolute in possession)**. However, in Northern Ireland a number of other forms of freehold (in addition to fee simple absolute in possession) are recognized (for further details, *see* FREEHOLD). An **estate in land** means a right of possession to that land (i.e. the ability to enjoy it). For the position in Scotland, *see* REAL RIGHTS (Scotland). *See also* LEASE; TERMS USED WHEN PROPERTY IS SUBJECT TO LEASE OR TENANCY.

leasehold enfranchisement *See* LEASEHOLD REFORM.

leaseholder *See* LEASEHOLD; LEASE; TERMS USED WHEN PROPERTY IS SUBJECT TO LEASE OR TENANCY.

leasehold reform Period of reform of the law concerning long *leasehold property (primarily long leaseholds of flats and houses) in England and Wales (for Northern Ireland and Scotland see below), commencing with the **Leasehold Reform Act 1967** and continuing to this day. Leasehold reform is primarily concerned with addressing longstanding concerns over the effectiveness of the traditional model of ownership of property in group occupation—say a block of flats or a housing estate—where a single *freeholder owns the *freehold whilst the individual unit owners each purchase a long leasehold and normally pay a ground *rent to the freeholder who remains responsible for maintenance of the common parts of the property. This situation sets up an immediate potential conflict between both sets of parties given that the interests of the freeholder and the leaseholders are not always the same. Leasehold reform has attempted to address this through a series of measures introduced in several pieces of *legislation:

- The **Leasehold Reform Act 1967** introduced the right for the owners of leasehold houses to buy the freehold *reversion from their landlord whether the landlord intended to sell or not (the right of **leasehold enfranchisement**). It also gave the owners of leasehold houses the right to extend their leases by fifty years; however, it did not apply to flats, as their freeholds cannot easily be separately identified.
- The **Landlord and Tenant Act 1987** introduced the opportunity for the leaseholders of flats to be given first refusal to buy the freehold collectively from the landlord, but only if the landlord intended to dispose of it.
- The **Leasehold Reform Housing and Urban Development Act 1993** effectively gave the leaseholders of flats the same rights as those granted to the leaseholders of houses: namely, that they could buy the landlord's freehold collectively with the other leaseholders (usually through establishing a company), whether the landlord wished to sell or not. Under this Act, individual leaseholders of flats also have the right to a new lease for ninety

years at a *peppercorn rent to replace their existing lease (and this is therefore of use to lessees whose leases are due to end relatively soon).

- The most significant recent reform was the introduction of *commonhold under the **Commonhold and Leasehold Reform Act 2002** (*see* COMMONHOLD for further discussion).

In December 2017 the *Law Commission announced its intention to further review commonhold and long-leasehold law in England and Wales. Similar provisions for long leaseholders to buy their freehold, or extend their lease, also exist in Northern Ireland, although commonhold is restricted to England and Wales. In Scotland, in 2015 many ultra-long leases of dwelling houses were automatically converted into outright ownership under the law. It is also difficult, under the law, to create new long leases of dwelling houses. For the position in relation to flats in Scotland, *see* TENEMENT FLAT.

lease terms *See* TERM.

LEC (Levy Exemption Certificate) *See* CLIMATE CHANGE LEVY.

Legionella Bacteria which can be found in poorly maintained water systems and which can cause **Legionellosis**, which can result in sufferers becoming seriously ill, and in severe cases can result in death. *See also* HEALTH AND SAFETY EXECUTIVE; HEALTH AND SAFETY EXECUTIVE FOR NORTHERN IRELAND.

legislation Law made through agreement by the elected representatives of the *Houses of Parliament, the *National Assembly for Wales, the *Scottish Parliament, or the *Northern Ireland Assembly. Laws can be established either directly by approval, through a series of votes, by the elected representatives (i.e. *Acts of Parliament or *Acts of Assembly, also known as **primary legislation**), or can be issued directly by ministers of the respective governments, under powers given to them through the relevant Act of Parliament (i.e. *delegated legislation or **secondary legislation**). Legislation in the United Kingdom can also incorporate that derived from the European Union, although the referendum decision of the United Kingdom on 23 June 2016 to leave the European Union (EU) will reduce the influence of European Union law once the UK leaves the EU. Law in the United Kingdom can be established through legislation (known as *statute law) and by *case law. *See also* REGULATION.

((⊕)) SEE WEB LINKS

- Website of the United Kingdom Parliament: includes information on how laws are made, and also the progress of legislation currently before Parliament.

legume A plant of the family *Fabaceae* or *Leguminosae* of significant importance to agriculture, especially *organic farming, due to its ability to fix atmospheric nitrogen and make it available to the plant as *nitrate, achieved by the symbiotic bacterium *Rhizobium* in nodules on the roots. Thus it is not necessary to apply nitrate *fertilizer to leguminous plants, and they can be grown to enhance soil fertility, essential in organic farming. There are two main groups of legume grown in agriculture: pulses grown for their seed, such as peas, *beans, or *soya beans, and others grown for forage, such as *clovers, *sainfoin, or lucerne, also known as *alfalfa. Pulses, high in *protein, are an important ingredient of *concentrated animal feeds.

LEP *See* LOCAL ENTERPRISE PARTNERSHIP.

LERAP *See* LOCAL ENVIRONMENT RISK ASSESSMENT FOR PESTICIDES.

lessee Person who has the right to use/occupy a property for a limited period of time under the terms of a *lease. *See also* TENANT; TERMS USED WHEN PROPERTY IS SUBJECT TO LEASE OR TENANCY.

Less Favoured Area (LFA) An area where farming is disadvantaged by geography, topography, or climate, and in which farmers are eligible for compensation from the *European Union for the extra costs incurred or *income foregone. Subsequent reforms of the *Common Agricultual Policy designated these areas as 'areas with natural or other specific constraints'. Established in 1975 (Council Directive 75/268/EEC) as a means to provide support to mountainous and hill farming areas, the LFAs in England, for example, are subdivided into two further areas. The more environmentally challenging areas within the LFA are classed as Severely Disadvantaged Areas (SDA), while the remainder are classified as Disadvantaged Areas (DA). This distinction determines eligibility for particular support payments and environmental schemes.

lessor Person who has granted permission for a *lessee to use/occupy the lessor's property. *See also* FREEHOLDER; LEASE; LEASEHOLDER; TERMS USED WHEN PROPERTY IS SUBJECT TO LEASE OR TENANCY.

levy An amount of money collected on the sale of farm produce to fund the *levy bodies. For example, when a farmer sells an animal for slaughter or a tonne of grain, a certain sum is deducted from the sale price automatically to raise the funds necessary to finance the levy bodies.

levy bodies Organizations, funded by *statutory *levy from the *agriculture and *horticulture industries, with the objective of making these industries more competitive and sustainable. They undertake research and development, knowledge transfer, and marketing and promotion, and provide market information. There are several levy bodies in the United Kingdom, representing different sectors.

   Agricultural and Horticultural Development Board: six bodies that came together with a central administration funded by a levy on the relevant produce sold, and rebranded in 2015:

   AHDB Pork: formerly BPEX, covering pig meat in England.
   AHDB Dairy: formerly DairyCo, covering milk production in Great Britain.
   AHDB Beef and Lamb: formerly EBLEX, covering beef and lamb in England.
   AHDB Horticulture: formerly the Horticultural Development Company, covering commercial horticulture in Great Britain.

AHDB Cereals and Oilseeds: formerly the Home-Grown Cereals Authority (HGCA), covering cereals and oilseeds in the United Kingdom.

AHDB Potatoes: formerly the Potato Council, covering potatoes in Great Britain.

These bodies conduct research into the production and marketing in each relevant area and promote consumption of the produce. For example, AHDB Cereals and Oilseeds grows trials of varieties of cereals and oilseeds, the information from which is used to publish a list of recommended varieties. Advertisements may be commissioned in the press or on television to promote consumption of particular meats. The central organization and management of these bodies is amalgamated into the *Agricultural and Horticultural Development Board, which is independent of commercial industry and Government. In addition, there are similar bodies in Wales, Scotland, and Northern Ireland:

Hybu Cig Cymru Meat Promotion: covering lamb, beef and pork in Wales.

Quality Meat Scotland: covering lamb, beef, and pork in Scotland.

Livestock and Meat Commission for Northern Ireland: covering beef and lamb in Northern Ireland.

(((⊕))) SEE WEB LINKS

- Website of the Agriculture and Horticulture Development Board, with links to the websites of its six divisions.
- Website of Hybu Cig Cymru Meat Promotion Wales.
- Website of Quality Meat Scotland.
- Website of Livestock and Meat Commission for Northern Ireland.
- Website of the British Wool Marketing Board.

Levy Exemption Certificate (LEC) *See* Climate Change Levy.

ley *Grass, sometimes in combination with other plants such as *clover, grown for a period of time in *rotation with other crops. Traditionally, a grass/clover ley was grown for two years for *grazing or hay/silage as a one- or two-year break in an arable rotation to help build fertility and reduce *weed, *disease, and *pest pressures, a practice widely carried out on organic farms.

LFA *See* Less Favoured Area.

liability An obligation to transfer something of economic value (typically cash or other *assets) to another party in return for, typically, a loan, or previous purchase of goods or services from that party. In simple terms this is what one party owes another party.

licence 1. Personal right to enter land for some defined reason; for example, to graze livestock on the land (*see* GRAZING/MOWING LICENCE). Under English law there are two types of licence: a **bare licence**, where no consideration (in effect, payment) is made, and a **contractual licence**, where there is some form of consideration given. For example, a friend temporarily sleeping on someone else's sofa would be a bare *licensee, whereas the visitor to a sporting event who had paid for a ticket would be a contractual licensee. In practical terms it may sometimes be difficult to see the difference between a *tenancy (or lease) and a licence. The *courts provide some guidance here through *case law, with a tenancy being generally identified as being for a *term and conferring upon the *tenant *exclusive possession, usually in return for the payment of *rent or some other consideration. (For further details on the difference, in English law, between a tenancy and a licence, *see* EXCLUSIVE POSSESSION.) Under Scottish law, whilst generally exclusive possession is also considered necessary for a lease to exist as opposed to a licence, the courts have interpreted this more flexibly, and the law around licences is not as well developed as under English law. **2.** Formal permission (usually from a public authority) enabling the holder of the licence to do something; for example, a *felling licence gives permission to cut down trees.

licensee Someone who holds a *licence. A *licensor grants a *licence to a licensee; for example, permission for the licensee to enter onto the licensor's land for a specified purpose.

licensor Someone who grants a *licence to a *licensee; for example, for that licensee to enter onto the licensor's land for a specified purpose.

lichen A symbiotic association between *algae and *fungi. Lichens are commonly encrusted on trunks of *trees, exposed rock, or *dry stone walls. The fungal component is able to absorb water and minerals, and the algae contribute organic nutrition through *photosynthesis. Slow growing and sensitive to pollution, they can provide useful natural monitors of the proximity and scale of pollution sources.

lien In certain circumstances, the right of one party to hold the goods of another until some kind of obligation is undertaken, or payment which is owed is made. For example, a shoe repair shop may keep a pair of shoes they have been reheeling until the bill for the repair is paid by the person who owns the shoes.

lifecycle costing The process of considering the whole cost of a product or service over its life (including manufacturing, purchase, running costs, maintenance, and disposal), often at the stage of **procuring** (or buying) it. Lifecycle costing can operate at a number of levels; for example, looking at overall financial costs, or other forms of cost such as those relating to carbon. For example, at a financial level a lifecycle costing approach could be used to compare the overall cost of two products in order to make a purchasing decision. For instance, if comparing two computer printers, Printer A may be cheaper to purchase initially than Printer B; however, printer A may have more expensive running costs (as its cartridges may be more expensive and print less pages) and may be more expensive to dispose of once it needs replacing. Therefore, on a lifecycle costing basis Printer B may have lower costs overall, and therefore may be the better printer to purchase. A lifecycle costing approach to purchasing is often incorporated into an organization's *environmental management system.

life estate *See* LIFE TENANT.

liferent, proper A **proper liferent** is a subordinate *real right in Scotland which allows the holder of it the right to use, for their lifetime, property owned by someone else. The person holding the right is known as the **liferenter** (if male) or **liferentrix** (if female). The owner is called the **fiar**, and the property which is subject to the liferent is referred to as the **fee**. A proper liferent is created by *deed and registered at the *Land Register of Scotland. It is also possible in Scotland to create an **improper (or trust) liferent**; this is not a subordinate real right, but is instead a *trust where the *beneficiary has the right to use the property for their lifetime. For the position in England, Wales, and Northern Ireland (where there are differences from Scotland), *see* LIFE TENANT.

life tenant Person (also known as a **tenant for life**), in England, Wales, and Northern Ireland who has an interest in land for their lifetime under a *trust; so, for example, Person A may have a life interest in land, with the **remainder** (the interest in land left when the holder of the previous interest dies) going to Person B. Person B may then have that interest either in fee simple (*see* FREEHOLD), or for their life with the remainder going to Person C, and so on. The life tenant is entitled to enjoy the property for their lifetime (by, for example, living in it) and is then said to have a **life interest in possession**.

   Under the nineteenth century **Settled Land Acts** land held for a life tenant was referred to as **settled land**. The life tenant has significant control (for example, the right to sell the property and manage it), so the **trustees** have only a relatively limited supervisory role. However, since 1997 it has not been possible to create new trusts of settled land in England and Wales (although previously existing ones continue to exist) and new trusts of this nature are now **trusts of land**. The legal principles of trusts of land apply to all trusts involving land, and this therefore means that the trustees now have the powers of the absolute owner to sell the land, although the beneficiaries generally have the right to occupy the property.

   Historically, a **life estate** could exist as an *estate in land in England and Wales limited to the lifetime of the holder. As a consequence of the **Law of Property Act 1925** the number of estates in land is now, however, limited to two in England and Wales: the fee simple absolute in possession (see *freehold) and the estate for a term of years absolute (*leasehold). Consequently, what would have been a life estate prior to this can now only be established in England and Wales as a trust (as discussed above). In Northern Ireland, however, a life estate can still exist as an estate in land, effectively as per the pre-1920s position in England and Wales.

   See also LIFERENT, PROPER.

lime 1. A deciduous tree of the *Tilia* family. *See also* TREE. **2.** A citrus fruit grown in Mediterranean climates. **3.** An alkaline mineral added to soil to reduce acidity. *See also* LIMING.

limestone pavement Outcrops of limestone largely comprised of calcium or magnesium carbonate, and as it is weakly soluble in water, over time rainwater permeates into the rock and erodes lines of weakness, creating a network of channels. The formed outcrops provide a **priority habitat** for a range of characteristic vegetation.

liming The spreading of an alkaline mineral to soil to reduce acidity. The ideal *pH of soil is around 7.0 for arable crops and 6.5 for grassland. If the soil has a pH lower than this, i.e. is acid, productivity is reduced. Crushed chalk or limestone, made up primarily of calcium carbonate, is applied to the soil to increase the pH which increases microbial activity and improves the uptake of nutrients by the crop. *See also* FERTILIZER.

limitation Rules setting out time periods by which actions in the civil law *courts must be taken. Whilst claims for breaches of *contract, and *tort, must normally be made within six years in the courts of England, Wales, and Northern Ireland, there are significant variations from this time period in some situations. In Scotland, a similar system exists, but many of the time periods are different from the rest of the *United Kingdom, and limitation mainly concerns personal injury, with *prescription, in effect, indirectly addressing time limits in other areas of potential claim by extinguishing the right or obligation in question. *See also* ADVERSE POSSESSION.

limited company Company structure where the liability of the owners is limited. Like *limited liability partnerships (LLP), limited liability companies are separate legal entities, and therefore the liabilities of the company are those of the company, rather than personally those of the owners (unlike in a *sole trader or traditional *partnership situation where each of the partners, or the sole trader, has personal liability for the debts of the business). In the United Kingdom, the liability of the owners of the company is **limited by shares** or **limited by guarantee**. Where liability is limited by shares, the liability of the owners of the company is limited to the value of their shareholdings. In the case of a company limited by guarantee, the members draw up a memorandum (agreement) guaranteeing that, in the event of the company being wound up, they will contribute a nominal amount (their guarantee) to cover the debts (if any) of the company, hence their liability is limited to this amount. This structure is more commonly used for non-profit companies (for example, clubs, societies, or community projects). As with LLPs, limited companies are required to register, and file *accounts (which become publicly available) with the government in the *United Kingdom (at **Companies House**). For some small businesses this can be off-putting, and they still prefer to retain a more private traditional partnership structure. Limited companies are identified by having the words 'limited' (Ltd) (for a private company) or 'public limited company' (plc) (for a public company which can offer shares to the public) after their name.

limited duration tenancy (LDT) *See* SCOTTISH AGRICULTURAL TENANCIES.

limited liability partnership (LLP) Business structure operating in the *United Kingdom where the members are able to limit their personal liability for the debts of the business. Traditional *partnerships in the United Kingdom are well understood, and many businesses still operate with a traditional partnership structure. However, within a traditional partnership, each of the partners has unlimited personal liability for any debts of the business (and potentially for any claims made against it, for example, if the partnership were to be sued); therefore, if things go wrong within the business it can be personally financially catastrophic for each of the partners. However, a limited liability partnership (LLP) structure provides the opportunity for each of the members (who are broadly equivalent to partners) to limit their personal liability. This is achieved because an LLP is a separate legal entity in its own right, unlike in a traditional partnership where the partners and the partnership are all the same legal body (though it should be noted that in Scottish law a partnership is a separate legal entity; however, its partners still have unlimited personal liability). Because an LLP is a separate legal entity, any borrowing, for example (and therefore any debts that flow from this), is made by the LLP itself, rather than by the members personally. LLPs have become a popular structure for professional services businesses, such as solicitors, accountants, and surveyors, where the members wish to limit their liability for any potential *negligence claims, and these businesses are identified by having LLP at the end of their business name. LLPs are required to register, and file information regularly, with the government via **Companies House**, and this then becomes a matter of public record. This can be off-putting for some small businesses that still prefer to retain a more private traditional partnership structure. *See also* LIMITED COMPANY.

Linking Environment and Farming (LEAF) A registered charity that aims to promote *sustainable agriculture and to help farmers improve both business and environmental performance. One approach has been the promotion of the LEAF marque to identify foodstuffs grown on LEAF participating farms. It has also given a focus to raising awareness and a better understanding of farming with the general public, which includes an established annual programme of **Open Farm Sundays (OFS)**, where farms are open to the public.

(((⊕))) SEE WEB LINKS
• LEAF website, providing details of what the organization can do for farmers, including the LEAF Marque and OFS farm details.

linseed A plant of the *Linaceae* family primarily grown for its oil, although another strain is grown for its fibre to make linen, when it is known as flax. The oil, very high in Omega 3 fatty acids, is available in food grades, and for industrial purposes, such as wood treatment. As a farm crop it is sown in spring as a *break crop and harvested in a similar way to *oilseed rape. The flowers are blue, opening in the morning and closing in the afternoon. *See also* CROP.

lintel A concrete, steel, stone, or wood support above a window or door. *See also* APPENDIX 2 (ILLUSTRATED BUILDING TERMS).

liquidated damages *See* DAMAGES.

listed building Building or structure in England, Wales, Scotland, or Northern Ireland that is included on a national list of buildings of special architectural or historical interest. Listed buildings in England and Wales are classified as Grade I, Grade II* or Grade II. Grade I buildings are of exceptional interest; Grade II* are particularly important buildings of more than special interest; and Grade II buildings are of special interest. A similar classification system exists in Scotland, with buildings being grouped into Category A (national or international importance), Category B (regional or more than local importance), and Category C (local importance). In Northern Ireland, listed buildings are divided into four grades (A, B+, B1, and B2).

Buildings can be classified on the basis of their own merit as an individual building or because of the contribution that they make to the architectural or historic interest of a group of buildings; for example, a Georgian terrace. They may also be listed because of the desirability of protecting something fixed to the building or its *curtilage; for example, carvings.

In practice, the process of listing a building is undertaken by *Historic England, *Cadw in Wales, *Historic Environment Scotland, and the Northern Ireland Department for Communities Historic Environment Division (advised by the Northern Irish *Historic Buildings Council). The main reasons why buildings are likely to be listed are:

- Age and rarity.
- Architectural or historic interest (for example, being a particularly good example of a particular style of architecture).
- Close historical association (for example, to a significant person or event).

Once a building is listed it becomes an offence to demolish it or alter it without first obtaining *listed building consent from the *local planning authority. Permission for demolition is only very rarely given, and if a local planning authority (or relevant government minister) is concerned that a non-listed building is about to be demolished, but in their view should be listed, they can serve a *Building Preservation Notice to temporarily halt the work prior to a formal listing application being made.

Listed buildings can be altered, subject to obtaining listed building consent, and where appropriate, *planning permission. However, permission is granted subject to strict controls which often require the use of traditional building methods and materials in accordance with the age of the building. Planning policy in connection with listed buildings is set out in the *National Planning Policy Framework and *Planning Practice Guidance in England (and the planning policy equivalents in Wales, Scotland, and Northern Ireland). Historic England (and its equivalents in Wales, Scotland, and Northern Ireland) also produce detailed guidance on working with, and listing, buildings. Listed buildings in England are included within the National Planning Policy Framework's definition of a *heritage asset. The planning protection relating to listed buildings includes both the building and its curtilage, and can also extend to its *setting (which can cover a larger physical area than curtilage).

See also CERTIFICATE OF IMMUNITY FROM LISTING.

listed building consent Approval required from the *local planning authority for any works of demolition, alteration, and/or extension to a *listed building which may affect its character as a building of special architectural or historic interest. It is an offence in the United Kingdom to carry out such work without obtaining listed building consent. The range of alterations which may require listed building consent is broader than those needing *planning permission, as planning permission only applies to things falling within the definition of *development (which does not include most internal alterations), whereas listed building consent requirements apply to internal and external changes.

The listed building consent and planning permission systems run in parallel, and therefore the same works can require planning permission and listed building consent. However, it is also possible for listed building consent to be needed where no application for planning permission is required, for example, because the works proposed do not fall within the definition of development, or benefit from the grant of blanket planning permission through *permitted development rights (although these rights themselves can be restricted for listed buildings).

The requirement for listed building consent applies to the building or structure described within the listing entry, and also to objects or structures fixed to it. It can also include buildings or structures within the *curtilage of the listed building (i.e. the area of, and around, it), even if not fixed to the listed building itself. In practice this can sometimes mean that it is difficult to determine whether or not works require listed building consent, because it is not always clear where the curtilage of a listed building physically stops. There is consequently an extensive body of *case law where the courts have assessed the factors that determine the extent of the curtilage of listed buildings. A range of aspects have been considered relevant, including age, layout, history, ownership, function, and whether a building or structure is related to the main listed building; however, the picture is complex, and it is often not easy to determine the situation on the ground.

A number of reforms to try to address the complexities of the listed building consent regime were introduced in England through the **Enterprise and Regulatory Reform Act 2013**. These include the ability to apply to the local planning authority for a **Certificate of Lawfulness of Proposed Works** to provide confirmation as to whether proposed works would be legal without listed building consent. If the local planning authority refuses to issue such a certificate, then an application for listed building consent must be made for the proposed work. The aim of this reform is to provide a greater degree of certainty as to whether or not listed building consent is required; for example, in relation to changes to buildings or structures that are close to the main listed building and therefore where it is not easy for the potential applicant to determine whether or not they would be considered to be within the curtilage (and therefore require a listed building consent application).

In addition, *Historic England is now able to state within a listed building entry whether attached or curtilage structures are protected by the listing, and also to exclude specific objects that are fixed to a building from listed building consent requirements. They are also now able to state within the listing that some feature or part of a listed building is not of special interest in relation to listed building consent. However, whilst these provisions apply to listed building entries made (or amended) after 2013, for buildings and structures with older listings the previous system continues.

Further measures introduced in England in 2013 also now make it possible for local planning authorities to issue **Local Listed Building Consent Orders**. These can grant blanket listed building consents for certain types of routine or minor work to identified listed buildings within a local area (subject to the works meeting certain conditions or criteria). This means that relevant works meeting the criteria set out in the order do not require individual listed building consent applications. It is also possible for national government in England to make **Listed Building Consent Orders**, which are essentially the same as Local Listed Building Consent Orders but cover specified works to identified types of listed building at a national or cross-*local planning authority level, rather than in a specific local area.

In addition, it is also now possible in England for the owner(s) of a listed building or buildings to agree a **Listed Building Heritage Partnership Agreement** with the local planning authority. This sets out a management plan (covering a period of up to twenty-five years) for the building(s), and grants listed building consent for those works specified within the agreement (although not for demolition works). This means that the owner does not then have to submit successive individual listed building consent applications to carry out the works specified in the agreement (though they may still need to apply for planning permission).

Whilst the general operation of the listed building consent system is the same throughout the United Kingdom, most of the reforms introduced by the Enterprise and Regulatory Reform Act 2013 apply only in England. However, in 2015, legislation was introduced in Scotland allowing *Historic Environment Scotland to state that particular objects or structures fixed to, or within the curtilage of, the listed building, or parts or features of it, are not of architectural or historic interest, thereby excluding them from the protection of the listed building consent regime. In Northern Ireland the **Planning Act (Northern Ireland) 2011** included a provision making it possible for a person wishing to carry out works to a listed building to make a formal application to their local planning authority to determine whether or not listed building consent is required. Additionally, there are proposals in Wales to introduce Heritage Partnership Agreements.

See also SETTING; CERTIFICATE OF IMMUNITY FROM LISTING.

Listed Building Consent Order *See* LISTED BUILDING CONSENT.

listed building consent enforcement notice *See* ENFORCEMENT, PLANNING.

Listed Building Heritage Partnership Agreement *See* LISTED BUILDING CONSENT.

litter 1. The offspring of an animal that has multiple births. So a *pig might have a litter of up to ten *piglets, or a dog a litter of five puppies. **2.** Material such as *straw or *wood shavings used as animal bedding. *See also* BEDDING.

livery Facility to rent to keep *horses. Livery yards or stables can vary considerably in size, standards, and costs. They range from full livery where all cleaning, feeding, and so on are paid for, through to the cheapest option, which is DIY livery where the horse owner normally undertakes all the associated care work. A working livery is another type of arrangement more common at a riding school, where the costs are discounted in return for the horse being made additional use of for riding lessons.

livestock Domesticated animals such as *cattle and *sheep kept on a farm primarily for *milk, meat, or fibre production. A livestock farm is where livestock production is the only or predominant enterprise.

Livestock and Meat Commission for Northern Ireland *See* LEVY BODIES.

liveweight The weight of an animal before it has been slaughtered. The price paid for an animal for slaughter may be determined either on a liveweight or deadweight basis, the latter being the weight of the carcass after slaughter. *See also* KILLING OUT PERCENTAGE.

living wage A suggested *United Kingdom minimum wage calculated independently of government by the **Living Wage Foundation** and based on the cost of living. The living wage is paid voluntarily by some employers; however, it is not the same as the **National Living Wage**, which is the minimum rate of pay which most employees of twenty-five years and over are entitled to by law in the *United Kingdom, and which is set by the government. For further details on the National Living Wage, *see* MINIMUM WAGE.

LLP *See* LIMITED LIABILITY PARTNERSHIP.

Local Access Forum (LAF) Created under the *Countryside and Rights of Way Act to provide a channel for local people, both owners and users of land, to advise their appointing authority on the improvement of public access to land for the purpose of open-air recreation and enjoyment of the area. The appointing authority might be, for example, a county council or a *National Park Authority. The LAFs themselves are made up of appointed members that are representative of users and landowners.

local authority Elected and administrative body delivering government functions at a local level in the United Kingdom. In some areas of England, local authority governance is delivered in a two-tier system with each area being covered by a single **county council** and several smaller **district councils**. In other areas of England (and in the whole of Scotland, Wales, and Northern Ireland), there is a single-tier **unitary authority** structure with a single authority delivering all the local authority functions in an area (although in some parts of Northern Ireland they are known as district councils). All local authorities comprise elected councillors and an administrative staff, and they deliver a range of services from waste and recycling through to sports and leisure facilities. Apart from in Northern Ireland, elected *parish councils, *town councils, or *community councils also exist in some areas, sitting underneath local authorities in the local government hierarchy, and serving smaller areas. *See also* LOCALISM; LOCAL PLANNING AUTHORITY.

local development documents *See* LOCAL PLAN.

local development framework *See* LOCAL PLAN.

Local Development Order *See* PERMITTED DEVELOPMENT.

local development plan *See* LOCAL PLAN.

Local Enterprise Partnership (LEP) Body deciding on priorities for economic investment in roads, buildings, and facilities within an area of England and typically consisting of representatives from businesses, local authorities, and educational institutions. The LEPs were established as part of the *localism agenda and have taken on many of the responsibilities of the former *Regional Development Agencies. There are thirty-nine LEPs in total, covering the whole of England. In addition to having responsibilities for *enterprise zones, they are also able to bid for central government and European Union funds to support economic development and infrastructure in their area.

Local Environment Risk Assessment for Pesticides (LERAP) Certain plant protection products require a *buffer zone of specified width between the *crop to which the product is applied using a horizontal boom *sprayer or *broadcast air-assisted sprayer and the nearest *watercourse. Such a requirement will be stated on the product container and application guidance. If it is intended to reduce the width of the buffer zone, for example, because the product is to be applied at a lower dose rate than that recommended, then a LERAP must be conducted and all details recorded.

local green space *Planning designation, applying in England and introduced in the National Planning Policy Framework (NPPF) (*see* PLANNING POLICIES), which allows local communities, through their *local plan or *neighbourhood plan, to designate areas they consider to be of particular importance to them with the effect of restricting future *development of those areas. The NPPF sets out a number of justifications for the introduction of a local green space designation, including that the area is valued for its recreational value, wildlife, historic significance, or beauty.

localism Philosophy underpinning many of the 2010-elected Conservative/ Liberal Democrat coalition government's policies (and also those of the subsequent Conservative government), particularly in England. The central objective of localism is that all governance is done at the lowest possible level, i.e. the level nearest those people most directly affected by decisions. In order to achieve localism, government policies focus around *decentralization, which involves moving additional elements of decision making away from central government to other bodies such as *local authorities. The objective of doing this is that it improves efficiency by reducing bureaucracy, as well as ensuring that the most appropriate services are delivered locally, as provision is guided by service-users rather than centrally. Whilst the themes of localism and decentralization ran throughout the coalition and Conservative governments' policies, the main direct delivery mechanism was the **Localism Act 2011**. This introduced a number of measures including *neighbourhood planning, a set of *community rights, and changes to the provision of *social housing.

Localism Act 2011 *See* LOCALISM.

local land charge A restriction affecting a specific piece of land for the benefit of a *local authority or other government body. In England and Wales each local authority has a **Local Land Charges Register** relating to properties in their area. Examples of local land charges include *Tree Preservation Orders, *planning permissions, *listed building designations, or the existence of a *Conservation Area. In England the government is in the process of moving this information to a single central searchable register which will be owned by the *Land Registry, and there are similar proposals for Wales. Currently, when a property is in the process of being sold, a **Local Authority Search**, or **Local Land Charges Search**, is usually carried out to ascertain the information that is held on the Local Land Charges Register. In Scotland, similar information is provided by the respective local authority in a **Property Enquiry Certificate**. In Northern Ireland a central **Statutory Charges Register** exists (*see* LAND REGISTRY), although some information is also held by local councils and Northern Ireland government departments. *See also* CONVEYANCING; LAND CHARGE.

Local Landscape Policy Area (LLPA) *Planning designation used in Northern Ireland for features and areas which adjoin settlements and which are considered to be worthy of protection because of their amenity value, *landscape quality, or local significance. The reasons used by *local planning authorities to designate such areas may include the presence of heritage sites (such as *listed buildings or *scheduled monuments), river banks or shorelines, attractive views, or areas of natural conservation interest (for example, woodland areas). Where local authorities have designated such areas, the local development plan (*see* LOCAL PLAN) will have specific *planning policies in relation to them. *See also* SPECIAL COUNTRYSIDE AREA.

Local Listed Building Consent Order *See* LISTED BUILDING CONSENT.

Local Nature Partnership The partnerships compromise representatives from statutory and non-statutory organizations whose aim is to work at a strategic scale to improve the multiple benefits for nature, people, and the economy that can be gained from good management of land. Furthermore, to become local champions influencing decision making, in particular, through working closely with *local authorities, *Local Enterprise Partnerships (LEPs) and **Health and Wellbeing Boards**. *See also* ECOSYSTEM SERVICES.

Local Nature Reserve (LNR) Sites more commonly of wildlife interest, but can be for their geological features, that are established by *local authorities as a statutory designation under the *National Parks and Access to the Countryside Act 1949. Normally owned by local authorities, but not uncommonly managed by a local county *Wildlife Trust. *See also* NATIONAL NATURE RESERVE; SITES OF SPECIAL SCIENTIFIC INTEREST; LOCAL WILDLIFE SITES.

local place plans (Scotland) *See* LOCAL PLAN.

local plan Document, or set of documents, also known as a local development plan, detailing local *planning policies within an area. The local plan is prepared by the *local planning authority, and applications for *planning permission and *listed building consent within that area are then assessed against it, the principle being that permission/consent should be granted for those applications that accord with the relevant policies, unless any *material consideration dictates otherwise.

The local plan in essence interprets national (and regional, where it exists) planning policy at a local level, in addition to identifying sites where future applications for planning permission are likely to be looked upon favourably (most typically for housing and business uses).

As development needs and national/regional planning policy change over time, local plans are regularly reviewed and rewritten (the frequency varies, though this is typically done every four or five years). In preparing the local plan, the local planning authority is required to carry out a process of assessing future local development needs (say for housing), and identifying sites where such needs could potentially be met during the period covered by the plan. The process of developing the plan also involves consultation with the public and a range of public bodies with an interest in the plan (for example, those relating to flood risk management, or highways, and neighbouring local planning authorities). Once the local planning authority has a draft plan in place, it is then submitted to an inspecting body for examination (in England and Wales the *Planning Inspectorate; in Scotland, *Scottish Government Ministers who will appoint a **Reporter** to examine the plan; and in Northern Ireland, the *Planning Appeals Commission). The inspections are carried out in order to ensure that the local plan policies are consistent with national (and regional where it applies) planning policies, and that the processes used to prepare the plan, and its contents, comply with legal requirements (in the event that they did not, developers who had had applications refused could then challenge the legal basis of the decision through the courts). Following the report of the inspection body, the local authority may then make further changes in response, following which the plan is then 'adopted' formally by the local planning authority. At this point, all applications must then be assessed against the newly adopted local plan.

Whilst the general role, and process of preparing local plans, is the same throughout the United Kingdom, there are some differences in the structures of the documents themselves.

- In England, the **National Planning Policy Framework (NPPF)** sets out national planning policy. All of the local planning policies for an area may be found in its local plan. Alternatively, the local planning authority may work with other neighbouring authorities to produce a strategic planning document covering several authorities, with each authority's local plan then only having policies which are specific to their local area. Following the introduction of the *neighbourhood planning system in England, there may also be a number of smaller and more localized (often at *parish level), neighbourhood plans sitting underneath a local plan, and against which applications in the relevant parish are also assessed. The combination of the local plan, any relevant strategic planning documents developed jointly with other authorities, and any made neighbourhood plans, is referred to as the **development plan** for an area and applications within that area are assessed against it. There may additionally be *supplementary planning documents sitting underneath the local plan, providing further detail about specific aspects; for example, designs of buildings within the local plan area (though these do not formally form part of the development plan). Prior to the introduction of the NPPF, local planning authorities were encouraged to have a set of documents within a **Local Development Framework**, which included a **Core Strategy** which set out the main policies for the area whilst the site allocations document may have been separate. This terminology continues within some local planning authorities.
- In Wales, overarching national planning policy is set out in **Planning Policy Wales**, with further detailed information on specific issues such as housing and tourism set out in **Technical Advice Notes**. Local planning authorities are required to produce a local development plan which has regard to national policy. (Local development plans replace the previous **unitary development plans** which served the same purpose as local development plans.) The local planning authorities may also, as in England, produce supplementary planning guidance to provide more detail on some aspects of the local development plan.
- In Scotland, national policy is set out in the **National Planning Framework** and **Scottish Planning Policy** documents. The National Planning Framework sets out the Scottish Government's economic objectives and how they impact at a strategic level on development and land use, with the Scottish Planning Policy more specifically focused on planning matters. Local planning authorities are expected to give significant weight to these documents when preparing their development plan documents. Development plans can consist of up to three parts: a local development plan which sets out policies and site allocations for an individual council area, a **strategic development plan** (which is required for Aberdeen, Dundee, Edinburgh, and Glasgow and addresses region-wide issues which cross council boundary areas), and supplementary guidance which can provide more information about policies within the local plan; for example, design guidance. There are, however, proposals to remove strategic development plans from the system and also amend the way in which information in supplementary guidance is dealt with within planning policy. In addition, there are plans to introduce **local place**

plans (which would be prepared by community bodies in a similar way to neighbourhood plans in England; *see* NEIGHBOURHOOD PLANNING).

- In Northern Ireland, until 2015 the planning system operated on a centralized basis with the majority of planning functions and decisions being carried out by the **Northern Ireland Department of the Environment**, which had powers to make development plans for any area of Northern Ireland. These took a number of forms: **Area Plans** (which were similar in geographic scale and scope to local plans/local development plans in the other areas of the United Kingdom), **Local Plans** (which tended to be smaller in geographic area; for example, addressing issues in a town centre), and **Subject Plans** (which were focused around planning issues relating to a particular subject; for example, *houses in multiple occupation in a particular locality). Northern Ireland-wide planning policy was set out in **A Planning Strategy for Rural Northern Ireland** and a suite of documents, known as **Planning Policy Statements** (each of which addressed a specific aspect, such as natural heritage or access). Following the **Planning Act (Northern Ireland) 2011**, however, the majority of planning functions transferred, in 2015, to local authorities who are now required to prepare their own local development plans. Northern Ireland-wide policy is now contained in the **Strategic Planning Policy Statement for Northern Ireland (SPPS)**, which was published in 2015 and complements the **Regional Development Strategy 2035** document which sets out wider strategic objectives for Northern Ireland. Local planning authorities are expected to take into account the provisions of the SPPS when preparing their new local development plans. Once all local planning authorities have adopted their own new local development plan, the SPPS will entirely supersede the previous Planning Strategy for Rural Northern Ireland and Planning Policy Statements; however, until this is the case, transitional arrangements exist which retain a number of elements of these old Northern Ireland-wide policies (as detailed in the SPPS). The new local development plans comprise two elements: a **plan strategy** which sets out the council's strategic objectives in relation to the development of land in its district, and a **local policies plan** which sets out information about local site-specific designations and related policies. Supplementary planning guidance is also used in Northern Ireland.

See also PLANNING REFORM WALES; STRATEGIC HOUSING MARKET ASSESSMENT; STRATEGIC HOUSING LAND AVAILABILITY ASSESSMENT.

local planning authority Authority that manages the development control and planning *enforcement processes in a specific locality. The majority of local development control and enforcement aspects are managed by departments within either district or unitary authorities, with the exception of most *National Parks where the *National Park Authority is the local planning authority.

The local planning authority prepares a *local plan (or local development plan) for its area which sets out the local *planning policies against which applications for *planning permission and *listed building consent are assessed. The core principle followed is that permission/consent should be granted for those applications that accord with the relevant policies, unless any *material consideration dictates otherwise. The local planning authority employs *planning officers to administer the process of determining planning-related applications, including those for planning permission, listed building consent, express consent in relation to *advertising controls, and *prior approval applications in respect of *permitted development rights. The decisions in respect of these applications may be made either by the planning officers (under *delegated powers) or by the local authority's planning committee (comprising elected councillors).

Applicants have a right of *appeal against decisions. Appeal applications in England and Wales are administered at a national level by the *Planning Inspectorate, and in Northern Ireland by the *Planning Appeals Commission. In Scotland, appeals where the decision has been made by a planning committee are dealt with centrally by the **Scottish Government Planning and Environmental Appeals Division**, whereas those decided by a planning officer under delegated powers are considered by the local planning authority's **local review body** (which comprises a group of three or more elected members of the planning authority).

Local planning authorities also manage planning enforcement in their locality (i.e. where there has been a breach of planning *conditions, or development has occurred without the appropriate permissions or consents).

Whilst the majority of planning applications are dealt with by local planning authorities, most nationally significant infrastructure projects in England are managed under a system operated by the Planning Inspectorate on behalf of central government. In Wales, the Planning Inspectorate examines energy and harbour development applications, with other types of application being dealt with directly by *Welsh Government Ministers. In Northern Ireland, regionally significant developments are dealt with centrally by the **Northern Ireland Department for Infrastructure**. In Scotland, whilst national developments are dealt with by local planning authorities, *Scottish Government Ministers have powers to call in applications to determine them themselves.

The planning system in relation to waste and minerals operates on a parallel basis to other types of non-infrastructure planning; however, in areas which are not covered by a unitary authority, minerals and waste planning, and general local planning matters are often not managed by the same authority. Where a two-tier local government system of district councils and a county council exists, the minerals planning authority is normally the county council, with other local planning matters being managed by the district councils.

Whilst the local authority planning system described above is well established within England, Wales, and Scotland, until 2015 the system in Northern Ireland operated on a centralized basis with the majority of planning functions and decisions (on all scales of project) being carried out by central government through the Northern Ireland Department of the Environment. Following the **Planning Act (Northern Ireland) 2011**; however, in 2015 the majority of non-infrastructure planning functions transferred to local authorities, on a broadly parallel basis to that already in operation in England, Wales, and Scotland. For a more detailed discussion of the situation in Northern Ireland, *see* LOCAL PLAN.

See also NEIGHBOURHOOD PLAN; PLANNING REFORM WALES.

local policies plan (Northern Ireland) *See* LOCAL PLAN.

local review body (Scotland) *See* APPEAL, PLANNING; LOCAL PLANNING AUTHORITY.

Local Wildlife Sites (LWS) Sites that are recognized as of local importance for wildlife conservation, but are non-statutory designations. They do not have the same level of legislative protection of, for example, *Sites of Special Scientific Interest, but can be of material consideration in matters of *planning and *development. Local Wildlife Sites will contain significant *species and/or *habitats, and can also be referred to as **Sites of Interest for Nature Conservation (SINC)** or **Sites of Biological Importance (SBI)**. Sites are also known by different terms across the countries of the UK. Many sites are owned and managed by organizations such as local *Wildlife Trusts or the *Royal Society for the Protection of Birds.

loch *See* LAKE.

lodging The collapse or permanent displacement of the stems of an annual *crop, making *harvest more difficult. The taller older *cultivars are more likely to lodge, but as improvements have been made to the *harvest index of crops, i.e. the weight of the grain, there is an added risk that the stems are not strong enough to hold up the plant. Thus plant breeders have selected for the length and/or strength of the straw to improve standing power.

logs Units of timber commonly used in relation to the supply of *firewood, and generally sold by weight or volume. *See also* SAWLOGS; CORD; WOODFUEL.

loosebox A description of a type of *stable accommodation for a *horse that allows it some freedom of movement, and is therefore more comfortable than a stall which is more restrictive. To accommodate a horse, a loose-box should measure about 3.6 metres square.

lop and top *See* BRASH.

lotting fee *See* AUCTION.

lough *See* LAKE.

louping-ill A viral disease that affects the nervous system and is passed onto *grouse mainly by sheep *ticks (*Ixodes ricinus*). It has high levels of mortality in grouse chicks, and can subsequently cause significant adverse economic impacts on grouse-shooting interests. Improved *sheep management and *bracken control has reduced the problem in both sheep and grouse. *See also* PARASITE.

lowland farm A *farm in an area of the country that is relatively low in terms of height above sea level. The uplands are generally thought to be land above 300 metres above sea level, so a lowland farm is one with land below 300 metres above sea level, though there are variant definitions both in the UK and the EU. The term does not indicate the type of farming carried out. Arable production would primarily occur on lowland farms, but there are large areas of lowland UK which are permanent pasture, particularly in the wetter west of the country.

lucerne *See* ALFALFA.

lungworm A parasitic nematode worm found in the lungs of mammals, especially farm and domestic animals. Larvae are ingested and migrate from the intestine to the lungs, where they cause inflammation, difficulty in breathing, and weight loss. The worms can be controlled by administration of a *wormer such as an *anthelmintic.

lupin A genus of flowering plants of the *Fabaceae* or *Leguminosae* family grown as a source of protein for animal feed. Sown in March or April, the crop can be cut for *silage or *harvested in a similar way to peas or *beans as a high-protein constituent of a *concentrated feed ration, often seen as a replacement for imported *soya bean. Although it is a *legume, to achieve the highest level of *nitrogen fixation the roots should be inoculated with extra *Rhizobium* bacteria. As a legume, it is an ideal *break crop in an arable rotation.

Lyme disease An infectious disease of humans caused by bacteria of the *Borrelia* genus. It is carried by *ticks of the *Ixodes* genus and is spread to humans by tick bite. The tick needs to be attached to the host for 36–48 hours before the disease is transmitted. Symptoms are a rash or area of red skin around the site of the tick bite that occurs around a week after transmission, headaches, fever, and feeling tired. If untreated, the symptoms may extend to joint pain, severe headaches with neck stiffness, and heart palpitations. These symptoms may last for six months, even after treatment, and may recur months or years later. Treatment is by means of a course of antibiotics. Deer are a common host of the tick, and incidence of the disease in the UK is increasing due to the increased population of deer. It is recommended that the body is covered, for example, by means of long trousers when walking in the countryside, especially through long grass or bracken which may contain the ticks.

lynchet 1. A strip or bank of earth formed in ancient times on the downhill side of a *ploughed field. *Soil movement down the slope creates positive lynchets, whereas the areas where the soil is reduced creates negative lynchets. The term is derived from 'lynch', which implies an agricultural terrace. **2.** A strip of unploughed land forming a temporary barrier between two fields.

MA *See* MILLENNIUM ECOSYSTEM ASSESSMENT.

machair A rare *grassland *habitat which occurs on the coast of north and west Scotland and western Ireland. The term machair is Gaelic and means a fertile low-lying grassy plain. It is a distinctive *landscape that can provide a habitat for extensive and diverse *flora which in turn supports an associated invertebrate *fauna.

macrocell *See* TELECOMMUNICATIONS EQUIPMENT.

mad cow disease *See* BOVINE SPONGIFORM ENCEPHALOPATHY.

MAFF *See* MINISTRY OF AGRICULTURE, FISHERIES AND FOOD.

MAI *See* MEAN ANNUAL INCREMENT.

maize A *cereal *crop *Zea mays*, it is also known as *corn in many parts of the world and is normally grown with the purpose of making *silage for use as an animal feed. However, it can also be grown for *grain or *biogas production. Where silage is the purpose, the crop is grown and harvested at the end of the season when the crop moisture content is approximately 65%. Advances in variety development and the warming UK climate has allowed for grain maize to be grown more reliably in the UK. With the numbers of *farms installing biogas plants maize grown as a fuel for energy production is also on the increase. *See also* RENEWABLE ENERGY.

malting The process of turning grain into malt which is then used in brewing. The grain is usually *barley for the brewing of beer or whisky, but may be other cereals such as *wheat. The grain is soaked in water to encourage germination, during which starch is turned into sugars. The process is stopped by heating and drying the germinated grain. Different types of malt are produced by varying the amount of heat used; the higher the temperature, the darker the malt, which is then used for darker beers such as stout. Malting barley is barley of the quality required for malting, such as low nitrogen content.

management objectives The aspirations of the managers and owners of an organization for the future of that organization. These objectives are then used as a management tool through their translation into **SMART (specific, measurable, achievable, realistic, and timely) targets** which relate to particular aspects of the operation of that organization, with specific members of staff having identified responsibilities for meeting particular targets.

manege Constructed outdoor or indoor riding arena providing an all-weather and well drained surface for schooling *horses and other activities. Most maneges for private use are 40 metres long x 20 metres wide, but those in commercial horse-riding establishments are often larger.

mange A skin disease of mammals caused by parasitic mites. It can affect cattle and is common in dogs and *foxes. **Sheep scab**, a very serious condition, is a form of mange. There are three types of mite: surface, burrowing, and **Psoroptic**, all causing intense irritation and thickening of the skin, hair loss, and loss of productivity. Treatment is by an *anthelmintic medicine either injected or poured on the skin.

mangel *See* FODDER BEET.

manger A trough-like container used to hold feed, commonly for *horses. It can be built into the *stable corner, attached to a door or gate, or simply placed on the floor. Mangers can be made of wood, metal, plastic, or rubber, and need to be shallow enough to allow the horse to feed, but deep enough to prevent feed being thrown out and wasted. Fixed mangers for feeding hay are usually installed at chest height to prevent a horse from catching its foot.

manorial rights Old legal rights (for example, for mines, some *minerals, *sporting, and to hold fairs and markets) which are still in existence today in England and Wales and which were originally those retained by lords of the manor when their land became *freehold in the early twentieth century. Lords of the manor were, historically, those noblemen given land (manors) by the king under the feudal system first established by the Normans. Whilst in some cases manorial rights in existence today have been inherited, they can also be bought and sold to third parties (although the land to which they originally related may have long since been sold to other people). Up until 2013, in England and Wales manorial rights did not have to be recorded on the register of *title of a property; however, they did remain in existence and were **overriding**, so they bound the later owners of the land to which they related (who may not have been aware of their existence). However, as a consequence of the **Land Registration Act 2002**, all such manorial rights had to be registered at the *Land Registry by 2013 or lost their overriding interest, and as a consequence will be lost when a relevant property is sold unless the rights are registered before the sale. The 2013 deadline led to a steep increase in the number of manorial rights being registered in England and Wales and some people becoming aware for the first time that their property is subject to these rights. At the time of writing, the *Law Commission is reviewing some elements of the law in connection with manorial rights in England and Wales.

manure *See* FARMYARD MANURE.

manure spreader *See* MUCK SPREADER.

maple (field, Norway) *See* TREE.

marketing mix A conceptual framework comprising elements which inform the decisions taken by marketing managers in shaping business ventures so that they meet their customers' needs effectively. Whilst there is some debate about the exact composition of the elements of the marketing mix, there is broad agreement that **product** (the physical item or service being offered), **price**, **promotion** (including advertising, public relations, or sponsorship), and **place** (comprising the whole distribution process from where something is produced to the point at which the customer gains access to it) are core aspects. These are often referred to as the **4 Ps**. Other ingredients are also sometimes included in the mix, particularly in relation to providing services (often expanding the list to **7 Ps**). These include **people** (particularly when staff have high levels of direct contact with customers), **process** (the customer's entire experience of interacting with a business, for example, from instructing a solicitor to carry out a task, through to receiving the final invoice), and **physical evidence** (evidence used by the customer to determine the quality of the service that they are buying, for example, the weight of paper used in a marketing brochure). It is argued that the customer's response to the marketing of a product or service will be the result of the way in which all the ingredients of the mix are combined and, therefore, that one cannot be looked at in isolation from the others. The marketing mix approach has been criticized by some (who instead often advocate a *customer relationship marketing approach) for not sufficiently addressing long-term customer relationships.

market rent The amount of rent that, in an open market, the *lessor of a property can reasonably expect to receive from a *lessee for the use of that property under the terms of an appropriate *lease. The *Royal Institution of Chartered Surveyors' (RICS) *Red Book (**RICS Valuation: Global Standards 2017**) defines it as 'The estimated amount for which an interest in real property should be leased on the valuation date between a willing lessor and willing lessee on appropriate lease terms in an arm's length transaction, after proper marketing and where the parties had each acted knowledgeably, prudently and without compulsion.' This is an internationally accepted definition as set out in the **International Valuation Standards 2017** published by the **International Valuation Standards Council (IVSC)**.

market segmentation *See* SEGMENTATION.

market value The amount that, in an open market, an asset or liability should exchange for if sold. The *Royal Institution of Chartered Surveyors' (RICS) *Red Book (**RICS Valuation: Global Standards 2017**) defines it as 'The estimated amount for which an asset or liability should exchange on the valuation date between a willing buyer and a willing seller in an arm's length transaction, after proper marketing and where the parties had each acted knowledgeably, prudently and without compulsion.' This is an internationally accepted definition as set out in the **International Valuation Standards 2017** published by the **International Valuation Standards Council (IVSC)**.

Whilst the IVSC definition of market value referred to above is used for a large proportion of valuations, in the *United Kingdom valuations for some purposes use different definitions (often set out in law). For example, valuations for *capital taxes (*Capital Gains Tax, *Inheritance Tax, *Stamp Duty Land Tax, and their Scottish and Welsh equivalents) must be undertaken using a slightly different definition of market value which is set out in *legislation (and which is supplemented by further *case law).

marriage value The additional element of value if two different property interests, or assets, are put together in comparison to the sum of the values of the two interests or assets held separately. For example, if the separate values of the *freehold and *leasehold of a property are added together it is likely to be less than the value of the same property as a freehold with *vacant possession (i.e. if the freeholder buys the leasehold and combines the two interests). Marriage value is also sometimes referred to as **synergistic value**.

marsh *See* WETLAND.

Martinmas *See* QUARTER DAYS.

mash In general terms, to mix ingredients, often with water for feed, as in bran mash for horses. To crush or beat something to a soft texture, as in mashed potato. More specifically, in brewing to mix *malted grain, usually *barley, with hot water in order for the sugars to dissolve to form the wort. The vessel in which this takes place is called the **mash tun**.

mast The production of a very large crop of *seeds, most commonly used to refer to beech and oak. This is a mechanism sometimes referred to as **masting**, to counter predation and aid the successful regeneration of new trees. A mast year is the specific year in which this production occurs. A significant consideration should be given to the timing of felling operations, and good mast years, where restocking is reliant on the availability of seeds to promote natural regeneration.

Master of Foxhounds (MFH) Responsible, or with others as a Joint Master of Foxhounds, for the conduct and good running of the *hunt. They operate the sporting activities of the hunt, maintain the kennels, and work with the professional huntsman. In many hunts an MFH may also actually hunt the pack of hounds. The term MFH applies to both men and women, and those so appointed can use these post-nominal initials with their name. Often, but not always, a master will also operate as **field master** on a hunting day, and will lead the mounted followers of the hunt and ensure that they ride where they are permitted, and do not obstruct the huntsman and hounds. The Masters of Foxhounds Association (MFHA) is an organization of current and past MFH, and is the governing body for *fox hunting. *See also* BEAGLING; STAG HUNTING.

material consideration Any factor which is relevant to the making of a decision in relation to a planning application. This can include a range of aspects, including *planning policy, responses made by *statutory consultees, or site-specific factors such a flood risk, or issues relating to access. It is up to the decision maker (normally the *planning officer or planning committee) to determine what weight to give each consideration (i.e. whether it is sufficiently problematic to refuse the application, or whether it can be addressed through the provision of a planning *condition). Because there is a large range of aspects which can potentially be material considerations in respect of any individual application, there is an extensive body of *case law where the courts have considered whether factors are material considerations or not; they have generally held that material considerations should relate to matters in the public interest, rather than in relation to private interests, such as the impact of a development on the value of a neighbouring property. *See also* PLANNING REFORM WALES.

MCPA A selective *herbicide, 2-methyl-4-chlorophenoxyacetic acid is used to control *broad-leaved weeds, including thistles and docks in *grassland, and cereal crops. It was one of the very first selective herbicides to be developed in the 1940s, and is selective because its action is based on plant hormones, auxins, in such a way that some are susceptible and others are not. *See also* PESTICIDES.

MDF (medium-density fibreboard) *See* PULPWOOD.

meadow A *grassland field that might often include wild flowers and is used as a grazing pasture, but more especially for making *hay. Meadows are an important and recognized *habitat which are becoming less common, particularly with the movement towards making *silage or converting to *arable cropping. *See also* WATER MEADOW.

mean annual increment The average rate of volume increase of a stand of trees over a given period of time, normally measured in cubic metres per hectare.

Meat and Livestock Commercial Services Ltd Company (formerly owned by the *Agriculture and Horticulture Development Board and now in the private sector) which provides data, equipment, and inspection services (including *carcass classification) to the meat and livestock sector.

mediation *See* ALTERNATIVE DISPUTE RESOLUTION.

medium-density fibreboard (MDF) *See* PULPWOOD.

mensuration The science of measurement used by all those in the timber trade. It uses and relies upon a series of standard mensuration tables and equations to accurately estimate volumes of timber, age, growth, and yields.

MFH *See* MASTER OF FOXHOUNDS.

Michaelmas Day *See* QUARTER DAYS.

microcell *See* TELECOMMUNICATIONS EQUIPMENT.

microchipping A system that involves the implantation of an RFID (Radio Frequency Identification) device just below the mane of the *horse. The microchip contains a unique identification number, and the horse's details can be accessed using a microchip reader. An EU regulation passed in 2009 strengthened the existing requirements of new *horse passports, and made microchipping compulsory.

Midsummer Day *See* QUARTER DAYS.

mildew A disease of plants caused by *fungi resulting in damage to leaf tissue. There are two main strains of significance to farming: *downy mildew, and powdery mildew. The impact of an infection can be severe in terms of yield loss, especially in cool, damp conditions, but the *disease can be treated by an application of *fungicide.

milk The white liquid produced by the mammary glands of female mammals to feed their young. *Cattle in particular have been domesticated to produce milk for human consumption, either as a liquid or for processing into a variety of *dairy products such as cream, butter, yoghurt, and ice cream. Although cow's milk is the most common, both *sheep and *goats are also used in milk production. The facility in which the cows are milked is known as the **milking parlour**, and the equipment as the **milking machine**. *See also* DAIRY; MILK FEVER; MILK QUOTA.

milk fever A condition of female animals, particularly *dairy cows, in the early stages of lactation. Also known as hypocalcaemia, it is caused by a low level of calcium in the bloodstream. *Heifers are rarely affected. It is a condition of older cows, as the calcium level of the body is syphoned into the *colostrum and milk for the new-born calf. Symptoms are unsteadiness on the legs and twitching of the limbs. If left untreated, the cow will lie flat, its heart beat will weaken, and death ensues. If caught early enough, treatment by injecting calcium borogluconate, perhaps with other minerals and dextrose, can bring a complete recovery. It is a similar condition to staggers or **hypomagnesaemia**, caused by a low magnesium level in the bloodstream.

Milk Marque A large-scale dairy farmers' cooperative founded in 1994 on the abolition of the statutory Milk Marketing Board. However, it was disbanded following an investigation by the Competition Commission into its monopolistic position in 1999, and was broken into three separate companies.

milk quota A measure introduced by the European Commission in 1984 in an attempt to reduce surplus production. Each *dairy farmer in the European Union was allocated a certain quota which limited the amount of milk he was allowed to produce. The quota could be traded, and at its peak and in extreme cases the milk quota was worth more than the farm. Milk quotas were finally abolished in 2015.

Millennium Ecosystem Assessment (MA) *See* NATIONAL ECOSYSTEM ASSESSMENT.

millstone A large stone used for grinding *wheat and other *grains to produce flour. Traditionally in a mill driven by wind or water, there would be a pair of stones, the bottom or **bedstone** remaining stationary, with the upper or **runner stone** rotating on a spindle. The grain is fed through a central hole in the runner stone and is ground between the stones, and the flour emerges from the outer edge. A millstone could be very heavy, hence the expression 'a millstone around the neck'.

minerals rights Legal rights of ownership of minerals. In *Great Britain, as a general principle the ownership of minerals lies with the land above them; however, it is also possible to separate that ownership, and hence minerals can be owned by a different party from the land. Mineral rights are also subject to *statutory controls; for example, the Crown has the right to gold and silver and mines of them are known as **Mines Royal**. Since the privatization of the coal industry in the 1990s, coal has been owned by the **Coal Authority** (a public body who have responsibility for unworked coal and past coal mining). Oil and gas are also owned by the Crown, and the UK Oil and Gas Authority (or in some situations, **Welsh** or **Scottish Ministers**) issue issues licences for exploring and exploiting them.

   In Northern Ireland, the rights to all minerals apart from gold and silver (owned by the Crown) and some other 'common' substances (including sand, gravel, and aggregates) are owned by the Northern Ireland government, who grant licences to prospect and mine them.

Minerals Safeguarding Area Area where *planning permission for non-minerals *development is unlikely to be granted because the location has been identified as having known deposits of minerals. The purpose of the designation is to ensure that new development does not restrict the ability of minerals to be mined at a future date, given that minerals are non-renewable and only sited in specific locations.

minimal tillage The means of establishing a crop using limited tillage operations. A *plough will not be used, and there will be only one or two passes of *cultivation machinery such as a *disc harrow or *power harrow. The extreme version of this, also known as zero tillage or direct drilling, is where the crop is sown directly into the stubble of the previous crop. A tillage train—a combination of cultivation tools pulled by a single tractor—may be used. Not all drills may be suitable for sowing crops in these conditions, as there is likely to be more trash or decaying plant material on the surface. The purpose is to disturb the soil as little as possible, relying on earthworms and microorganisms together with the root systems of previous crops to maintain a good soil structure. Where the soil becomes compacted, it will be necessary to use a deep cultivator to improve structure.

minimum wage The minimum hourly wage that can be paid to most employees over school leaving age, by law, in the *United Kingdom. The government sets a **National Minimum Wage (NMW)** rate (which is tiered for younger workers), and all employers must comply with this. Most employees are entitled to the National Minimum Wage rate, with a small list of exceptions, which includes young or new apprentices who are instead entitled to the **apprentice rate**. The National Minimum Wage applies to employees up to the age of 24, with those aged 25 and upwards being entitled to at least the higher **National Living Wage (NLW)** hourly rate.

Agricultural workers have had long-established government minimum wage rates (tiered for differing levels of responsibility) set by the *Agricultural Wages Boards (England and Wales having a joint board, with Northern Ireland and Scotland having separate ones). The Agricultural Wages Boards set rates on behalf of the respective governments (who also set out legal rights for agricultural workers in relation to a range of other employment areas including holiday, sick pay, and overtime in **Agricultural Wages Orders**).

Workers employed in England before 1 October 2013 are still covered by the Agricultural Wages Order (and must be paid the **Agricultural Minimum Wage** or the respective NMW/NLW, whichever is higher); however, those employed since then are subject to the same NMW/NLW rights as other workers.

However, in Wales the government continues to operate an AMW system with rates being set by the **Agricultural Advisory Panel for Wales** with workers being entitled to either the AMW or the NMW/NLW, whichever is higher. Similarly, Scotland and Northern Ireland also continue to operate their Agricultural Wages Boards which set AMW rates (with workers again entitled to the higher of the AMW or the NMW/NLW).

See also LIVING WAGE.

Ministry of Agriculture, Fisheries and Food The UK government department responsible for *agriculture and food safety that gave rise to concerns of conflict of interest with outbreaks of *bovine spongiform encephalopathy and later *foot and mouth disease in 2001. Food safety came under the responsibility of the new *Food Standards Agency, and the remaining parts of MAFF merged with environment the following year to create the *Department for Environment, Food and Rural Affairs (DEFRA).

Ministry of Housing, Communities and Local Government Government department (formerly known as the **Department for Communities and Local Government**) with responsibility for planning, housing, and community matters in England.

mink (*Neovision vision*) The American mink became widespread within the UK following escapes or deliberate releases from established mink farms from as far back as the 1950s. They share some characteristics with the otter for which they are sometimes mistaken, but are much smaller. They are an effective predator and are of concern because of their adverse impacts on native wildlife species, including some *waterfowl and iconic species such as the water vole (*arvicola amphibious*). Mink hunting using hounds adapted from otter hunting packs continues within the limitations of the **Hunting Act 2004**. *See also* HUNTING.

mintil *See* MINIMUM TILLAGE.

miscanthus (*Miscanthus fuscus*) A perennial crop grown for energy, also known as elephant grass. It can be harvested each year, having grown to around 3 metres tall and looking a little like bamboo. It originates in south Asia and started to be grown on UK farms to be co-fired with coal in power stations. That market has now declined due to the closure of many coal-fired power stations, but miscanthus can still be burnt as *biomass for heat or in small-scale electricity generation. *See also* ENERGY CROPS.

missives *See* CONVEYANCING.

mite A group of very small **arachnids** of which there are many different species, each with four pairs of legs when adult. Some live in the soil and have a beneficial action in breaking down organic matter to form humus. Others are parasitic on plants and animals and cause disease such as *mange and **sheep scab**.

mitigation Actions taken to reduce how serious, intense, or severe something is. *Climate change policies in the *United Kingdom are structured around mitigation policies, alongside *adaptation approaches. The mitigation elements of the policies comprise targets to reduce *greenhouse gas emissions (*see* CLIMATE CHANGE ACT 2008) in order to try to minimize climate change impacts, whilst adaptation policies look to change future mechanisms for the delivery of goods and services in order to adapt them to any changes in climate that result from climate change impacts that cannot be mitigated.

model clauses *See* AGRICULTURAL HOLDINGS ACT 1986.

modern limited duration tenancy (MLDT) *See* SCOTTISH AGRICULTURAL TENANCIES.

modulation A mechanism that was introduced and modified in reforms of the *Common Agricultural Policy to allow for transfer of monies from *European Union farm payments (Pillar 1) to *rural development programmes (Pillar 2). This is essentially to promote a better funding balance between agricultural support and the promotion of rural development. Individual EU member states have some flexibility in the proportions of modulated funding that they choose to implement and it is now referred to as 'transfers between pillars' by the European Commission.

molasses A thick liquid by-product of the refining of sugar cane or sugar beet into sugar. It is high in sugars but also contains minerals and trace elements. It is used as an ingredient for animal feeds to provide *carbohydrate, and can also be used as a soil conditioner to encourage microorganisms.

mole (*Talpa europaea*) The common mole has a widespread distribution throughout Britain. It has a characteristic almost black and velvety fur, and feeds predominantly on earthworms and insect larvae. It lives underground in a network of tunnels and chambers, and leaves distinctive mounds of *soil on the surface, called molehills. The molehills can cause a particular difficulty for farmers in that soil can contaminate *silage grass and spoil the stored silage. They can also cause injury to *livestock and damage to *farm machinery, and hence there has been a long-established industry to control mole numbers through trapping and poisoning.

mole plough An implement used to improve drainage, particularly of clay soils. It is not a true plough in that the soil is not inverted, but it has a single tine or blade that is pulled through the soil at an appropriate depth. At the bottom of the tine is a cylindrical foot or torpedo that creates an unlined passage through the soil for water to pass. Over time, that channel will collapse and the exercise will need to be repeated. In lighter soils where this is likely to happen more quickly, sometimes a porous pipe is inserted into the channel as a lining to keep it open. A mole plough may also be used to bury wires such as telephone or electricity to avoid digging a trench. *See also* FARM MACHINERY.

mollusc A large phylum of invertebrate animals, many of which have shells. There are about 85,000 species of mollusc, including bivalves such as oysters, scallops, and mussels, cephalopods such as squid and octopus, and gastropods such as whelks, limpets, snails, and slugs. From a farming perspective it is the last group that are the most important. Slugs are a pest of growing crops, eating holes in the stems and leaves of plants, particularly at the seedling stage. Control is by *molluscicide, mainly the application of **metaldehyde** pellets, but this is controversial because the chemical has been found in rivers in the UK, which has led to calls for restriction on use and more reliance on increased biological control. Snails are considered a pest more of gardens than of farms, but are also cultivated for human consumption and are considered a delicacy particularly in France.

molluscicide A chemical compound that is used to control molluscs and gastropod pests, specifically slugs and snails. Slugs in particular can cause significant damage to emerging crops, as they eat the leaves and stems. The most common molluscicide used in agriculture is metaldehyde, but environmental concerns have been raised, particularly over levels of the chemical found in watercourses. Certain *nematodes can offer levels of protection through biological control. *See also* PESTICIDES.

monocropping The growing of the same crop on the same land in repetitive years without any *rotation of cropping. It is not commonly practised or encouraged, as it can lead to a build-up of *pests and *diseases and a reduction in soil fertility and profitability.

monoculture The growing of a single crop over a large area, and usually contiguous fields. Unlike *monocropping, this may be part of a rotation of crops so that the large area may be sown to a different crop the following year. It is a common practice on larger farms or blocks of land that are contract farmed, to ease management in that the same activities, such as *spraying or *harvesting, can bring economies of scale.

moor An open *upland landscape, usually poorly drained, treeless, and with patches of different vegetation including *marsh, *heath, and unimproved grassland.

moorland *See* HEATH.

Moorland Association A membership organization representing the interests of *moorlands in England and Wales, with a stated objective to encourage the conservation of *heather moorland. It has a particular focus on the interests of those who own and manage *grouse moors.

(⊕) SEE WEB LINKS
• The website details its campaigning work and profiles its other activities.

mortar *See* CEMENT.

mortgage A **security interest** where one party (the **mortgagee**) makes a loan to another party (the **mortgagor**), usually for the payment of *interest and using the borrower's property as 'security' in return. If that loan is not being paid back when it should be, then the party loaning the money (usually a bank or building society) normally has the power to sell the property in order to recoup the money, plus any unpaid interest. In England, Wales, and Northern Ireland a mortgage will be recorded on a property's *title at the respective *Land Registry. Commonly, mortgages are used by people to buy their own home (which is therefore the 'security'); however, it is possible for people or businesses to borrow money for something other than the purchase of a home (say a new business enterprise) and for the bank or building society to use as security their home (assuming this itself is not heavily mortgaged) or another *asset. In Scotland, land law operates on different principles from those parts of the *United Kingdom operating under the English legal system, and in Scotland a **standard security** is broadly equivalent to a mortgage and accordingly is recorded on the *Land Register of Scotland.

mortgagee The party lending (usually money) to a *mortgagor. *See also* MORTGAGE.

mortgagor The party borrowing (usually money) from a *mortgagee. *See also* MORTGAGE.

mosaic virus One of a range of virus diseases of plants. Examples include tobacco mosaic virus and cucumber mosaic virus. Some can attack *cereals such as *barley yellow mosaic virus or soil-borne wheat mosaic virus. Control may be cultural by avoiding a build-up by rotation of crops or by controlling the vector of the virus, which is often an insect such as *aphids.

mouldboard The part of the *plough that inverts the soil.

mountain ash *See* TREE.

moveable property Property which can be moved (i.e. that which is not land or attached to land). The term is particularly used in Scottish law to describe all property other than *heritable property.

movement licence The licence required for the movement of any sheep, pigs, goats, or deer. To keep any farm animals, a farm holding must be registered and any movement must be recorded and be accompanied by the appropriate licence. This is a necessary precaution to prevent the spread of disease. The movement of horses and cattle is covered by the passport that each animal is required to have. In England, for example, the **Animal Reporting and Movements Service (ARAMS)** took over this function from *local authorities that still retain primary responsibility for enforcing the rules to protect animals during transportation, alongside veterinary inspectors from the *Animal and Plant Health Agency.

mow To cut grass or other crops, as in 'mow the lawn'. A mower is the implement used for mowing; it may be self-propelled, mounted on the *three-point hydraulic linkage of a tractor, or towed when it is driven by the tractor's *power take-off. At one time, mowers had reciprocating knives but are now mostly rotary with blades on a vertical spindle. *See also* FARM MACHINERY.

mowing licence *See* GRAZING/MOWING LICENCE.

MRICS *See* CHARTERED SURVEYOR.

muck heap A pile of *farmyard manure or other refuse. It is particularly common where animals, such as horses, are mucked out by hand so that the faeces together with the straw or sawdust bedding is removed from the stable or shed and stored in a heap before being applied to the soil. A large muck heap may be found in a field where the manure has been removed by mechanical means and piled up to rot down before subsequent spreading on the land.

muck spreader A farm implement for the spreading of *farmyard manure or muck. Like a trailer, it is towed behind a tractor and has one of two mechanisms. One works by moving chains taking the manure to the back of implement where it is spread by a single horizontal or several vertical rotating spindles with flails to aid the spread. The other is approximating to a cylindrical container with an opening on one side through which the muck is thrown by a horizontal rotating spindle with flails along the length of the machine. *See also* FARM MACHINERY.

mud Wet earth. *See also* SOIL.

mud fever A difficult and painful disease associated with a number of different causes, and affecting *horses' lower legs where the superficial layers of the skin of the legs become waterlogged and penetrated by bacteria. This is then encrusted with scabs which seal in the infection and can result in fever in severe cases. It can often be caused by wet and mud, and is normally more common in winter, though it can be more prevalent in some areas of the country than in others. In North America it is known as 'scratches'.

mulch A layer of material spread around newly planted trees to control weeds and to aid growing conditions. The process is described as **mulching** and can utilise thick polythene sheeting, old fertilizer bags, layers of bark, or manure.

mule 1. More usually in farming, it is a *crossbred *sheep, specifically a lowland flock ewe bred from an upland breed for hardiness and mothering qualities, and a lowland ram for prolificacy and carcass characteristics. The original mule is a cross between a Bluefaced Leicester ram on a Swaledale ewe, but other hill ewes have been used such as the Scottish Blackface or the Cheviot. The ram is usually a Bluefaced Leicester but is sometimes other breeds, notably the Border Leicester. *See also* SHEEP. **2.** The offspring of a male donkey and a female horse.

multifunctional countryside The provision of a number of benefits or functions from the *countryside, commonly in addition to the usually dominant *agriculture and its farming activities. These can include, for example, beautiful *landscapes, clean water, recreation including health and well-being services, abundant *wildlife, better *flood management, trees, and improved carbon storage. This is linked with trends towards a multifunctional nature of farming itself, and the broader range of policy measures to encourage a more diverse sector. *See also* DIVERSIFICATION; ECOSYSTEM SERVICES.

mustard A plant of the Brassica family grown for its seed, which is used as a condiment for food. However, it is grown more frequently as a *cover crop or green manure. It is very fast growing, and can be sown after the harvest of a *cash crop or even *broadcast into it before *harvest. It will quickly cover the ground, thus acting as a weed suppressant, will help to hold the *soil together to prevent soil erosion, and will take up nutrients, nitrates in particular, and thus minimize *leaching. As such, it may be grown alone or as a component of a mixture.

mutton The meat of mature sheep. A young sheep is known as a *lamb until it is a year old, and its meat is also described as such. A *hogget is a sheep between one and two years old, and the meat is also so termed. The meat from more mature sheep, commonly three to four years, is known as mutton and may be stronger in flavour. It may also be tougher so may be cooked for longer. Once very common, it is uncommon to find mutton for sale in certain countries, including the UK, as it is usually considered to be cheaper and of less eating quality than lamb. Sometimes, the age distinction may become blurred in the marketing of the meat.

myxomatosis Frequently shortened to 'myxy', it is a disease caused by the myxoma virus and is spread by blood-sucking insects. It affects only rabbits. It was introduced into Australia in 1951 in an effort to control its large rabbit populations, and within a few years had appeared in Europe, including the UK, where it has since spread widely. Infected rabbits develop skin tumours and blindness, and usually die within fourteen days. Hares are not normally susceptible to the disease, but can act as a vector.

National Assembly for Wales Elected assembly in Wales which is authorized by the United Kingdom Parliament in Westminster, under *devolution powers, to make laws relating to Wales. *See also* WELSH GOVERNMENT.

National Character Area (NCA) A provision of natural areas dividing England into 159 NCAs, each an appraisal of its distinctive *landscape, *biodiversity, geology, history, and socio-economic context. They are defined by *Natural England and used by its advisory staff as well as by the public and private sector in the planning process. Their main use is to provide a national resource to inform local planning policies, landscape assessments, and further opportunities to improve the natural environment. The NCAs more often overlap administrative boundaries and can provide a framework to facilitate cross-boundary working.

(⊕) SEE WEB LINKS
• Website outlining each numbered and mapped NCA, with linked details.

National Ecosystem Assessment (NEA) A comprehensive overview of the natural environment in the UK published in 2011, and providing a basis for estimating the value of natural resources. The UK NEA recognized four categories of *ecosystem services: provisioning, regulating, supporting, and cultural. These services include, for example, clean air, clean water, flood control, food, fibre, and natural resources. The categories mirrored the divisions within the **Millennium Ecosystem Assessment** (MA), a document published by the United Nations in 2005 providing a comprehensive appraisal of the global ecosystems and the services that they provide.

national envelope The amount of direct agricultural support payments allocated to farmers in each *European Union member state, also referred to as the 'financial envelope'. In order to promote greater subsidiarity, specific proportions of the national envelope can be transferred to various direct payments and to their individual *Rural Development Programmes.
See also COMMON AGRICULTURAL POLICY; GREENING MEASURES; MODULATION.

National Farmers' Union A membership organization founded in 1908 that looks after the interests of farmers and growers in England and Wales. It has around 55,000 members covering two thirds of the farmland in England and Wales. It is a political lobby representing the interests of its members at national, international, and local level, and also offers help and advice to members on issues that affect agriculture and horticulture.

(⊕) SEE WEB LINKS
• Website of the NFU, containing information on the structure and activities of the organization at national and local levels.

National Forest The National Forest Company is a government-led initiative set up in the early 1990s with an aim for significant woodland creation on an area of 200 square miles in the centre of England. The aims included linking existing *Ancient Woodland with the overall intent to promote *multifunctional woodland. *See also* COMMUNITY FOREST.

(⊕) SEE WEB LINKS
• Website of the National Forest, detailing its role, activities, and achievements.

National Gamekeepers' Organisation (NGO) The body that represents and assists the *gamekeepers of England and Wales, although it aims to promote gamekeeping UK-wide. It is particularly concerned with raising standards throughout the profession, and is also concerned with promoting *field sports in general. The interests of gamekeepers in Scotland are represented by the **Scottish Gamekeepers Association**.

(⊕) SEE WEB LINKS
• NGO website, including information on events and training programmes.

National Heritage List for England Official register of *listed buildings, *scheduled monuments, protected wrecks, *registered parks and gardens, and *registered battlefields in England.

(⊕) SEE WEB LINKS
• Website of the National Heritage List for England.

National Insurance Payments made by workers, and their employers, to government in the *United Kingdom in order to enable the worker to access government-funded benefits and pensions such as the State Pension and unemployment benefits. The rates of National Insurance contributions made by an individual (and their employer) are set by government as a percentage of the individual's salary. National Insurance payments are compulsory and are made in addition to *income tax. Individuals build up an entitlement to government payments such as the State Pension by contributing National Insurance payments over their working career. For example, an individual paying National Insurance for only fifteen years would have a lower level of State Pension on retirement than someone paying National Insurance for their whole career.

National Living Wage *See* LIVING WAGE; MINIMUM WAGE.

National Minimum Wage *See* MINIMUM WAGE.

National Monuments Record of Wales National collection of information about the historic environment in Wales.

(⊕) SEE WEB LINKS
• Coflein website: online database of the National Monuments Record of Wales, the national collection of information about the historic environment in Wales.

National Nature Reserve (NNR) Protected areas of some of the most important natural and semi-natural terrestrial and coastal ecosystems in Great Britain. In addition to management to benefit the site's particular *biodiversity, they can also provide special opportunities for scientific study, and provide appropriate public recreation. In the UK, NNRs are declared by the statutory country conservation agencies under the National Parks and Access to the Countryside Act 1949 and the Wildlife and Countryside Act 1981. In Northern Ireland, Nature Reserves are designated under the Amenity Lands Act (Northern Ireland) 1965. *See also* SITES OF SPECIAL SCIENTIFIC INTEREST; SPECIAL AREAS OF CONSERVATION; SPECIAL PROTECTION AREAS.

National Park A large designated area of natural, *semi-natural,or developed land. At an international level there is a range of different National Park categories to reflect aims and levels of protection. The fifteen National Parks of the UK are described as a Category V for areas managed mainly for *landscape protection and recreation. All of these National Parks are farmed at relatively low levels of intensity, and many are in the *uplands. *See also* AREA OF OUTSTANDING NATURAL BEAUTY; CONSERVATION BOARD; NATIONAL PARK AUTHORITY.

(⊕) SEE WEB LINKS
• Website providing a visitors' guide and some details of the fifteen National Parks in the UK.

National Park Authority (NPA) Organization with responsibility for the running of their respective *National Park. Each NPA in England and Wales is overseen by a Board comprising a mix of locally and nationally appointed members, and is guided by the responsibilities set out under the *National Parks and Access to the Countryside 1949 and the *Countryside and Rights of Way Act. These responsibilities include the conservation and enhancement of the natural beauty, wildlife, and cultural landscape, and the promotion of public understanding and enjoyment of their special qualities. The **Association of National Park Authorities** (ANPA) provides the NPAs of England, Wales, and Scotland a focus for collaboration. *See also* CONSERVATION BOARD; PLANNING AUTHORITY.

National Parks and Access to the Countryside Act 1949 The UK Act that created the National Parks Commission that evolved and merged ultimately into the *Countryside Agency and finally into *Natural England. It established powers to notify *Sites of Special Scientific Interest, declare *National Nature Reserves, and to allow *local authorities to establish *Local Nature Reserves. The Act also provided the framework to create *Areas of Outstanding Natural Beauty.

National Planning Framework (Scotland) *See* PLANNING POLICIES.

National Planning Policy Framework (England) *See* PLANNING POLICIES.

National Policy Statements (England) *See* PLANNING POLICIES.

National Scenic Area Designated areas created to give special protection to some of the best *landscapes in Scotland. Protection is mainly through the special development control measures, and planning policies for NSAs. NSAs are broadly equivalent to the *Areas of Outstanding Natural Beauty in England and Wales. *See also* NATIONAL PARK.

National Trail These are normally *footpaths that have the quality and character to be described as national. They allow long-distance journeys on foot, horseback, or bicycle, and are of significance to attract domestic and overseas tourists. There are fifteen National Trails in England and Wales, with equivalent trails in Scotland called **Scotland's Great Trails**. *See also* PUBLIC RIGHTS OF WAY.

(((()))) SEE WEB LINKS
• The National Trail website, providing detailed information on the routes.

National Trust The full but not commonly used title is the 'National Trust for Places of Historic Interest and Natural Beauty'. This better defines its purpose as the largest conservation organization, with a membership of over four million in England, Wales, and Northern Ireland, and with a separate and analogous organization operating in Scotland, the **National Trust for Scotland**. In addition to ownership and management of *heritage properties, the National Trust is one of the largest *landowners in the UK, many of these areas being in marginal *uplands, coastal, and other scenic areas.

National Union of Agricultural and Allied Workers The trade union of agricultural and allied workers founded in 1906 to represent its members. In 1982 it became the Agricultural Section of the Transport and General Workers' Union, which in 2007 amalgamated with another union to become Unite.

National Vegetation Classification A set of common standards to produce a systematic description and classification of GB plant communities which cover all natural and semi-natural habitats. It is maintained by the *Joint Nature Conservation Committee, which is significant because the NVC is used to apply aspects of statutory national and international site designations. *See also* SITES OF SPECIAL SCIENTIFIC INTEREST; SPECIAL AREAS OF CONSERVATION; SPECIAL PROTECTION AREAS.

natural capital The stock of biodiversity, soil, air, water, and all natural assets contained within, for example, a particular area, *catchment, or *habitat. This is primarily an economic definition to describe the derived benefits that humans obtain from natural capital, i.e. *ecosystem services.

Natural England (NE) Officially classified as non-departmental public body and under the responsibility of the *Department for Environment, Food and Rural Affairs (DEFRA), it is the government adviser for the natural environment in England, and its responsibilities include the promotion and protection of *biodiversity, *landscape, and *public access to the countryside. Its many functions include implementation of wildlife legislation and the designation of protected areas. It has direct or indirect responsibilities for the management of *Sites of Special Scientific Interest, *National Nature Reserves, *Special Areas of Conservation, and *Special Protection Areas. Its commonest association with the farming industry in England is via its delivery of *agri-environment schemes. Natural England was formed by the amalgamation of separate countryside agencies, including **English Nature**, the **Countryside Agency**, and the **Rural Development Service**. *See also* JOINT NATURE CONSERVATION COMMITTEE; NATURAL RESOURCES WALES; SCOTTISH NATURAL HERITAGE; NORTHERN IRELAND ENVIRONMENT AGENCY.

(((●))) SEE WEB LINKS
• Natural England website, providing detailed information on each of its listed main areas of responsibility.

natural environment Encompasses the natural occurrence of all living and non-living things, usually in reference to Earth or a defined part of it. Purely defined natural environments are becoming less common, and degrees of naturalness vary with minimal human modifications at one extreme, to agricultural *landscapes and the man-made surroundings of the **built environment** at the other.

natural heritage Encompasses the elements of *biodiversity and physical formations, usually within defined areas that can be designated sites or the wider countryside. The term 'heritage' describes what is inherited from past generations, maintained and passed on to future generations. Natural heritage can be used where its meaning is more easily understood and less scientific.

natural regeneration A method of restocking trees that is reliant on natural *seed germination and/or *coppice. This method is normally less costly than *planting of *transplants, and is a preferred choice for promoting wildlife and other environmental benefits. *See also* WILDWOOD; MAST.

natural resources Materials naturally occurring in the *environment and without intervention by humans. Ubiquitous natural resources, i.e. found everywhere, will include, for example, air or sunlight. Many natural resources are localized; for example, coal or fresh water. It can also be considered divided between *biotic or *abiotic resources. *See also* BIODIVERSITY; NATURAL CAPITAL.

Natural Resources Wales (NRW) A statutory body under the responsibility of the *Welsh Government, with roles that include regulatory and advisory functions on a wide range of environmental and natural resources issues. It was formed by the amalgamation of separate countryside agencies including

Countryside Council for Wales, Environment Agency Wales, and Forestry Commission Wales. *See also* GLASTIR; NATURAL ENGLAND; SCOTTISH NATURAL HERITAGE.

SEE WEB LINKS
• Website of Natural Resources Wales, detailing individual roles and operational functions.

nature conservation *See* CONSERVATION.

Nature Improvement Areas (NIA) A pilot initiative to create *ecological networks and connecting and creating *wildlife sites at a *landscape scale, and run by private and public partnerships. Funding from the *Department for Environment, Food and Rural Affairs (DEFRA) has now ceased, however, comparable initiatives and versions of local NIAs continue in existence.

navy bean (*Phaseolus vulgaris*) A small white bean also known as a **haricot bean** eaten as a vegetable. It is widely grown in the United States but not in the UK and is commonly processed and sold as **baked beans**.

NCA *See* NATIONAL CHARACTER AREA.

NEA *See* NATIONAL ECOSYSTEM ASSESSMENT.

negligence Situation when one party has, by act or omission, acted so carelessly that they have breached their *duty of care towards another party which has resulted in that party sustaining injury or loss. In general terms, a person is considered to be negligent if they have not acted in a way that it would be expected that a reasonable person would have acted. In the situation of a professional acting within the scope of their professional expertise, the courts expect them to act as a reasonably competent member of their profession—if not, they may be found to be negligent. The *tort (in English law; law of **delict** in Scots law) of negligence allows a party to go to the civil law *courts to claim *compensation (*damages) against another party who they consider has breached their duty of care towards them (even if there was no contractual relationship between the parties). *See also* PROFESSIONAL INDEMNITY INSURANCE.

neighbour consultation scheme *See* PERMITTED DEVELOPMENT.

neighbourhood area Geographic area, often a single parish, in England, which is the subject of a *neighbourhood plan and/or *neighbourhood development order. *See also* NEIGHBOURHOOD PLANNING.

neighbourhood development order Type of *permitted development order which relates to a designated *neighbourhood area in England (under the *neighbourhood planning regime). They are proposed by the relevant town or *parish council, or neighbourhood forum, and are made by the *local planning authority following approval in a referendum. They grant blanket *planning permission for certain types of development within the neighbourhood area.

neighbourhood forum *See* NEIGHBOURHOOD PLANNING.

neighbourhood plan Document sitting underneath a *local plan and comprising a set of *planning policies for a *neighbourhood area. It forms one element of the *neighbourhood planning rights system in England.

neighbourhood planning Rights introduced as part of the *localism agenda, and through the **Localism Act 2011**, which allow authorized community organizations in England (typically *parish councils and *town councils in rural areas) to draw up *planning policies and *permitted development rights for their locality, which, subject to approval in a local *referendum, become a formal part of the planning and *development control system. In practice, the rights exist in three forms, and a local community can exercise its rights to use any or all of them, individually or together:

- *Neighbourhood plans comprise a set of planning policies for the neighbourhood area. These policies must align with the *local plan, and be positive (i.e. specifying what development is to be permitted, rather than restricted). Applications for *planning permission in the neighbourhood area must then be decided in accordance with the neighbourhood plan and any other *material considerations.
- *Neighbourhood development orders are a local form of *permitted development order which can provide blanket planning permission for specified types of development within the neighbourhood area.
- *Community right to build orders operate on a similar basis to neighbourhood development orders, with the main difference being that they can be proposed by community organizations (whereas neighbourhood development orders must be proposed by the official neighbourhood planning body).

There are a number of different stages if a community wishes to exercise its neighbourhood planning rights:

- The *local planning authority must agree a **neighbourhood area**; in rural areas this is typically a single parish or town, though it may incorporate a number of parishes or towns. This area must be proposed by an official neighbourhood planning qualifying body; where they exist, this will be the parish council or town council. In areas without a parish council or town council, another body (for example, a residents' association) may apply to be officially recognized as a **neighbourhood forum** which is able to exercise the neighbourhood planning rights.
- The official neighbourhood planning body is then able to prepare a draft neighbourhood plan and/or permitted development right order. The process of doing this must involve consultation with the community, and the final documents must align with other local and national planning policies and be underpinned by evidence.
- The documents are then subject to review by an **independent examiner** who assesses whether they meet the relevant legal requirements (known as the **basic conditions**).
- Once the independent examiner is satisfied, the local authority then organizes a *referendum, during which the electorate in the neighbourhood area are asked whether they wish to approve the documents.

- If a majority of the votes received approve the plan or order, then the local planning authority make the plan or order and it becomes an official part of the planning system.

nematode A type of *roundworm, nematodes are the most numerous multicellular animals on earth. There is a very wide range, many living as soil microorganisms, some of which are beneficial to agriculture and others harmful. Amongst the former are species that can be used as biological control of a number of pests, such as slugs and **leatherjackets**. Amongst the latter are plant *parasites that can cause significant crop loss, particularly in vegetable crops. An example is the cereal cyst nematode. Control can be biological—for example, some brassicas produce chemicals that suppress nematodes—or chemical, by the use of a nematacide. *See also* PESTICIDES.

nest box A supplement for hole-nesting birds where there is a shortage of naturally occurring trees with suitable holes and cavities. Normally of wooden construction which does not produce condensation and can blend in, and placed in larger trees or a suitably located structure.

net present value *See* PAYBACK PERIOD.

Newcastle disease A highly contagious disease of birds caused by the virus *Avian Paramyxovirus*, also known simply as Newcastle disease virus. Symptoms include respiratory distress, tremors or paralysis, diarrhoea, lack of appetite, and loss of productivity. Different strains of the virus cause disease of varying severity, but the mortality rate is high in severe cases. It is spread by direct contact with the bodily fluids of infected birds, especially faeces, and indirectly by movement of people via their clothing, footwear, or vehicles. Wild birds, such as pigeons, can spread it to poultry units, where it may have severe consequences. It is a *notifiable disease, which means that it must be reported to the appropriate authorities, which may then slaughter all susceptible animals that may have come into contact, and impose movement restrictions over a wider area. The carcasses are incinerated, rendered, or buried, and the premises thoroughly disinfected. Prevention can be achieved by vaccination and by good *biosecurity.

New Forest pony *See* HORSE.

New Zealand rug A hardy rug designed to keep *horses or ponies warm and dry when turned out or living out in the winter. They are particularly useful for *clipped horses. Rugs are made of lined waterproof canvas or man-made fibre, and are attached by a variety of fittings and leg straps. The canvas type is heavier and is more robust, but the lighter nylon rugs are less likely to rub or rip.

NFFO *See* NON FOSSIL FUEL OBLIGATION.

NFU *See* NATIONAL FARMERS' UNION.

NGO *See* NATIONAL GAMEKEEPERS' ORGANISATION.

niche market A group of potential customers sharing a very specific set of characteristics and needs, who are therefore likely to find a particular product, service, or brand attractive, and can consequently be targeted in a marketing strategy. Where an organization uses a *segmentation approach to its marketing, a niche market will often be a single segment. *See also* DIFFERENTIATION.

night storage heaters Heaters which contain firebricks which are heated up (usually overnight) by an internal electrical heater, and then release the heat stored in the bricks throughout the day. *See also* DRY HEATING SYSTEM.

NIMBY Acronym of 'not in my back yard'. It is commonly used to refer to a person or group who objects to a new building or installation of some sort in their immediate neighbourhood, but does not object to comparable developments in another location. Nimbys can be local activists protecting local environments or communities, but the term is more usually used in an almost pejorative sense. *See also* CAMPAIGN FOR THE PROTECTION OF RURAL ENGLAND; PLANNING.

nitrate Salts of nitric acid. Nitrate is a major *nutrient for plant growth and is applied as a *fertilizer. *Ammonium nitrate manufactured in prills is the most common form, but sodium, potassium, and calcium nitrates are also used. *Urea is a carbamide rather than a nitrate, but may be applied as a fertilizer as it is broken down into nitrate in the soil. Nitrate is highly soluble, and thus, if not taken up by the plant, may be *leached from the soil in rainwater drainage. The presence of nitrate in *watercourses has led to the introduction of *Nitrate Vulnerable Zones.

Nitrates Directive A European Council Directive (91/676/EEC) to promote the use of *good agricultural practice for the protection of waters against pollution caused by nitrates from agricultural sources. The Directive set limits for the permissible concentration of nitrates in water, and allowed EU member states to target management measures countrywide or in areas designated as *Nitrate Vulnerable Zones. *See also* CATCHMENT SENSITIVE FARMING; WATER FRAMEWORK DIRECTIVE.

Nitrate Vulnerable Zones Areas of land that are designated as vulnerable to agricultural *nitrate pollution, and are a requirement in the UK in order to comply with the European Commission *Nitrates Directive. There are a number of rules and management activities that farmers in NVZs have to implement, which are also further re-enforced by *cross-compliance regulations. Previously to NVZs, Nitrate Sensitive Areas were introduced as part of a now ended *agri-environment measure, also aiming to reduce nitrate levels on targeted farmland.

(⊕) SEE WEB LINKS
• Government webpage detailing area maps and management requirements.

nitrogen A gas that forms 78% of the air by volume. It is a major *nutrient of plant growth in the form of *nitrate. Nitrogen fixation, nitrification, or

mineralization is the process whereby microorganisms convert nitrogen into
*ammonia which can then be taken up by plants. These microorganisms may be
free living or may be bacteria of the *rihizobium* genus which live in a symbiotic
relationship in nodules on the roots of leguminous plants such as *clovers or
pulses. Nitrogen may be lost from soils either by the *leaching of nitrate or by
denitrification, the opposite process to nitrification whereby nitrogen
compounds are broken down into gaseous nitrogen and lost to the air.

nitrogen cycle The cycle in which nitrogen is converted into various forms in
the atmosphere, in soils, and in water. There are five processes in the cycle:

- Nitrogen fixation, where atmospheric nitrogen is turned into ammonium
  compounds by the action of rhizobium bacteria in the root nodules of
  *leguminous plants.
- Nitrogen uptake by the growth of organisms; for example, the ammonium
  compounds absorbed by the roots of the plants in which the nitrogen fixation
  has taken place.
- Nitrogen mineralization, when organisms containing nitrogen compounds die
  and decompose, resulting in ammonium compounds.
- Nitrification is the process by which bacteria turn ammonium compounds into
  *nitrate. Ammonium ions are positively charged, and thus stick to *clay
  particles, but nitrate ions are negatively charged and can be more readily
  *leached from the soil.
- Denitrification is the process by which bacteria convert nitrate and nitrite into
  atmospheric nitrogen.

See also NITRATE VULNERABLE ZONES.

NNR *See* NATIONAL NATURE RESERVE.

nodule In general terms an abnormal growth, but in an agricultural context the
swellings on the roots of *leguminous plants in which live the *Rhizobium* bacteria
that fix atmospheric nitrogen, turning it into *nitrate that is available as a
*nutrient for the crop.

Non Fossil Fuel Obligation (NFFO) Former *renewable energy support
scheme that operated in England, Wales, and Northern Ireland, and was
designed to incentivize the increased production of renewable electricity. It
operated alongside the equivalent **Scottish Renewable Obligation**, and both
schemes were replaced by the *Renewables Obligation.

non-intervention A management term more frequently directed at *woodland;
it reflects a purposeful policy of no management to allow the *habitat to develop
with as minimal a level of human interference as possible. *See also* WILDWOOD.

non-market goods A particular reference to environmental goods or services
that are not easily traded in markets, and hence their economic values are not
easily identified. Such goods include *wildlife, *landscapes, and clean air and
water. Attempts to value non-market goods is an aim within the understanding

and development of *ecosystem services. *See also* NATURAL CAPITAL; VALUATION; VALUATION METHODS.

non-native species Plant and animal species, which can also be described as **alien species**, that have been brought into a country deliberately or by accident, and which otherwise would have been prevented by natural barriers. This might be particularly relevant to an island country such as the UK. A recognized difficulty is the categorization of species that have become long established to the point of being considered by some as native, such as the rabbit, believed to have been introduced into Britain by the Romans. Species that have or cause significant adverse impacts on, for example, native species or economically are referred to as **invasive non-native species**.

non-renewable Term used to describe a natural resource which, if used or lost, is, in effect, not replaceable because it develops as part of a cycle which is so long it is outside the time frame of human comprehension (one example being *fossil fuels, which develop as part of a cycle lasting millions of years). This contrasts with *renewable materials that have a much shorter growing cycle (such as timber), so, if used or lost, can be replaced through the growth of new trees relatively quickly. *See also* SUSTAINABLE DEVELOPMENT.

Northern Ireland Assembly Elected assembly in Northern Ireland which is authorized by the United Kingdom Parliament in Westminster, under *devolution powers, to make laws relating to Northern Ireland. *See also* NORTHERN IRELAND EXECUTIVE.

Northern Ireland Buildings Database Online database, managed by the *Northern Ireland Executive, detailing *listed buildings or those that have been surveyed with a view to potentially being listed. *See also* HISTORIC ENVIRONMENT RECORDS.

(((⊕))) SEE WEB LINKS
• Website of the Northern Ireland Buildings Database.
• Website of the Northern Ireland Sites and Monuments Records, managed by the Northern Ireland Executive.

Northern Ireland Countryside Management Scheme *See* AGRI-ENVIRONMENT SCHEME.

Northern Ireland Environment Agency Northern Irish government agency responsible for the natural environment within Northern Ireland. Its responsibilities include environmental regulation, water and waste management, the marine environment, *biodiversity and *conservation, and landscape. *See also* ENVIRONMENT AGENCY; SCOTTISH ENVIRONMENT PROTECTION AGENCY; NATURAL RESOURCES WALES.

Northern Ireland Executive Executive of the *devolved government for Northern Ireland, exercising authority on behalf of the **Northern Ireland Assembly**. The Northern Ireland Executive comprises a First Minister and

Deputy First Minister and other Ministers, each of whom is nominated to take charge of a particular government department. The Ministers are nominated by political parties in accordance with the proportion of seats that they hold in the Northern Ireland Assembly (with the exception of the Minister of Justice, who is appointed by a cross-community Assembly vote). The Northern Ireland Assembly is the elected body that makes laws for Northern Ireland in devolved areas, and holds the Northern Ireland Executive to account. Elected members of the Assembly are identified by the letters MLA after their name (i.e. Member of the Legislative Assembly). The Northern Ireland Assembly has the right to pass laws in a number of areas including *agriculture; health and social services; education; employment and skills; housing; economic development; local government; environmental issues including *planning; transport; culture and sport; justice and policing; social security and pensions; and child support. For those areas where the Northern Ireland Executive does not have the right to make laws, the United Kingdom Parliament and Government is the main national law-making body.

Northern Irish residential tenancies Forms of *tenancy for occupation of rented residential property in Northern Ireland (referred to as **private tenancies**).

- **Protected** or **statutory tenancies** (also formerly referred to as **regulated** or **restricted** tenancies). These mainly commenced prior to 2007. They are referred to as protected tenancies when the original *contract of the tenancy is still in operation, and statutory tenancies if the original contract has ended (possibly because it was for a *fixed term, or alternatively, the original tenant has died and the tenancy has passed on by *succession). Protected or statutory tenancies are subject to *rent control, with the rent generally being set by the *Rent Officer. The tenants also have *security of tenure, with the landlord only able to obtain possession via a *court order if one of the *grounds for possession apply. There are also *succession rights.
- **Other tenancies** are mainly a matter for agreement between the landlord and tenant. The landlord does not need to demonstrate grounds for possession to regain *vacant possession of the property at, or after, the end of the fixed term, or after the first six months of a *periodic tenancy, and provisions, for example, for *rent review are agreed within the tenancy agreement. All tenants, however, have a legal right to a *rent book.

notice to quit Notice, served by either a *landlord or a *tenant, on the other party giving notice that they wish to end a *tenancy on a particular date.

notice to remedy breach Notice issued by one party to a *contract to the other party to the contract informing them that they must fix a *breach of contract or *breach of covenant.

notice to treat *See* COMPULSORY PURCHASE.

notifiable disease Any contagious disease of animals that the authorities in many countries determine must be reported so that control or eradication can take place. In the UK, such reports must be made to the Animal and Plant Health Agency, an agency of the *Department for Environment, Food and Rural Affairs, which will then take the appropriate action, including movement restrictions. For some diseases this will be a compulsory slaughter policy for which compensation may be paid. In the UK there are around forty notifiable diseases of animals, including *foot and mouth disease, *bovine tuberculosis, *Newcastle disease, avian flu, and rabies.

NPPF (National Planning Policy Framework) *See* PLANNING POLICIES.

NRW *See* NATURAL RESOURCES WALES.

nuisance *See* PRIVATE NUISANCE; STATUTORY NUISANCE.

nursery A business or site devoted to the production of plants. The nursery trade expanded from the early eighteenth century onwards to supply forestry plants as well as the significant demand for woody shrubs, mainly hawthorn, for the surge in new *hedgerow planting. Nurseries supplying forestry plants are normally separate commercial enterprises, but some estates also operate their own nurseries. *See also* TRANSPLANTS; WHIPS; SAPLINGS; STANDARDS.

nurse species Used as part of a mix of tree *planting to promote the growth of the desired main crop. Nurse species or nurse **crop** grow faster and can help to suppress ground vegetation and protect other species from extremes of weather. Often, *conifers may be used to promote height and straighter stems of planted *broadleaves. In agriculture, some *annual crops such as *oats can be used to aid establishment of *perennial crops by suppressing *weed growth or by providing protection from wind or *erosion.

nutrient Foodstuff that plants need for survival. There are two main types of nutrient groupings: macro and micro. Macro nutrients are needed by the plant in large amounts, these being *nitrogen, *phosphate, potassium, and sulphur. Micro nutrients are needed by the plant in very small amounts, examples of these being manganese, copper, and boron. Macro nutrients are usually supplied to the plant via *fertilizer products that are broadcast onto the growing *crop. Micro nutrients are generally mixed with *water and are applied as a foliar application via a crop *sprayer. *See also* FARMYARD MANURE; NITROGEN; SOIL.

NVC *See* NATIONAL VEGETATION CLASSIFICATION.

NVZ *See* NITRATE VULNERABLE ZONE.

oak (English or pedunculate, sessile, Turkey) *See* TREE.

oats (*Avena sativa*) A *cereal grown for its *grain, used for human consumption and as an ingredient of animal feeds; horses in particular. It may be sown in autumn or spring and harvested in late summer. Described as a health food, oats have high levels of certain nutrients such as magnesium, and the fibre content slows down digestion, preventing spikes in blood sugar levels. Oatmeal or rolled oats are eaten in many forms, such as porridge or biscuits.

obligation A course of action or a duty (either to do, or not to do, something) by which a party is legally bound. Obligations forming part of a *contract are known as **contractual obligations**. In the context of a *lease they are also sometimes referred to as *covenants. *See also* SECTION 106 PLANNING OBLIGATION AGREEMENT.

obsolescence The features or condition of a piece of equipment or a property (say land and buildings) which mean that it may not be as well suited for current uses as a new piece of equipment or property. For example, the layout of a Victorian university building may not be as well suited to house modern laboratories as a new purpose-built building; therefore there is said to be a degree of obsolescence in the Victorian building. Similarly, an old building may not be as energy efficient as newly built equivalents. Obsolescence may have some impacts on the value of properties (as they may be less attractive to purchasers); therefore when modern buildings are being designed one of the objectives of the architects is often to allow the building layouts to be easily modifiable in future, should they need to be, in order to try to minimize future obsolescence.

occupier's loss payment *See* COMPULSORY PURCHASE.

oestrus A regularly recurring period of sexual receptivity in female mammals when ovulation takes place and intercourse may result in fertilization of the ovum or ova and thus pregnancy. At this time, the female is said to be in season or on heat.

oilseed A *crop that is grown for the oil within its seed. The main oilseed crops are *soya, *oilseed rape, sunflowers, and **olives**. The oil content varies from around 20% for soya beans to over 40% for sunflowers and oilseed rape.

oilseed rape (*Brassica napus*) A plant of the *brassica family grown for the oil content within its seed. It can be sown in autumn or spring, but most is planted in late August in Europe and harvested in July. The oil is used mostly in cooking, though there are some industrial uses. It is also processed for use as a *biodiesel. The meal left after the oil has been extracted is used as an animal feed. *See also* CROPS.

on-the-hoof Literally of livestock still standing, i.e. yet to be slaughtered. The distinction is similar to the terms *liveweight and *deadweight. From this, it has come to mean taking action whilst on the move or without adequate preparation.

open access Within the context of land use, this normally refers to a type of public access to specific categories of land in England and Wales. This was enforced by the *Countryside and Rights of Way Act 2000, and although initially resisted by many *landowners it has now become embedded. This right to access is also commonly referred to as 'right to roam' or 'freedom to roam'. The land types include those categorized as mountain, *moor, *heath, *downland, and registered *common land. Rights are restricted to certain uses and do not, for example, include horse riding, cycling, or camping. Landowners are permitted to apply for certain temporary exemptions to restrict or close access. *See also* PUBLIC RIGHTS OF WAY.

(⊕) SEE WEB LINKS
• Government website detailing types, rights, and responsibilities, of public access in the UK.

open season The periods of the year when *shooting, *angling, and *hunting are permitted. The system of seasons around individual *game species has developed over time in a mixture of complex traditions and laws. Variations will occur between localities and species, and are also commonly restricted on Sundays and Christmas Day. Most rules are legislated, but others are a matter of custom. For example, there is a prescribed *fox hunting season, but it is legal to kill a fox throughout the year as it is classed as vermin. In the UK, for example *rabbits and *pigeons and other species of vermin can be shot or otherwise killed the whole year round. In other words there is an open season on these species. Many other species of *game, *deer, and fish that are permitted to be shot or fished have controlled periods of the year when they are protected, i.e. the *close season. This is to allow sustainable populations to breed and flourish. (See tables on pp. 296 and 297 for open season dates for deer, and game and waterfowl in the UK.)

(⊕) SEE WEB LINKS
• Government website guidance on wild bird protection and licences.

opportunity cost The costs attached to undertaking one course of action as a consequence of losing the opportunity to gain the benefits of taking another course of action. In financial terms, this can be measured as the financial cost attached to using money for a particular project, thereby meaning that it cannot be used to generate income another way. An example would be if a landowner

Open season for deer in the UK

Species/sex	England and Wales	Scotland	Northern Ireland
Red			
Stag	1 Aug to 30 April	1 July to 20 Oct	1 Aug to 30 April
Hind	1 Nov to 31 Mar	21 Oct to 15 Feb	1 Nov to 31 Mar
Fallow			
Buck	1 Aug to 30 April	1 Aug to 30 April	1 Aug to 30 April
Doe	1 Nov to 31 Mar	21 Oct to 15 Feb	1 Nov to 31 Mar
Roe			
Buck	1 April to 31 Oct	1 April to 20 Oct	
Doe	1 Nov to 31 Mar	21 Oct to 31 Mar	
Sika			
Stag	1 Aug to 30 April	1 July to 20 Oct	1 Aug to 30 April
Hind	1 Nov to 31 Mar	21 Oct to 15 Feb	1 Nov to 31 Mar
Red/Sika hybrid			
Stag	1 Aug to 30 April	1 July to 20 Oct	1 Aug to 30 April
Hind	1 Nov to 31 Mar	21 Oct to 15 Feb	1 Nov to 31 Mar
Chinese Water Deer			
Buck	1 Nov to 31 Mar		
Doe	1 Nov to 31 Mar		
Muntjac	Found only in England and Wales. There is no closed season for this species, but it is recommended that when culling females, immature or heavily pregnant does are selected to avoid leaving dependent young.		

had funded a *development project through using their own cash instead of borrowing the money to undertake the work. Although the landowner would not be paying interest on a loan, they would also not be receiving the income (in the form of interest) that they would have had had they left their money in the bank. Hence there is an opportunity cost to using their own money to finance the project.

orchard A purposively planted area of *trees grown and maintained for fruit or nut production. The earliest planting of trees was probably in orchards in the Stone Age, but they were often planted in the Middle Ages.

order *See* COURT ORDER; DELEGATED LEGISLATION.

organic Relating to or derived from living matter. In *agriculture it describes a system where manufactured compounds such as inorganic *fertilizers or

Open season for game and waterfowl in the UK

Species	England and Wales	Scotland	Northern Ireland
Pheasant	1 Oct to 1 Feb	1 Oct to 1 Feb	1 Oct to 31 Jan
Partridge	1 Sept to 1 Feb	1 Sept to 1 Feb	1 Sept to 31 Jan
Red grouse	12 Aug to 10 Dec	12 Aug to 10 Dec	12 Aug to 30 Nov
Black grouse	20 Aug to 10 Dec Somerset, Devon, New Forest 1 Sept to 10 Dec	20 Aug to 10 Dec	
Ptarmigan		12 Aug to 10 Dec	
Duck and goose inland	1 Sept to 31 Jan	1 Sept to 31 Jan	1 Sept to 31 Jan
Duck and goose below mean high water	1 Sept to 20 Feb	1 Sept to 20 Feb	1 Sept to 31 Jan
Common snipe	12 Aug to 31 Jan	12 Aug to 31 Jan	1 Sept to 31 Jan
Jack snipe	Protected	Protected	1 Sept to 31 Jan
Woodcock	1 Oct to 31 Jan	1 Sept to 31 Jan	1 Oct to 31 Jan
Golden plover	1 Sept to 31 Jan	1 Sept to 31 Jan	1 Sept to 31 Jan
Coot/moorhen	1 Sept to 31 Jan	1 Sept to 31 Jan	Protected
Brown hare	There is no closed season for hares in England and Wales, though there has been a campaign to introduce one. However, it is illegal to sell hares for meat during the breeding season between 1 March and 31 July. In Scotland the open season for hares is 1 October to 31 January, and in Northern Ireland it is 12 August to 31 January.		

*pesticides are not used. Organic farms are usually mixed, having arable and *livestock enterprises. *Soil fertility is maintained or enhanced by a *rotation of crops that may include *legumes, such as *clovers or pulses, grazing livestock, and *farmyard manure, *compost, or other organic fertilizers. Certain naturally occurring products may be used, and pests controlled by cultivation or biological means. Organic farms may be certified by a number of organizations such as the *Soil Association.

Organic Entry Level Stewardship *See* ENVIRONMENTAL STEWARDSHIP SCHEME.

Approved UK Organic Certification Bodies

Organic Farmers & Growers CIC
Organic Food Federation
Soil Association Certification Ltd
Biodynamic Association Certification
Irish Organic Association
Organic Trust Limited
Quality Welsh Food Certification Ltd
Organic Farmers & Growers (Scotland) Ltd

osier Willows (*salix sp*) that are grown and *coppiced on a one- or two-year cycle. The most common species of willow used for osier growing is *Salix viminalis* (osier willow). Osiers provide the **rods**, **wands**, or **withes** that are used in particular for making baskets, hence the term **basket willow coppice**. The growing of osiers is specialized, and has more in common with an agricultural, rather than forestry, industry activity.

OSR *See* OILSEED RAPE.

outdoor activities These include a wide range of outdoor pursuits and sports, covering *countryside, *woodland, water, and coastal environments. They range from visiting attractions such as historic buildings, nature reserves, and *farms; undertaking activities such as *field sports, water sports, walking and riding; and more informal social activities such as visiting a country pub or picnicking.

outline planning permission Type of planning permission granted in England, Wales, and Northern Ireland which establishes the general principle that a proposed development is acceptable in planning terms, but where full details of the proposed development have not been submitted at the point of application to the local planning authority. Those matters which were not included within the original application are known as *reserved matters, and can include some details of access to and within the site, appearance and layout of buildings or places in the development, landscaping, and the scale of buildings relative to their surroundings. Before the development can proceed, the details of the reserved matters must be agreed by the local planning authority via a reserved matters application. *See also* FULL PLANNING PERMISSION; PLANNING REFORM WALES.

outwintering A system where livestock are kept outdoors during the winter months rather than being brought into a shed. In the uplands, cattle and sheep will usually be brought down from the hills to *in-bye land for the winter. Dairy cows and beef cattle being finished for slaughter will usually be housed in the colder and wetter months to ensure that productivity is maintained. Beef cows and *store cattle may be kept out on suitably sheltered yet free-draining pasture or even on winter *fodder such as *stubble turnips. Store lambs may also be outwintered, as may pregnant *ewes, either if they are to lamb in late spring outdoors or before they are housed for lambing.

overstorey The trees forming the upper canopy of the *woodland. When *clear felling and promoting *natural regeneration, the retention of an overstorey can be used to control sunlight onto the woodland floor and hence help suppress weed growth as well as provide possible additional *mast.

owner-occupier Term used to describe someone who is occupying (for example, living in) a property of which they have legal ownership, as opposed to a *tenant who occupies a property which is owned by another party.

ox (*pl.* **oxen**) A member of the *cattle species used as a draught animal, i.e. to pull a plough or cart. Usually a castrated male or *bullock.

P

package treatment plant *See* SEWAGE TREATMENT PLANT.

paddock A small enclosed area of land, usually pasture often used for horses. *See* FIELD.

paddy A level field or area of agricultural land in which *rice is grown, particularly in southern and eastern Asia. The land is flooded to a depth of 10–15 cm before seedlings are planted in early summer, and the water level is maintained for three quarters of the growing season before harvest in autumn.

pannage The practice of permitting *pigs into *forest areas at certain times to feed on acorns, beech mast, or other nuts. Historically, local people enjoyed rights to this access on *common land or in **Royal Forests**. It is rarely carried out now in most areas, but is continued in the New Forest in southern England where it is also known as 'common of mast'. It is also recognized as an important part of forest ecology. *See also* COMMONER.

paperboard *See* PULPWOOD.

paraquat A non-selective *herbicide that acts as a defoliant, killing off leaves and stems but with little effect on roots. It was very widely used as a defoliant to clear land of all green material, but has now largely been superseded by *glyphosate. It is highly toxic to mammals but rapidly becomes inactive in the soil. Paraquat is now no longer authorized, or its use is restricted, in some countries.

parasite An organism that lives on or in another organism, deriving nutrients from its host. The relationship is not reciprocal in that the host gains nothing in return; indeed many parasites have a detrimental impact on the host that may result in death. They may carry disease; for example, malaria spread by mosquitos. The action of a parasite is described as parasitic, whilst a substance used to control parasites is called a **parasiticide**. An **ectoparasite** lives on the outside of its host such as on the skin; for example, **fleas**, **ticks**, and **lice**. An **endoparasite** lives within its host; for example, **intestinal worms**.

parish council Elected body of councillors serving an individual parish, and being the first level of local government in England. Parish councils generally exist in rural areas and often serve an individual village and its surrounding area. They are often known as *town councils if serving a town, but may also choose to call themselves a **neighbourhood council**, **community council**, or

village council. Parish councils are normally the officially recognized body in relation to *neighbourhood planning. *See also* LOCAL AUTHORITY; LOCALISM.

park An area of natural or *semi-natural land with a primary objective of recreation and enjoyment for humans, or for protection of *biodiversity. Parks show considerable variation in the emphases of their objectives, size, and levels of protection. For example, the relatively small **urban** and **country parks** provide accessible recreational and amenity green space to residents of towns and cities. Significantly larger areas are set aside with higher levels of protection, for example, for *landscape or for wildlife conservation. **Deer parks**, often dating back centuries, are large enclosed semi-natural areas harbouring wild deer used as a food and sporting resource, and are often associated with large rural estates. *See also* NATIONAL PARK.

Parry's Valuation and Investment Tables Collection of tables, originally produced in 1913 and updated several times since, used by valuers for, amongst other things, taking account of *discounting when undertaking valuations of property interests where an income is produced over a sustained period of time. Tables included within the publication include **present value of £1; years' purchase (present value of £1 per annum)**; and **years' purchase in perpetuity (present value of £1 per annum in perpetuity)** (for further information, *see* DISCOUNTED CASH FLOW). Parry's also contains other information such as conversion tables. In addition to the printed publication there are now several online and computer program-based calculators based on the tables in Parry's.

partial budget A management tool to allow alternative courses of action to be evaluated financially. Thus only the costs and benefits that alter with the proposed choices under consideration are included in the budget so that the impact of incremental changes can be specifically assessed.

partnership A business owned by two or more individuals where the partners have personal liability for the debts of the business. The assets of the business are owned by the partners (either in, or outside, the partnership), and for taxation purposes in the *United Kingdom. Each of the partners are treated in the same way as individuals, so each pays *Income Tax on their share of the profits of the business, rather than the business itself paying *Corporation Tax. Farms in the UK are commonly run in partnerships (often with several family members being partners), and partnership structures are very well established (being governed, in the absence of any written agreement between the partners, by the **Partnership Act 1890**). In a partnership, each of the partners has unlimited personal liability for any debts of the business (and potentially for any claims made against it; for example, if the partnership were to be sued). Therefore, if things go wrong within the business it can be personally financially catastrophic for each of the partners. Whilst partnerships must hold *employers' liability insurance in the UK, and will also normally hold public liability insurance and other insurances, concerns over the unlimited liability of each of the partners in a partnership have resulted in some businesses altering their structure from a traditional partnership to a *limited liability partnership. *See also* SOLE TRADER; LIMITED COMPANY.

partridge The two significant species in the UK are the grey partridge (*Perdix perdix*), which is native, and the red-legged (*Alectoris rufa*) or French partridge, which was first introduced from continental Europe in the eighteenth century. The grey partridge is a small to medium-sized *game bird restricted predominantly to lowland parts of the UK. Once very common and widespread, it has now declined and is the focus of conservation initiatives to increase its population. The red-legged partridge is slightly larger than the grey partridge, and it can be readily hand-reared and released for *driven shooting. It is now by far the commonest species of partridge in the UK. *See also* GAME; GAME MANAGEMENT; SEASON; SHOOTING.

pasteurize Describes the process of destroying pathogenic bacteria in foodstuffs by heating and then cooling rapidly. Variations in the temperatures used and the lengths of duration are important factors in achieving results, as well as consideration given to retaining the food's nutrients, flavouring, colour, and texture. It is most commonly referred to in its use in the production process of *milk for human consumption. Some people prefer to consume raw or unpasteurized milk and *dairy products that have not gone through this process and therefore may contain harmful germs. Raw milk is also sold as green top milk, not to be confused with the green colour-coded lid of semi-skimmed.

pasture Land covered with *grass or similar plants used for grazing livestock. It may be temporary as part of a *rotation of crops, but is more likely to be permanent. The adjective 'pastoral' refers to the countryside, especially grassland, and, in romantic culture, can apply to the life of a shepherd looking after his flock. A **pastoralist** is a cattle or sheep farmer. Grassland that is described as permanent pasture is not part of a *rotation of crops, and is never *ploughed up or otherwise changed. It is often on land that is not suitable for cultivation or arable cropping. It may not be as productive as a *ley or a *sward that has been specifically sown, but may contain a much wider range of plants including herbs and wild flowers. It is used for *grazing, *hay, and/or *silage.

PAT *See* PORTABLE APPLIANCE TESTING.

payback period An estimate of how long an investment will take to pay for itself. The payback method is one of the main methods of appraising investment into capital items or new projects (**capital investment appraisal**), and is the most straightforward. In simple terms:

- Company A invests in some new machinery which costs £2,000.
- The 'net cash flow' is then determined for each year. This is essentially the difference between the cash generated and/or spent in connection with the project, and the instalments that will need to be paid to pay it back, and when they need to be paid (this may have to be estimated).

- It is estimated that the estimated 'net cash flow' for each of the first few years is year 1 (£400); year 2 (£600); year 3 (£1,000), and year 4 (£1,000) . . .
- So, the payback period is calculated:
  - Year 1: £400
  - Year 2: £400 + £600 = £1,000
  - Year 3: £400 + £600 + £1,000 = £2,000
  - Year 4: £400 + £600 + £1,000 + £1,000 = £3,000
- Given that the machinery cost £2,000 to purchase, the original investment will have been paid back at the end of year 3, so it has a payback of three years.

This method, whilst simple, does not necessarily indicate whether the project is a good idea in the long term, as it does not indicate the cash flow after the initial period and therefore does not indicate whether it is necessarily a better investment in the longer term than something which might take slightly longer to pay back initially. It also does not reflect the timings of when cash is received in connection with the project; receiving £1 now is worth more than receiving £1 in five years' time, because £1 received now will receive five years' more interest. Money received now is also considered to be less risky as market conditions, and so on, will be better understood now than those which may apply in several years' time.

Because of these weaknesses, a number of other approaches are also used when appraising capital investments. These include:

- **Rate of return (accounting)**: a measure of the profit of a project (usually the average profit earned by the project over its lifetime measured before interest and taxation) against the capital invested (either initially or over the lifetime of the project), which is usually given as a percentage. So:

$$\frac{\text{average profit per year}}{\text{capital invested}} \times 100 = \text{rate of return}$$

In simple terms, if Farmer A had invested £150,000 in a diversification project which was producing an average profit of £20,000 per year, the rate of return would be:

$$\frac{£20,000}{£150,000} \times 100 = 13.33\%$$

As with all methods of investment appraisal, this approach has strengths and weaknesses. Although it is relatively easy to calculate, like payback periods, it does not take into account the value of receiving money sooner (because of additional interest), and in addition requires the length of the project's life to be determined (in order to calculate the average profit).

- **Discounted payback**: one of a series of approaches which use *discounted cash flows to underpin them; in this case to give a discounted payback period (i.e. the point at which the initial investment is recouped, taking into account that cash that is received today may be worth more than cash received in future, because interest has not been earned on the money received at a later date for as long). It still, however, requires an estimate to be made of the amount and timing of instalments, and also of a suitable interest rate to use, and, as with payback periods, ignores net cash flows received after the payback period.

- **Net present value (NPV)**, like the discounted payback system, takes into account the value of receiving money sooner and is based on similar principles. In essence, initially each year's net cash flow figures are multiplied by a discount factor (figures are available from sources such as *Parry's Valuation and Investment Tables), which take account of the fact that in order to receive a set sum of money at a given point in the future it would not be necessary to invest the whole of that sum initially, as the sum received at the later date would comprise both the original investment sum and the interest received on it. (For further information, *see* DISCOUNTING.) After each year's net cash flows have been multiplied by the relevant **discount factor**, this produces a series of present values (i.e. today's value of the right to receive that sum of money at a future date). These are then added together and compared to the initial cost of the investment in the project, producing a net present value. If the total NPV is positive, then it may be worth proceeding with the project:

Year 1 net cash flow × discount factor   = present value
Year 2 net cash flow × discount factor   = present value
Year 3 net cash flow × discount factor   = present value
Year 4 net cash flow × discount factor   = present value
Total present value (present values of years 1–4) = Figure A
Less: initial cost   Figure B
Equals: Net Present Value   Figure C

So, if Figure A (the total present value) was £180,000 and the initial cost of the project (Figure B) was £150,000, this would give a Net Present Value of £30,000, which would suggest the project was potentially worth proceeding with. This approach allows NPVs between different projects to be compared, allowing a determination to be made about which one may be the better investment. However, it does require the selection of an appropriate *interest rate in order to determine which set of discount factors to use.

- **Internal rate of return (IRR)** is a method very similar to the NPV method, but it produces an estimate of the rate of return that would have to apply in order for the NPV to equal the cost of the project. This then allows the IRR to be compared to the interest rate that would have to be paid on any borrowings required to fund the project. Estimating the IRR is done by choosing two different discount factors and then running through an NPV exercise for each. Ideally, one discount factor will generate a positive NPV, and the other a negative (if not, the exercise would need to be repeated until this is the case). Once a positive and a negative NPV have been generated, then a further calculation known as a **linear interpolation** is undertaken to assess at what point between the two discount factors selected the NPV would be equal to the cost of the project (i.e. the internal rate of return). This is a more complex exercise for evaluating potential investments than using the NPV approach alone.

payment in kind Paying for goods or a service by providing goods or a service in return, rather than by paying money. For example, Person A mows Person B's lawn, and Person B cleans Person A's kitchen in return. It is sometimes used in agriculture, where, for example, a farm worker is provided with a rent-free house by his or her employer in lieu of part of what their wages would otherwise be (although it does not necessarily mean that the farm worker can avoid paying tax on that element of their wages which is comprised of the rent-free house).

peat Partially decomposed vegetative remains formed in layers over thousands of years in waterlogged conditions that exclude oxygen. Commercially extracted and used in the gardening and horticultural industries, peatland is now increasingly recognized as of significant importance for *flood prevention and *carbon sequestration, as well as an important *habitat. *See also* BLANKET BOG.

pedology The study of *soils, including their physical and chemical properties, their formation, and their distribution. The characteristics of the soil, including microorganisms and their biological activity, are important aspects of pedology, particularly in relation to *agriculture. A soil scientist may be called a pedologist.

pellet A small compressed mass of substance. **1.** Often tubular in shape, and frequently used to describe animal feed with the ingredients mixed and then compressed into a pellet for storage and ease of use. Also describes wood shavings or sawdust that is dried and compressed into a pellet for a biomass fuel for stoves or boilers. **2.** In ornithology, a ball of undigested matter that some bird species, such a owls, regurgitate.

peppercorn rent A nominal rent that is very substantially below *market rent. For example a piece of land may be let at a peppercorn rent of £3 per year. Peppercorn rents are sometimes used for *leases with very long terms when the *freeholder essentially wishes to offer the lease rent free, but in order to reinforce that it is a lease, and not a *freehold that is being offered to the *lessee, a peppercorn rent is included within the lease agreement.

perch Old unit of measurement, used in the United Kingdom, equal to 5½ yards (approximately 5 metres). A perch is also known as a rod or pole. *See also* CHAIN.

perennial A noun to describe a plant that lives for more than two years, as distinct from a plant that persists for only one year which is described as an *annual, or for two years and described as a *biennial. For example, Italian ryegrass is an annual or biennial and is thus used in short-term grass *leys, whilst perennial ryegrass is used in longer-term leys and *permanent pasture. It can also be used as an adjective to describe systems of perennial *agriculture whereby several crops can be grown on the same land over the year, or perennial *irrigation in which land can be irrigated at any time.

period *See* PERIODIC TENANCY.

periodic tenancy Under English law, a *tenancy which may continue indefinitely. A periodic tenancy effectively automatically renews at the end of each **period**. The period is either defined in the *tenancy agreement, or, in the absence of such an agreement, is normally determined by reference to the period which is used to calculate the *rent. Periodic tenancies can therefore be, for example, weekly, monthly, quarterly, or annually (i.e. a **tenancy from year to year**). Because a periodic tenancy automatically renews at the end of each period (each of which is of a defined length), a periodic tenancy still meets the requirement that a tenancy should have a defined duration, even though the total number of periods may not be known. (For further information on the characteristics of a tenancy, *see* EXCLUSIVE POSSESSION.) A tenancy can exist either as a periodic tenancy from the outset (if no fixed term is ever agreed, therefore preventing the tenancy from being a *fixed-term tenancy) or can arise at the end of a fixed-term tenancy because the *landlord continues to accept rent beyond the end of the fixed term. *See also* TENANCY AT WILL.

peri-urban *See* URBAN FRINGE.

permanent pasture *Grassland that is not part of a *rotation of crops, that is never *ploughed up or otherwise changed. It is often on land that is not suitable for cultivation or arable cropping. It may not be as productive as a *ley or a sward that has been specifically sown, but may contain a much wide range of plants including herbs and wild flowers. It is used for *grazing or *hay and/or *silage. *See also* PASTURE.

permissive access Where permission is given for people to access private land as a linear route or a defined area. This type of access is normally for walking, cycling, or horse riding, and is not a public right as provided through *Public Rights of Way or *open access. As with legislated access, the landowner has a responsibility and a duty of care to those using it, and in some cases may reach agreement with the *highways authority for its management. Some examples of permissive access have been integrated as an option within certain *agri-environment schemes. Unlike Public Rights of Way and open access, this type of access is not normally shown on **Ordnance Survey** maps, but may rely on more local publicity and signage.

permitted development Type of *development which, providing it meets certain criteria, does not normally require an application for *planning permission to be made to a *local planning authority, because it is of a type which has been granted blanket planning permission through a permitted development order. The permitted development system exists in order to avoid the planning system in the United Kingdom becoming overwhelmed by large numbers of planning applications for developments which are unlikely to cause any significant issues. Permitted development rights are most commonly set out by central government in *General Permitted Development Orders (GPDO), which grant blanket planning permission for certain types of development. England, Wales, Scotland, and Northern Ireland each have their own GPDO, and they are frequently amended. In addition to detailing the types of development which are permitted, GPDOs also set out criteria which each development must meet in order to be covered by the GPDO (which may include factors such as height or size). For example, the GPDOs each grant permission for extensions to domestic properties; but only those of certain sizes and layouts. Extensions which are larger than the size permitted by the relevant GPDO are still subject to the requirement to submit a planning application. As well as granting permission for physical construction, the GPDOs also grant permission for some *changes of use.

Although the GPDOs are the main system used to grant permitted development rights, there are also a number of other mechanisms which can be used to grant permitted rights in more localized areas. These include:

- **Local development orders**. These can be made by local planning authorities in England and Wales, and grant planning permission for specific types of development in their local area (over and above those granted by GPDOs).
- **Neighbourhood development orders**. These cover designated *neighbourhood areas in England (under the *neighbourhood planning regime). They are proposed by the relevant *town council, *parish council, or neighbourhood forum covering the area, and are made by the local planning authority.
- **Community right to build orders**. These also sit under the neighbourhood planning regime in England, and grant permission for development specified in the order. They differ from neighbourhood development orders in that they can be prepared by community organizations.

Although permitted development does not require an application for planning permission, some types of permitted development are subject to the requirement to obtain **prior approval** (also known as **prior notification**) from the *local planning authority in relation to specified elements of the development before it can proceed. Prior approval applications are designed to be less complex than applications for planning permission. The matters which the local authority must consider in a prior approval application are set out in the relevant permitted development order, and are specific to each type of development. They are generally more limited than in an application for planning permission, but may include factors such as contamination or flood risk.

Householders in England who wish to rely on permitted development rights to undertake some larger extensions must also notify the local authority in advance under the **neighbour consultation scheme**. Once the local authority has been notified they will then consult adjoining neighbours, and if they raise objections, the local planning authority will then consider whether or not to allow the work to proceed.

Whilst permitted development rights are often of significant benefit to developers, they can be restricted in certain areas (for example, *National Parks). In addition, *article 4 directions can also be used to restrict permitted development rights in specified areas and are often used to control development more tightly in *Conservation Areas. Permitted development rights can also be removed through a *condition attached to a planning permission.

Developments which benefit from permitted development rights may also require other forms of permission or consent, including *listed building consent, *building regulations approval, *advertising control consents, *environmental permits/licences, or permissions relating to trees; for example those covered by *Tree Preservation Orders.

See also CERTIFICATE OF LAWFULNESS OF PROPOSED USE OR DEVELOPMENT.

personal bar In Scottish law, a legal principle which is broadly equivalent to *estoppel in English law.

personal protective equipment (PPE) Equipment that must be provided to workers free of charge, under health and safety law in the *United Kingdom, in order to enable them to safely carry out potentially hazardous tasks at work without harm to themselves, if the risk cannot be controlled in another way. Examples may include safety goggles, hard hats, high-visibility clothing, or gloves. *See also* HEALTH AND SAFETY AT WORK ACT 1974; RISK ASSESSMENT.

pest Any organism that is detrimental to others. In a land management context, this would more specifically relate to an organism which impacts on, for example, *crops, *livestock, or *game, and leads to loss of production through competition for sunlight, water, or nutrients, through direct action such as feeding, or through spreading disease. Common agricultural pests include animals such as rabbits and pigeons, *weeds, *bacteria, *fungi, viruses, and *parasites such as aphids and ticks.

PEST analysis *See* PESTLE ANALYSIS.

pesticide Any substance, preparation, or organism used to control or render harmless unwanted species of plants and animals. In most cases a pesticide contains an active ingredient or ingredients, the role of which is to control the target species identified. Pesticides are not solely used for plant protection in the arable sector, but also within the *livestock sector, to aid control of mites and parasites. Pesticides can be subdivided into four main categories based upon their activity. Contact: these do not penetrate tissue and are not transported or translocated (moved internally around the plant), and control is at the point of contact. Systemic: these are either ingested or absorbed by the target and are transported and translocated around the host to various kill sites. Residual: this persists within the environment and provides control of the target species over a relatively short period of time. Persistent: this acts as a residual, but over a longer period of time.

The main types of pesticide used are *herbicides to control unwanted plants (weeds); *fungicides to control plant diseases, viral as well as fungal; *insecticides to control insect pests; *rodenticides to control rodents (rats, mice, moles, rabbits); nematicides to control *nematodes; *molluscicides to control *molluscs; and acaricides to control *mites and *parasites. Adjuvants (surfactant, sticker) are mixed with pesticides to enhance the pesticide performance, and although some of these may not be pesticides, they are usually described in the same way.

Pesticides are generally applied as a liquid, where the pesticide product is mixed with *water and applied via a *sprayer. Other application methods are granules, dusts, smoke, and gas. The main influences that affect the performance of pesticides are the influence of the *crop, including *seedbed, the influence of the spray quality, weather conditions, choice of pesticide product, and application timing. *See also* SPRAYER.

PESTLE analysis Tool used in business and marketing which involves an organization analysing the influence on it of external factors which it may not be able to control. The acronym PESTLE refers to the 'political', 'economic', 'social', 'technological', 'legal', and 'ecological' environments. The last two aspects are sometimes omitted in PEST, or STEP analysis (although less so recently with the increasing prominence of the *sustainability agenda). The process can be used in a number of ways, including for assessing risks (for example, the impacts of new *legislation, or economic instability), or for identifying new business opportunities which are likely to be presented by changing external factors, such as the invention of new technologies. PESTLE analysis is often used in combination with *SWOT analysis.

pH A measure of the acidity or alkalinity of an aqueous solution on a scale of 0–14. Pure water has a pH of 7, which is neutral, with a reading below being acidic, and above being alkaline. When applied to *soils, plants can grow within a range, some thriving in slightly more acidic conditions than others, but most thrive in pH 5–7. Outside this range, *nutrients become unavailable to the plant; for example, calcium and magnesium in acid soils, and manganese, copper, zinc, and cobalt in alkaline soils. High levels of *organic matter tend to have a lower pH, so if soils are alkaline, increasing the organic matter may help, whilst application of *lime can increase the pH of acidic soils.

pheasant (*Phasianus colchicus*) A large, long-tailed *game bird, native to Asia, that was thought to have been initially introduced into Britain 2,000 years ago. There is some debate over various reintroductions, but it appears likely that they were familiar by the fifteenth century. They have since interbred with a variety of other introduced pheasant species. During the twentieth century the pheasant became an increasingly significant game bird, in part due to its ready adaptability to intensive hand-rearing and subsequent release, predominantly for *driven shooting. It is now the commonest and best-known game bird both in the UK and in other countries worldwide. These levels of intensity of pheasant rearing and releasing are the cause of some concern on the impacts which they might have on native *biodiversity.

phosphate A chemical compound that contains phosphorus. A nutrient essential for all life, it is one of the three primary nutrients for plant growth (the others being *potash and *nitrate) and is thus applied as a *fertilizer, either on its own or in combination with the other two. It is a component of nucleic acid and is thus vital for cell division and plant growth. There are continuing concerns that global supplies are limited and might only last another century. There are also concerns at damaging levels of phosphate in watercourses, particularly in the context of the *Water Framework Directive.

photosynthesis The process by which plants, algae, and some bacteria capture energy from the sun and create *carbohydrates. **Chlorophyll**, the green pigment found in most plants, absorbs energy from the sun which allows carbon dioxide and water to combine to create carbohydrate and oxygen which is released into the atmosphere. Other pigments, such as **carotenoids**, perform a similar function. Photosynthesis is fundamental to life, as it creates food from solar energy, and by absorbing carbon dioxide and emitting oxygen it maintains balance in the atmosphere.

photovoltaic *See* SOLAR POWER.

picocell *See* TELECOMMUNICATIONS EQUIPMENT.

piebald *See* HORSE.

pig An animal of the *Sus* family, domesticated from wild boar and farmed for its meat. A female pig is a **sow**, a male pig is a **boar**, and the young are called **piglets**. The building in which pigs are kept is a **piggery**, and someone who looks after pigs is called a **pigman**. The pig is an omnivore, but rations fed on the farm are based on *cereals and a *protein source such as *soya bean. Pigs reach sexual maturity at six to eight months of age and, with a gestation period of 115 days, are capable of having more than two *litters of piglets each year. A litter might be as many as ten piglets, so a single *sow can produce up to twenty-five piglets per annum. A **weaner** is a young pig that has been weaned off its mother's milk and is transferred from a rearing to a finishing unit. A **gilt** is a young female that has yet to give birth. Whilst the traditional breeds of pigs are still kept on specialist farms, most pigs today are the result of extensive selective breeding. The meat from pigs is used in a variety of ways. Pork is the fresh meat and may be sold as such or processed into sausages. The meat may be cured, when it becomes bacon, gammon, or ham. Pigs are also known as **hogs** in certain parts of the world, notably North America, and the term 'hog roast' is more generally used as the term to barbecue a whole pig, usually on a spit.

Breeds of pig in the UK

Large White	Tamworth
Large Black	Welsh
British Saddleback	British Lop
Berkshire	Oxford Sandy and Black
Gloucester Old Spot	

pigeon The UK's commonest pigeon is the wood pigeon (*Columba palumbus*), which is of particular relevance because of the potential harm that it can cause to agricultural *crops. They can cause significant damage to some *field vegetables and young crops, and this has created a diverse industry supplying preventive measures such as the gas-operated bird-scarer. The wood pigeon also provides shooting opportunities and a high-quality meat source.

piggery *See* PIG.

piglet *See* PIG.

pigman *See* PIG.

PII *See* PROFESSIONAL INDEMNITY INSURANCE.

pine (Scots, Corsican, lodgepole) *See* TREE.

pitchfork A hand tool with a long handle and projecting tines or *prongs used for moving material such as *bales or *manure. The name derives from the original use of pitching **sheaves** of *grain.

plagio-climax *See* CLIMAX COMMUNITY.

plan-led system Term describing a planning system which operates on the principle that decisions in relation to planning applications should be made on the basis of the degree to which they align with the *planning policies set out in local, regional, or national development plans. England, Wales, Scotland, and Northern Ireland all have development control systems which are based on the plan-led system model. For a more detailed discussion on the structure of the planning system in the United Kingdom, *see* LOCAL PLAN; LOCAL PLANNING AUTHORITY; PLANNING OFFICER.

planning System (also often referred to as town and country planning) of controlling changes in the use of, and/or physical changes to, land and property. England, Wales, Scotland, and Northern Ireland all operate a *plan-led system, whereby applicants must apply for planning permission before undertaking *development (unless it is *permitted development). Development can include both physical changes (for example, the construction of a new building) or a *change of use in the land or building. Applications are submitted to the *local planning authority, which is responsible for preparing a *local plan (or local development plan) for its area which sets out the local *planning policies against which applications for *planning permission are assessed. The core principle followed is that permission/consent should be granted for those applications that accord with the relevant policies, unless any *material consideration dictates otherwise. Waste and minerals planning operates on a parallel basis to other non-infrastructure planning matters; however, in areas with a two-tier system of local authorities, waste and minerals planning is often dealt with by the county council, with other planning matters being managed by the district council (for further details, *see* LOCAL PLANNING AUTHORITY). *Infrastructure planning in the United Kingdom also operates on a plan-led system; however, in England, Wales, and Northern Ireland it is managed on a more centralized basis, with policy making and decisions made on a national or regional basis. *See also* APPEAL, PLANNING; ENFORCEMENT, PLANNING; LOCAL PLANNING AUTHORITY; NEIGHBOURHOOD PLANNING; PLANNING DESIGNATION; PLANNING REFORM WALES.

planning agreement (Northern Ireland) *See* SECTION 106 PLANNING OBLIGATION AGREEMENT (ENGLAND AND WALES).

planning appeal *See* APPEAL, PLANNING.

Planning Appeals Commission Public body operating in Northern Ireland that decides on appeals in relation to planning and environmental decisions made by Northern Ireland government departments and *local authorities. In addition to deciding on *planning appeals, the Commission also carries out independent examinations of *local development plans, and determines appeals in relation to environmental matters, for example, in relation to the refusal of licences relating to waste management. Decisions are made either by a single **Commissioner** or a panel of Commissioners who are public appointees qualified in town planning or related disciplines. Whilst the Commission is funded by the Northern Ireland government, it is independent of any government department.

planning application *See* PLANNING PERMISSION.

planning authority *See* LOCAL PLANNING AUTHORITY.

planning circular Document issued by a government minister which provides detailed guidance on how a change in planning policy or legislation is to be implemented in practice.

planning committee *See* DELEGATED POWERS; PLANNING OFFICER.

planning conditions *See* CONDITIONS, PLANNING.

planning contravention notice *See* ENFORCEMENT, PLANNING.

planning designation Classification applied to a site or area which impacts on the nature of any *development which can be carried out in that location, and may also require a developer to obtain additional permissions, over and above those required in a non-designated area. There is a large range of designations which impact on planning decisions (and which are specifically addressed in *planning policy). These include *listed buildings, *scheduled monuments, *Conservation Areas, *National Parks, *Areas of Outstanding Natural Beauty, *Green Belt, *registered parks and gardens, *Sites of Special Scientific Interest, *Tree Preservation Orders, and *local green spaces. These may be designated at a national level (for example, listed buildings), or by *local planning authorities (for example, local green spaces). The impact of each designation varies, depending on the legal framework under which it is established. For example, the significance of the presence of a designation as a registered park or garden is that it becomes a *material consideration in any planning decision relating to the site, which must be guided by national and local planning policy in relation to it as a registered park or garden. Work on a listed building, however, must not only be in accordance with planning policy in relation to listed buildings, but is also subject to additional requirements to obtain *listed building consent, over and above any *planning permission. Sites or areas which are subject to planning designations also often have restricted *permitted development rights. In addition, other regulations, outside the planning system, can also often apply to designated sites, and in particular to those with statutory wildlife designations which are also often protected through wildlife and nature protection laws. *See also* HERITAGE ASSET.

planning gain *See* SECTION 106 PLANNING OBLIGATION AGREEMENT.

planning inquiry *See* APPEAL, PLANNING.

Planning Inspectorate Public body which deals with *planning *appeals, national *infrastructure planning applications, and examinations of *local plans in England and Wales. It is sponsored by the *Ministry of Housing, Communities and Local Government and the *Welsh Government.

planning obligation *See* SECTION 106 PLANNING OBLIGATION AGREEMENT.

planning officer An official of a *local planning authority who deals with planning applications. A planning officer will review an application in the context of relevant local and national *planning policy, and will judge whether or not the application complies with policy. They will usually visit the site of the proposal and will also invite comments from *statutory consultees. These consultees will vary, depending on the type of application, and may include *highways authorities (for issues relating to roads and traffic), environmental authorities (for example, the *Environment Agency in England, or its Welsh, Scottish, and Northern Irish equivalents), or heritage bodies (for example, *Historic England, and its equivalents). The planning officer will also ensure that the application is advertised so that those likely to be affected by it are aware, and have an opportunity to comment if they wish. Having considered all aspects of the application (including planning policy and any other *material considerations), the planning officer will either prepare a report for the local authority's planning committee (comprising elected councillors), recommending what decision should be made in respect of the application, or, alternatively, for a straightforward application where the planning officer has been given *delegated powers to make a decision by the planning committee, they will make the decision themselves. Following either a delegated decision, or a decision by the planning committee, the planning officer will prepare and issue a *decision notice confirming whether the application is given permission (and any relevant *conditions) or is refused (and if so, why). *See also* CONSERVATION OFFICER.

planning permission Form of permission granted for the *development of land or buildings (either physical works or *changes of use). Development is defined within planning *legislation in the United Kingdom, and anything falling within the definition requires planning permission. Permission can be granted either for a specific project (via a *planning application made to the appropriate planning authority), or through *permitted development rights (which give blanket planning permissions for certain types of development, subject to their meeting certain conditions or criteria). For most development projects, applications for planning permission are made to the *local planning authority. However, for larger/national or regional infrastructure projects, applications are normally considered by central government. In England, most nationally significant infrastructure projects are managed under a system operated by the *Planning Inspectorate on behalf of central government. In Wales, the Planning Inspectorate examines energy and harbour development applications, with other types of application being dealt with directly by *Welsh Government Ministers. In Northern Ireland, regionally significant developments are dealt with centrally by the **Northern Ireland Department for Infrastructure**. In Scotland, whilst national developments are dealt with by local planning authorities, *Scottish Government Ministers have powers to call in applications to determine them themselves. *See also* FULL PLANNING PERMISSION; DETAILED PLANNING PERMISSION; OUTLINE PLANNING PERMISSION; PLANNING PERMISSION IN PRINCIPLE.

planning permission in principle 1. In Scotland, a term (broadly equivalent to *outline planning permission in England, Wales and Northern Ireland) for *planning permission which establishes the general principle that a proposed development is acceptable in planning terms, but where only limited details of the proposed development are submitted at the point of application (at minimum, a location/land ownership plan and the point of access from a road). If permission is granted it will be done so with planning conditions attached to it which must then be approved through a subsequent **approval of matters specified in conditions application**. *See also* DETAILED PLANNING PERMISSION. **2.** In England, a type of *planning permission introduced by the **Housing and Planning Act 2016**, which enables *local planning authorities to grant planning permission for housing (subject to the further approval of a *technical details consent) through the allocation of a site within a *brownfield planning register or via a specific application to the local planning authority. Planning permission in principle grants permission in general terms for housing development on a site; however, in order for a development to proceed, a further application must be made to the local planning authority for technical details consent (which provides the precise information about the development). Once granted, this then gives the site planning permission so that the development may proceed (subject to any planning *conditions). *See also* FULL PLANNING PERMISSION; OUTLINE PLANNING PERMISSION; PLANNING REFORM WALES.

planning policies Statements setting out the principles against which applications for *planning permission or other forms of planning-related consent (for example, *listed building consent) will be assessed. Planning in the United Kingdom operates on a *plan-led system basis, with the principle being that permission/consent should be granted for those applications that accord with the relevant policies, unless any *material consideration dictates otherwise. Planning policies in the United Kingdom operate in a hierarchy. Central government sets out policies at a national and/or regional level, with local planning authorities setting out the policies for their area in their *local plans. In some areas, policies may also then exist at a *neighbourhood planning level. All of the layers of planning policy align with each other, hence local and neighbourhood planning policies (where they exist) essentially interpret national/regional planning policy objectives as they apply to an individual locality.

Whilst central governments in England, Wales, Scotland, and Northern Ireland each have their own planning policies, they are all based around the principle that the core objective of the planning system is to deliver *sustainable development.

In England, national planning policy is set out in the **National Planning Policy Framework (NPPF)**, with supporting detail on how NPPF principles are to be applied by local planning authorities set out in the online **Planning Practice Guidance**. Planning policy in respect of national infrastructure projects is set out in **National Policy Statements**.

In Wales, national planning policy is set out in **Planning Policy Wales**, with further detailed information on specific issues such as housing and tourism set out in **Technical Advice Notes**.

In Scotland, national policy is set out in the **National Planning Framework** and **Scottish Planning Policy** documents. In addition, the Scottish Government also produces separate policy documents on architecture and design, as well as a set of **Planning Advice Notes**, with guidance on the Scottish Government website providing further details on specific areas of policy.

Northern Ireland-wide policy is contained in the **Strategic Planning Policy Statement for Northern Ireland (SPPS)**, which was published in 2015 and complements the **Regional Development Strategy 2035** document which sets out wider strategic objectives for Northern Ireland. Until 2015, the planning system operated on a centralized basis, with the majority of planning functions and decisions being carried out by the **Northern Ireland Department of the Environment**, which had powers to make development plans for any area of Northern Ireland. For a transitional period (i.e. until local planning authorities have published their own local planning policies), some elements of the pre-2015 Northern Ireland-wide policies are retained; this includes policy and guidance provided in **Planning Policy Statements** and **Development Control Advice Notes**.

In addition to these national/regional policy documents, government ministers may also issue *planning circulars, or policy clarification letters, which provide further detail as to the interpretation of specific elements of national policy and planning *legislation. Each government's Chief Planner may also issue letters to assist in the interpretation of policy or legislation. For a more detailed discussion of the different planning policies and structures in England, Wales, Scotland, and Northern Ireland, *see* LOCAL PLAN.

Planning Policy Guidance *See* PLANNING POLICIES.

Planning Policy Statements 1. Superseded collection of documents comprising statements of national *planning policy in England. Planning Policy Statements **(PPS)** and their predecessors, **Planning Policy Guidance notes (PPG)**, detailed national planning policy in England until they were all cancelled by the publication, in 2012, of the **National Planning Policy Framework**, which replaced them. **2.** Partially superseded statements of Northern Ireland-wide planning policy. Following the introduction of reforms to the planning system in Northern Ireland in 2015 (*see* LOCAL PLAN), and the publication of the **Strategic**

Planning Policy Statement for Northern Ireland, some of the Northern Irish Planning Policy Statements were cancelled. However, others have been retained for a transitional period whilst the *local planning authorities prepare new local planning policies.

Planning Policy Wales *See* PLANNING POLICIES.

Planning Practice Guidance *See* PLANNING POLICIES.

Planning Reform Wales Current proposals from the *Law Commission and the *Welsh Government to produce a new **Planning Code** which would include all *planning *legislation relating to Wales (therefore effectively replacing existing legislation including the Town and Country Planning Act 1990 and the Planning (Listed Building and Conservation Areas) Act 1990), as well as the supporting secondary legislation. The impetus behind this is the highly complex structure of current planning legislation in Wales. This has arisen in part because until Welsh devolution, planning legislation generally covered both England and Wales. Following devolution, amendments to this legislation have not always applied in the same way in England and Wales, which has led to a complex structure of legislation in Wales. Proposals include amendments to legislation in relation to the definitions of *development and *material considerations, and also to some of the planning application processes (including *outline planning permission).

Planning Strategy for Rural Northern Ireland *See* PLANNING POLICIES.

planning viability A determination of whether a potential *development is likely to be financially viable, i.e. show a reasonable return for the developer once the costs of the project (including that of ensuring an appropriate site value for the landowner), and the risks attached to it, are taken into account. One of the key planning-related variables affecting the viability of a potential development is likely to be the cost of any planning obligations, such as ones included in a *section 106 planning obligation agreement (for example requiring *affordable housing to be built as part of a housing development). If a *local planning authority requires a high proportion of affordable housing on a housing development as an obligation of granting planning permission, then this is likely to reduce the potential *profit for the developer (or mean that the developer can only afford to pay less for the land which may make the proposal unattractive to the landowner). In this situation, a developer may argue that this would make the potential development so unprofitable that they would not go ahead with it. The developer may then put forward arguments to the local planning authority, based on the *residual valuation method, proposing the number of affordable houses they consider they can provide on the site whilst still making sufficient profit to make the project worthwhile. The local planning authority may then decide to reduce the number of affordable houses they are planning on requiring, or they may consider that it is not an unreasonable requirement and grant the permission subject to the provision of the original number of affordable houses, leaving the developer to decide whether or not to go ahead with the overall development. *See also* DEVELOPMENT APPRAISAL.

plan strategy (Northern Ireland) *See* LOCAL PLAN.

plantation *Woodland that has been specifically established through *planting, either on *felled woodland sites or other lower value land. Plantations in the UK that are of substantial size are commonly made with faster growing *non-native conifers, rather than native *broadleaf species. This was particularly prevalent after the First World War, and was led by the newly created *Forestry Commission as well as private forestry interests.

planting Trees are normally established through *natural regeneration or by the planting of young trees, usually one to three years old, transplanted from a *nursery. Planting requires consideration of species choice, siting of the *plantation, purpose, and how and when planting is undertaken. Consideration needs also to be given to appropriate ground preparation, pre- and post-planting weed control, and protection from mammalian pest species such as rabbits and deer. **Re-planting** and **re-stocking** can refer to planting of trees on existing woodland sites and replacement of failed newly planted trees. In systems of *continuous cover forestry management, **underplanting** of seedlings under the canopy is undertaken to eventually replace the older trees. **Enrichment** refers to the measure of planting trees at wider spacings among existing but degraded and low-quality stands; for example, areas of predominant *scrub. *See also* NATURAL REGENERATION.

plough An agricultural implement for breaking up and inverting soil, used in ancient times to bring virgin land into production. It was pulled by horses or oxen, but today is usually mounted on the *three-point hydraulic linkage of a tractor. It has one or many *furrows made up on *mouldboards that turn over the soil to bury any weeds or the residue of a previous crop. In this respect it differs from other *tillage implements that loosen or disturb the soil without inversion. It can be used to break up *compaction in the soil, but may also cause it in heavy soils at the base of the furrow, known as a plough pan. As ploughing is a slow process using a lot of energy, its use has been in decline, especially as it can cause damage to the soil microbiology. Someone who ploughs is a ploughman.

ploughman *See* PLOUGH.

plough pan *See* PLOUGH.

poaching 1. The illegal taking or killing of animals, including protected wild species as well as *game species. It is a criminal offence to hunt, shoot, or fish game species without the permission of the owner or lessee of the sporting rights of a property. **2.** The compaction of soil by the hooves of *livestock. Where the soil is damp or wet, animals, particularly heavy animals such as *cattle or *horses, may cause compaction, with the aggregation of soil particles under the pressure of the weight of the animal. This may occur especially where animals congregate; for example, round a *feeder or water *trough. If the poaching is significant, it will constitute a breach of *cross-compliance.

podzol An infertile acidic *soil having a pale layer from which minerals have been *leached and a lower dark layer containing *organic matter. Podzols are typically found under temperate *coniferous or *broadleaved forest. They may occur at some height above sea level, particularly under coniferous woodland where the breakdown of the plant material such as needles causes the soil to be acidic. Podzols are widespread throughout Scotland, generally associated with semi-natural *heath or coarse *grassland vegetation and coniferous woodland where *aerobic conditions prevail and water can percolate freely through the upper part of the profile.

point source pollution Pollution, usually coming from a single identifiable source, and which often has an impact immediately at the location of that source. Point source pollution can usually be traced back to the original source relatively easily, and contrasts with *diffuse pollution which is likely to be cumulative in nature, coming from multiple sources, and is therefore less easy to trace. A leak of chemicals from a factory into a watercourse which kills fish immediately in the local area would be an example of point source pollution, as opposed to high nitrogen levels in a watercourse causing *eutrophication which is likely to be the consequence of pollution from a large number of low-level sources, including, for example, *runoff of nitrogen from *fertilizer use on agricultural land over a large area.

point-to-point(er) In the UK, an amateur form of **steeplechasing**, i.e. horseracing over fences. Each point-to-point in the UK is run by a **hunt** or **hunt club** in order to raise funds, so there is an extensive use of volunteers at these events. The courses are commonly situated on private farmland. To be eligible to run in a point-to-point race a horse normally needs to have had a minimum of four days hunting in order to obtain a hunter certificate, which is usually issued by a *Master of Foxhounds. *See also* FOX HUNTING.

policy An official plan to address a particular situation such as agreed by a business, organization, or government(s). For example, it may be EU and member governments' policy to promote environmentally friendly farming, but it is the individual schemes and other measures that detail and implement the policy. *See also* WHITE PAPER.

Policy Commission on the Future of Farming and Food *See* CURRY REPORT.

pollard (pollarding) A tree that is cut about 2–3 metres above ground level and allowed to grow again. Similar in principle to *coppice, indeed most trees that coppice will also pollard, but affords protection for the new shoots which are above the reach of browsing animals. The technique is described as pollarding, which also increases the longevity of the tree itself.

pollination The transfer of pollen from the male reproductive organ of a flower to the stigma or female reproductive organ of the same or another flower. Pollination is the prerequisite for fertilization and the development of seeds.

Self-pollination is where pollen and stigma are from the same flower or plant, whereas cross-pollination requires different plants. There is a wide variety of different vectors to aid pollination, including wind, water, birds, mammals, and insects. The value of pollinating insects including bees to some agricultural crops is becoming more recognized, for example, in the *National Ecosystem Assessment.

polluter pays principle Principle, first coming to prominence in the *United Nations' 1992 Rio Declaration on Environment and Development which establishes that a person, or body, which has caused or allowed some form of pollution to happen, should bear the costs of cleaning it up (as opposed to another party, such as a government, doing so).

pollution prevention and control permit Form of permission from a government authorized body in Scotland or Northern Ireland (normally the *Scottish Environment Protection Agency, the *Northern Ireland Environment Agency, or a *local authority) to carry out some form of activity that may impact on the environment or human health. *Regulations set out the situations where a pollution prevention and control permit must be applied for, and the permits are normally issued with conditions which must be complied with, and which are designed to reduce the risk to the environment and health.

The pollution prevention and control permit system in Scotland and Northern Ireland operates alongside *waste management licences, whilst in England and Wales the types of activity covered by pollution prevention and control permits and waste management licences are both covered by the *environmental permit regime.

polytunnel An elongated arched tunnel constructed of a frame, usually metal, with a clear polythene covering. Like a temporary greenhouse, they offer shelter from wind, rain, and extremes of temperature for the growing of fruit and vegetables. They are large enough for people and machines to work within them. *Irrigation will be required to ensure that the plants have enough water, most commonly trickle irrigation. The use of polytunnels has transformed the growing of strawberries in the UK, for example, protecting the fruit and extending the season by several months. However, they are controversial as they are thought by some to be an eyesore in the landscape.

pond A smaller area of water than a *lake, which can be natural or man-made. It may have a relatively shallow depth of water and contain aquatic vegetation, and can be an important wildlife *habitat, especially within farmland with limited other standing water resources. Flight ponds are constructed or managed for duck shooting. *See also* DEW POND; WILDFOWLING.

ponding 1. Condition of *soil where small *water-filled indentations, or hollows, appear on the soil surface as a consequence of soil *compaction caused by the feet of *livestock which prevents water on the surface from draining into the soil (the small hollows being caused by the pressure of the feet of the livestock). Soil that is damaged in this way is said to be exhibiting *poaching. Soil that is *waterlogged can be more prone to ponding and poaching, but it can also be caused in soils which are naturally more free-draining if livestock stocking densities are too high or if livestock are left on the same area of land for too long. Additionally, it is more likely to be found in the areas of a field where livestock tend to gather in large numbers; for example, gateways, or areas around feeders. **2.** Condition which can occur on flat roofs where rainwater sits on the surface of the roof, rather than draining into guttering. It can increase the risk of structural problems within the roof.

pony *See* HORSE.

poplar (white, black, hybrid, Lombardy) *See* TREE.

population A group of individuals of a given species in a given area at a given time. Normally a term used in reference to humans and wild *fauna, but not to farm *livestock.

portable appliance testing (PAT) Term used, in the United Kingdom, to describe the inspection and testing of portable electrical equipment and appliances in order to ensure that they are safe. *See also* ELECTRICAL SAFETY REGULATIONS.

portal frame building A common type of *farm building, comprising a steel framework or steel-reinforced concrete columns and rafters. These form the portal frames and allow for wide-span enclosures ideally suited for a range of *agricultural purposes including housing *livestock and *grain storage.

Porter's five forces model Model, developed by business strategist Michael Porter, which conceptualizes the factors, or forces, which influence the competitive environment in which an organization operates and which can be used to inform strategic business planning. The five forces identified in the model are:

- The bargaining power of suppliers.
- The bargaining power of buyers.
- The likelihood of new entrants coming into the market.
- The likelihood of replacement products and services becoming available and potentially making existing products or services less attractive.
- The intensity of rivalry between existing competitors (in relation to their relative size, number, and market share).

possession The control and use of property (i.e. treating it as one's own) including, in the case of a *landlord, the right to receive any *rent. *See also* ADVERSE POSSESSION; EXCLUSIVE POSSESSION; VACANT POSSESSION.

possessory freehold *See* FREEHOLD; TITLE.

possessory leasehold *See* LEASEHOLD; TITLE.

post and rail *See* FENCE.

potash A chemical compound that contains potassium. A nutrient essential for life, it is one of the three primary nutrients for plant growth (the others being *potash and *nitrate). It is thus applied as a *fertilizer, either on its own or in combination with these other two. It is abundant, found in many forms, and is mined from rock deposits.

potato (*Solanum tuberosum*) A crop widely grown for the starchy tubers it produces. Originating in South America, it was brought back to Europe in the late sixteenth century. Seed potatoes are planted in spring and lifted from summer through to autumn. The earth is banked up along the rows, leaving furrows or trenches between them to protect new growth and to ensure that light does not reach the growing tubers. A tuber is a swollen undergrown stem which will *chit to produce new shoots. Potatoes are susceptible to a number of diseases, notably the fungal potato *blight which caused the great **Potato Famine** in Ireland in the middle of the nineteenth century. Control requires applications of *fungicide. A **potato harvester** is a machine that lifts the potatoes from the ground at harvest.

Potato Council *See* LEVY BODIES.

potentially exempt transfer (PET) *See* INHERITANCE TAX.

poultry Domesticated birds that are kept primarily for their meat and eggs but also for their feathers. Common species of poultry include chickens, turkeys, ducks, and geese. Chickens are reared for their eggs, when they are called layers, or for their meat, when they may be called *broilers. Broilers are usually kept in buildings and have been selectively bred to be very fast growing with a high feed conversion rate. Layers may be kept in cages but due to the animal welfare lobby in many countries, most are now *free range, which means they have outdoor access and can move around. Turkeys are mainly kept for meat, especially for certain feasts such as Christmas or Thanksgiving in the USA. Geese are an alternative to turkey for feasts in some cultures. Ducks are also mainly reared for meat, though there is a market for duck eggs.

power take-off (PTO) A mechanism on the back (and sometimes the front) of tractors and other vehicles that can provide power for a static, trailed, or mounted implement. It is a horizontal spindle to which a shaft can be attached that rotates to power the moving parts of the implement. *See also* FARM MACHINERY.

PPE *See* PERSONAL PROTECTIVE EQUIPMENT.

PPG *See* PLANNING POLICY STATEMENTS.

PPS *See* PLANNING POLICY STATEMENTS.

prairie A large area of temperate *grassland, typically in North America, similar to the *steppes of Asia and other parts of the world. Originally the prairies of the USA and Canada that included the Great Plains of grassland roamed by buffalo,

Breeds of poultry in the UK

Chicken	Turkey	Geese	Duck
Araucana	Blue	African	Abacot Ranger
Brahma	Bourbon Red	American Buff	Aylesbury
Dorking	Bronze	Brecon Buff	Bali
Ixworth	Buff	Buff Back/ Grey Back	Black East Indian
Legbar	Crimson Dawn or Black-winged Bronze	Chinese	Blue Swedish
Leghorn		Czech	Campbell
Marans	Crollwitzer (Pied)	Embden	Cayuga
Modern Game	Harvey Speckled	Pilgrim	Crested
Old English Game	Narragansett	Pomeranian	Hook Bill
Orpington	Nebraskan	Roman	Indian Runner
Rhode Island Red	Norfolk Black	Sebastopol	Magpie
Scots Grey	Slate	Steinbacher	Muscovy
Sebright	White	Toulouse	Orpington
Sussex		West of England	Pekin
Vorwerk			Rouen
Welsummer			Saxony
			Silver Appleyard
			Silver Bantam

p

they are now mostly in arable cultivation, growing mostly cereals and oilseeds. The prairies are prone to drought, during which the soils are at risk of wind erosion, notably a period during the 1930s known as the Dust Bowl.

precautionary principle Principle, first coming to prominence, in the *United Nations' 1992 Rio Declaration on Environment and Development, which establishes that international governments should make decisions on matters which may affect the environment on a precautionary basis, i.e. if there is a possibility that something may be causing, or may cause (in the case of a new development), damage to the environment, not waiting until there is full scientific evidence of this before refusing to permit its use. The reasoning behind this stems from the principle that whilst laboratory tests and trials can allow some level of understanding of whether a new innovation is likely to cause damage to the wider environment if released into it, the only way to obtain scientific certainty that this will be the case, is to release it into the wider environment and monitor its effects. The precautionary principle urges caution in taking this type of approach, particularly if there is a risk that irreversible harm may be caused. Similarly, if there is a suspicion that something already released into the wider environment may be causing harm to it, then the precautionary principle urges governments to take a precautionary approach to the stage at which they consider there is sufficient evidence to refuse to continue to permit its use, as the longer that governments wait for a body of evidence to build, the greater the risk of damage to the environment. Whilst this principle is widely followed by governments throughout the world, there is an element of subjectivity in its interpretation (as, in the absence of specific international agreements such as those relating to *climate change, it requires governments to take their own views of the weight of scientific evidence that they consider sufficient in order to determine that something is, or is not, causing damage to the environment). Therefore, different governments may take different views of the same products or innovations, with some countries permitting their release into the wider environment, and others not.

precedent *See* CASE LAW.

precision farming A system of arable farming that aims to concentrate resources to gain the optimum results. The first requirement is an accurate digital map showing areas of the field where, for example, *soil types change, and where yields are particularly good or bad, or where specific weeds grow or other pests exist. Satellite mapping using *Global Positioning Systems and computerized systems on *tractors and *combine harvesters create these maps and then use them to adjust inputs. For example, the seed rate might be varied across the field, more *fertilizer applied to areas of the field where the yield potential is high, or herbicides sprayed only where specific weeds are growing. The machinery used, including *drills, *sprayers, and fertilizer spreaders must be capable of responding to the message to adjust activity accordingly. Significant reductions in input costs can be achieved using this system.

predator Animal that kills and feeds upon another animal. Predator control is a common term used especially in *game management where the control of one species that preys upon a *game species is common practice. It can be controversial where protected species such as birds of prey are killed illegally in order to protect *game birds. *See also* GAMEKEEPER.

pre-emergent A chemical, usually a *herbicide, that protects the crop, literally before the emergence of a plant from the ground. The action is residual in that it controls the target weeds over a period of time rather than by direct contact, remaining in the soil and killing the weeds as they germinate and emerge. They are widely used to control grass weeds such as *blackgrass or *wild oats in cereal crops.

premium 1. Capital (or lump) sum sometimes paid at the beginning of a *lease or *tenancy, in addition to the *rent, by the *lessee to the *lessor usually to reflect a reduced *rent during the *term of the tenancy. A premium was sometimes known historically in Scotland as a **grassum**. **2.** Sum paid to an insurer by the insured for an insurance policy. *See also* AUCTION; COMMODITIES.

prescription 1. In England, Wales, and Northern Ireland, the acquiring of a new right, say an *easement or a *profit à prendre*, by long, uninterrupted use over a period of time (normally twenty years for an easement and thirty years for a *profit à prendre*). Although there are some similarities with *adverse possession, prescription does not require that the action that has resulted in the new right is one that is directly in conflict with the owner's rights. **2.** Prescription in Scotland can be either **negative** (the loss of a right over time) or **positive** (the creation of a right over time). Negative prescription can apply to a broad range of things, including, for example, a debt that is not paid which can be eventually extinguished if the person who is owed the money does not take action. It can also include rights such as **servitudes** (the Scottish equivalent of an easement) which can be extinguished if they are not exercised for the relevant period of time set out in *statute. The ownership of land cannot, however, be lost by prescription (so the owner of a house cannot lose ownership of it by negative prescription, even if *squatters have been living in it for a lengthy time). Ownership of land can, however, be acquired by positive prescription. For this to apply, the applicant must have possessed the land openly, peaceably, and without challenge, for a period of time set out in *statute (at the time of writing ten years), and they must also have had the *title of the land transferred to them by someone who it transpires did not themselves have good title, so, in effect, was not actually the owner. Therefore, whilst prescription in this context shares some similarities with the principle of *adverse possession which operates in England, Wales, and Northern Ireland, prescription in this context has a narrower application, as in adverse possession it is the possession (for the correct period of time) that is fundamentally the only element that is required. As in the rest of the United Kingdom, servitudes (the equivalent of easements) can also be created by prescription in Scotland. At the time of writing, a Bill was proceeding through the *Scottish Parliament to amend some elements of the law of prescription following a Scottish *Law Commission review. *See also* LIMITATION.

present value *See* DISCOUNTING.

primary production Production that comes directly from raw materials as opposed to the further processing of a primary product. Thus, growing crops is primary production as it uses the raw materials of soil, seed, water, and nutrients to produce food. Primary products can then be utilized or processed into the secondary products used further up the food chain; for example, milk processed into *dairy products.

prior approval *See* PERMITTED DEVELOPMENT.

private nuisance An action or situation resulting in interference with another's legitimate rights to use and enjoy their land and related rights. For example, someone may continually be playing loud music which affects their neighbour's right to enjoy their own property. Other potential examples of nuisance include excessive dust, smells, or fumes, or actions causing physical damage to a property. Private nuisance is a *tort (and a **delict** in Scotland) and therefore a party who has rights in the affected property (for example, the owner or a *lessee) may go to the *courts to try to remedy the situation. The courts are able to award *damages (i.e. *compensation) or issue an *injunction (or its Scottish equivalent, an **interdict**) requiring the party causing the nuisance to stop. The courts may also require the carrying out of remedial works. Members of the public also have similar protections under the law of **public nuisance**. *See also* RIGHT TO LIGHT; STATUTORY NUISANCE.

Private Rented Housing Panel *See* FIRST-TIER TRIBUNAL FOR SCOTLAND (HOUSING AND PROPERTY CHAMBER).

private residence relief *See* CAPITAL GAINS TAX.

private residential tenancy *See* SCOTTISH RESIDENTIAL TENANCIES.

probate, grant of Document issued by the courts in England, Wales, and Northern Ireland (the equivalent process in Scotland is called **confirmation**) confirming that a will is valid so that the executors of that will can obtain the legal right to deal with the deceased person's **estate** (for example, their *real property, money, and possessions). A grant of probate is not normally required for smaller estates, nor if all of the estate is passing to a spouse; however, for other estates the executors of the will need to have a grant of probate before they can **administer the estate**, i.e. deal with matters such as collecting the assets, paying any debts, paying any *Inheritance Tax, and distributing the estate to the **beneficiaries** (those people entitled to it). If someone dies without a will, then the equivalent process to applying for a grant of probate is applying for a **grant of letters of administration**.

probate valuation *See* INHERITANCE TAX VALUATION.

productive capacity *See* AGRICULTURAL HOLDINGS ACT 1986.

professional indemnity insurance (PII) Insurance which is designed to protect the insured professional (or their firm) and their clients from a substantial financial loss in the event of a claim being made by a third party because there has been a breach of professional duty by the insured in the course of carrying out their professional activities. In addition to protecting the financial position of the professional (or their firm) if a claim is made against them (perhaps because they have given poor or negligent advice), it is also designed to protect a third party who is entitled to *compensation, from a position where the professional (or firm) is unable to pay the claim because it does not have sufficient financial resources itself. A number of professional bodies (for example, the *Royal Institution of Chartered Surveyors) require their membership to ensure that their professional work is covered by professional indemnity insurance, and failure to do so will result in disciplinary action being taken by the professional body.

professional statement Document comprising guidance, good practice, and in a professional context, often specific expectations and requirements of a professional in relation to an area of work. In the context of the *Royal Institution of Chartered Surveyors (RICS), the term 'professional statement' has a specific meaning and is one of the five categories of professional guidance produced by the RICS; namely, *international standards, professional statements, *codes of practice, *guidance notes, and *information papers. The term 'professional statement', in the RICS context, covers a number of different documents including **practice statements**, *Red Book professional standards, global valuation practice statements, regulatory rules, RICS Rules of Conduct, and government codes of practice. Under the bye-laws and regulations of the RICS, its members must comply with RICS professional statements and international standards. Codes of practice and guidance notes provide advice and 'best practice' guidance to practitioners (often in relation to procedures), whilst information papers provide information and explanations to RICS members on various matters relating to surveying (although do not recommend or advise on professional procedures). Although the different types of guidance have slightly different statuses, all practitioners are expected to familiarize themselves with new or updated professional guidance, and the RICS reiterates that if it is alleged that a surveyor has been professionally *negligent, it is likely that the courts will take into account the professional guidance produced by the RICS in deciding whether or not the surveyor has acted with reasonable competence.

profit The amount of money (after all costs have been taken into account) made as a consequence of the trading activities of a business. *See* PROFIT AND LOSS ACCOUNT.

profit and loss account A financial statement (also known as an **income statement**) showing what *profit or loss an organization has made over a period of time. A simple profit and loss account for a *sole trader or *partnership essentially shows:

- The sales and purchases of goods traded (*see* trading account), plus an allowance for the value of trading goods held in store at the beginning and end of the trading period.
- Expenses of running the business (for example, electricity, insurance. etc.).
- *Depreciation of business *assets.
- Any change in the allowance for doubtful debts (i.e. where money owed to the business or organization is unlikely to be paid back), and any bad debts written off.
- Any *interest on loans held by the business.
- Salaries of the employees of the organization.

This then gives the **net profit** (or loss) for the financial period.

In a sole trader or partnership situation, from this net profit (assuming there is one), any **drawings** (in effect the business owner's equivalent of their salary if they were an employee) are then deducted. However, they may also leave some of the profit within the business; this is known as **retained earnings** and may be shown in a **statement of retained earnings**.

For *limited companies, the same principles are essentially followed; however, once the net profit is determined, *Corporation Tax is then deducted. This is because a limited company is a separate legal entity and so pays tax in its own right; whereas in a sole trader or partnership situation the business and its owners are not separate legal entities (apart from in the case of partnerships in Scotland, although this is not treated as relevant for the purposes of Income Tax); hence the sole trader, or partner, pays *Income Tax instead of Corporation Tax (for further information *see* PARTNERSHIP). The financial statements of a limited company will also show what **dividends** (effectively, shares of profit) were paid to the *shareholders, and what element of the profits was retained in the business (these are shown in a separate **statement of changes in equity**, which also shows any other changes relating to shareholdings; for example, if the company has issued new shares or cancelled any).

profit à prendre A legal right allowing one or more parties to take something specific from land owned by someone else; for example, to collect wood from the land to use for fuel. A *profit à prendre* is a legal *interest in the land and is either created by a formal written *deed or by long use of the right (*prescription). There are several types of *profit à prendre* including **profit à pasture** (the right to graze pasture), **profit à pannage** (the right to turn pigs out to eat acorns etc.), **profit à piscary** (the right to catch fish and take them away), and **profit à estover** (the right to collect wood). Whilst there are some similarities in practical effect between a *profit à prendre* and a *licence (which may, for example, give permission for one person to put their livestock on another person's land in order to graze the grass for a specified period), a licence does not create any rights in the land itself, so can later be withdrawn and does not need to be created by deed.

profit averaging *See* INCOME TAX.

profits method Valuation method used to determine the *rental value of properties in situations where their value is likely to be closely related to the *profit that can be generated from using the property for a trade use which the property is specifically fitted out for, and therefore where it would be difficult to use the *comparative method owing to a lack of comparable properties. This approach is typically used for facilities such as hotels or petrol stations, and is often used to assess the **rateable value** of such properties. (For further information on rateable value, *see* BUSINESS RATES.) In essence, the profits method operates as follows:

> Gross earnings of the business
> − Purchases
> = Gross profit
> − Working expenses (except rent)
> = Net profit (fair maintainable operating profit)
> − Allowance for tenant's risk, their work, and interest on their capital
> = Divisible balance.

Once the net profit is determined, an allowance is made from it for the tenant's work in the business (taking care that this has not already been included in the net profit figure), together with a sum to recognize the risk they have in running the business, and also an allowance for interest (*see* INTEREST RATE) on the capital (often money) the tenant has invested in the business. This leaves a **divisible balance** which is then divided between the *landlord and the *tenant (the landlord's portion representing the annual *rent). This method is dependent on the valuer using appropriate figures to assess the profit, and hence is used only by specialist valuers. Whilst the actual *accounts of the business currently operating in the property may be useful for the valuer to see in helping to determine these figures, the valuation is actually based on a hypothetical profit that a reasonably efficient operator could be expected to make in the property. Basing the valuation on the actual business operating in the property may be unfair to either the landlord or tenant, as the actual business may be operated by exceptionally successful and astute business people, or, alternatively, very ineffective business people. If the current business is exceptionally successful it would be unfair for the landlord to receive the benefit of their success in an overly high rent; however, the opposite is also true, as the landlord should not receive an unreasonably low rent if the actual business operating from the premises is badly run and makes a lower profit than other business people could reasonably make by operating the same type of business from the same premises.

pro indivisio *See* TENANCY IN COMMON.

prong The projecting tines of a *pitchfork or the whole tool. The front loader of a tractor may be fitted with a single tine for the handling of big round *bales of *hay, *straw, or *silage.

proprietary estoppel *See* ESTOPPEL.

protected species Certain species of *flora and *fauna that are provided with levels of protection through legislation. Some legislation may protect species at specified times of the year, or, for example, limit numbers and methods of killing or taking of species. Protection of species in Great Britain, for example, is contained within the *Wildlife and Countryside Act 1981, and specifically within its scheduled lists. Subsequent regulations have added to and amended this, including separate legislation within the devolved countries.

protected tenancy *See* RENT ACT 1977; TENANCY; SCOTTISH RESIDENTIAL TENANCIES; NORTHERN IRISH RESIDENTIAL TENANCIES.

Protection from Eviction Act 1977 *Legislation which protects the occupiers of residential properties in England and Wales from someone wishing to regain vacant possession (for example, the *landlord) attempting to **evict** (remove) them by unlawful means or harassment (for example, by cutting off the water or electricity supply in order to force the *tenants to leave). It is an offence to do this, as whilst it may appear that the tenants have left voluntarily, it is, in effect, the actions of the landlord that have forced them to leave. The Act also prevents landlords from using *forfeiture for residential properties with someone living in them, and provides that if a *landlord or *licensor wishes to recover vacant possession from a residential occupier they must do so through the *courts. Parallel provisions protecting the occupiers of residential property from harassment or unlawful eviction also exist in Scotland and Northern Ireland.

protection from eviction legislation Laws in place in the *United Kingdom which make it an offence for someone who wishes to regain vacant possession of a residential property trying to do this by using unlawful means or harassment of the occupiers. In England and Wales this is addressed by the *Protection from Eviction Act 1977, with separate legislation in Scotland and Northern Ireland.

protein An organic compound made up of amino acids that is one of the essential components of living tissue, especially structural as in muscle, or enzymes and antibodies. Thus protein is an important part of the diet of animals, coming from animal sources such as meat or fish, or from plant sources high in protein such as the *legumes, *soya bean, peas, *beans, and *clovers. A protein crop is one grown for its high protein content.

protocol, United Nations An agreement between member states at the *United Nations. They are often used as an instrument to implement a convention, for example, by dealing with ancillary matters, or interpretation of clauses of the convention, or by establishing specific obligations in order to deliver the convention's objectives. *See also* KYOTO PROTOCOL.

provenance A reference to the variety of slightly different characteristics in a *species related to its geographical origin. Where a species occurs over a wide geographical area, these characteristics can often relate to variations in environmental conditions. These can be a significant factor in the choice and success of seed or young trees, for example. In general forestry use the term 'provenance' can be used to describe the source of seed and the trees grown from them.

provisioning services One of the four broad categories of *ecosystem services identified within the *National Ecosystem Assessment. Provisioning services describe the products or energy supplied and utilized by humans such as food, fibre, fuel, and fresh water. Unlike other categories of ecosystem services, these are more commonly traded, and economic values attributed.

PROW *See* PUBLIC RIGHTS OF WAY.

pruning In reference to forestry use, it is the removal of branches up to 4–5 metres above ground, i.e. above the height where *brashing would normally be carried out. The intent would be to limit the knots in any future timber, but it is a labour-intensive and costly operation and is not carried out routinely.

PTO *See* POWER TAKE-OFF.

public access *See* OPEN ACCESS; PUBLIC RIGHTS OF WAY.

public footpath A path over which the right to travel is permitted and is legally recognized. Public footpaths occur in urban areas but are perhaps more commonly recognized in the *countryside. Definitive maps of the network are maintained by *local authorities, and these are used in derived mapping documents such as the **Ordnance Survey**. Access is usually restricted to walkers and, for example, a pushchair or a dog on a lead, but access by horse riders or cyclists is not normally allowed. *See also* PUBLIC RIGHTS OF WAY; NATIONAL TRAIL.

public goods A product or service that is provided without profit to the general public. It can be by an individual, organization or government, but crucially its use by one individual does not limit its availability to be used by others. Common examples would include clean air, views, *parks, and *Public Rights of Way, as well as goods such as knowledge and security. *See also* ECOSYSTEM SERVICES; NON-MARKET GOODS.

public liability insurance Type of insurance held by businesses and organizations that interact with the general public in some way (as a supplier of goods or services, or where the public are entering or crossing their property) and which covers the financial liabilities of a business or organization that is legally required to pay *compensation, and legal costs, if a member of the public dies, or is injured, or made ill, or has their property damaged, as a consequence of the member of the public's interaction with the business or organization. Employees of the business or organization itself who are similarly affected would be covered by their employers' liability insurance. *See also* PROFESSIONAL INDEMNITY INSURANCE.

public path creation order Legal document prepared by a *local authority in England or Wales creating a new public footpath, bridleway, or restricted byway (*see* PUBLIC RIGHTS OF WAY). When such an authority wishes to create an order it will initially publish, and publicize, a proposed order and will invite any objections or representations to be made concerning it. If no such objections are received then the authority may confirm the order. If objections are received, and are not subsequently withdrawn, then the proposed order must go to the Secretary of State for Housing Communities and Local Government, or the Welsh Ministers (*see* WELSH GOVERNMENT). A **public path diversion order** is created in the same way and is used for changing the route of a public footpath, bridleway, or restricted byway, whilst a **public path extinguishment order** is used to stop up a public footpath, bridleway, or restricted byway.

public path diversion order *See* PUBLIC PATH CREATION ORDER; PUBLIC RIGHTS OF WAY.

public path extinguishment order *See* PUBLIC PATH CREATION ORDER; PUBLIC RIGHTS OF WAY.

Public Rights of Way (PROW) Technically described as highways, although usually far from a vision of a metalled trunk road, they are more likely to be an unsurfaced *public footpath or *bridleway though farmland. Other categories of PROW which have defined access characteristics include *Byways Open to All Traffic, *restricted byways, and *green lanes. Landowners may own the sub-soil of the rights of way, but the responsibility of maintaining the surface is at the public expense, although these rights and responsibilities can vary. A series of public path orders can be used by *local authorities in dealing with landowners under the **Highways Act 1980**. These include a **Public Path Creation Order** to create a new route where there is a demonstrable need, although this power is not widely used. Where a **Public Path Diversion Order** is used to divert a route, then consideration must be given to its impact on the general public. These may also be of a temporary nature; for example, to enable development work to be safely undertaken. A **Public Path Extinguishment Order** is also not common and is usually opposed by access organizations such as the *Ramblers Association. They may be used in conjunction with created or diverted routes that bring increased or equivalent public benefits. *See also* NATIONAL PARKS AND ACCESS TO THE COUNTRYSIDE ACT 1949; COUNTRYSIDE AND RIGHTS OF WAY ACT 2000; OPEN ACCESS.

((⊕)) SEE WEB LINKS
• Government website detailing types, rights and responsibilities of PROW in the UK.

puddling 1. Lining, repair, or maintenance of the clay lining of a *pond, reservoir, or canal. A thick layer of clay is laid down and then compacted by some means. This might be by using the bucket of an **excavator**, or a traditional method using the trampling effect of *cattle or *sheep. **2.** Tillage of flooded *rice paddies using *livestock or *tractor to pull a weighted *harrow.

pulpwood Low-value small-diameter *conifer and *broadleaved *roundwood, as well as sawmill offcuts and so on, that are fragmented and enter the manufacturing process for paper production or as sheet material for *chipboard and *fibreboard. *Medium-density fibreboard (MDF) is a type of fibreboard that can be produced in a dry process using heat and pressure. These products are used widely by the construction industry and in furniture.

purchaser Person or organization buying something. In the case of property, the purchaser buys the property from a *vendor.

Q

quad bike A four-wheeled vehicle similar to a motorbike without a cab or covering, also known as an all-terrain vehicle (ATV). They are very versatile machines that can cover difficult terrain and have a low impact on the ground. Often used by those looking after livestock such as shepherds or gamekeepers to move around the farm and carry small loads. They may be equipped with a small spreader or *sprayer for the application of *pesticides such as slug pellets. They may also be used as recreational vehicles to travel or race around a track or across open country.

quadrat A plot of land used in ecology or agronomy to establish the distribution of the organism under study. It is usually a square, such as a square metre, and is replicated across a field. It might be used, for example, to estimate the population of *pests such as *aphids to determine whether a threshold has been reached that justifies the use of a *pesticide.

quail (*Coturnix coturnix*) The European or common quail is a migrant *game bird, and the UK represents its northern range. It has stocky body and is smaller than the *partridge. It has a distinctive call, though is more commonly heard than observed. It is a listed species under the Wildlife and Countryside Act 1981, and is therefore protected.

qualified freehold *See* FREEHOLD; TITLE.

qualified leasehold *See* LEASEHOLD; TITLE.

Quality Meat Scotland *See* LEVY BODIES.

quarantine To isolate or the isolation of animals or plants that have or may be carrying disease. It is used to prevent the spread of infectious diseases or to determine whether the target is or might become infectious. For example, a dog being brought into a country may have a period of quarantine to ensure that it does not have rabies.

quarry species An individual species that is the target to shoot or catch. It may include, for example, all species described as *game and *waterfowl, as well as including hunted species such as *fox and *mink. *See also* GAME BIRD; HUNTING; SHOOTING; STALKING.

quarter days Days which are commonly used as the date when payment of *rent is due. A *lease or *tenancy agreement may specify that annual rent, for example, is paid on a particular quarter day. Quarter days in England and Wales are:

- 25 March (Lady Day).
- 24 June (Midsummer's Day).
- 29 September (Michaelmas Day).
- 25 December (Christmas Day).

In Northern Ireland, quarter days (also referred to as **gale days**) are:

- 1 February.
- 1 May.
- 1 August.
- 1 November.

In Scotland, different dates and names are also used for the quarter days:

- Candlemas (28 February under modern law, although formerly 2 February, which still sometimes applies under old agreements).
- Whitsunday (28 May under modern law, although formerly 15 May).
- Lammas (28 August under modern law, although formerly 1 August).
- Martinmas (28 November under modern law, although formerly 11 November).

quiet enjoyment *See* COVENANT.

quota In the agricultural sector this term has been more commonly associated with one of the intervention tools used by the *European Union to control *milk production. Milk quotas were attached to individual *farms, and they effectively placed a cap on the quantity of milk that could be sold by the *farmer unless a levy was paid. Milk quotas were introduced in 1984, but as part of *Common Agricultural Policy reforms were withdrawn in 2015.

q

rabbit (*Oryctolagus cuniculus*) The European rabbit is a common and distinctive mammal, up to 20 inches long and with long ears. The males are termed bucks, the females are does. They are not a true native breed and were introduced into the UK as a food source by the Romans and subsequently by the Normans. As with the *pigeon they have a particular relevance because of the potential damage caused to agricultural *crops and *trees. They are susceptible to a virus called **myxomatosis**, which is a disease affecting the eyes and brain, and was introduced to the wild rabbit population to reduce numbers.

raddle A harness worn by a *ram which holds a coloured crayon that marks the wool on the hindquarters of a *ewe when the ram serves the ewe. The colour of the crayon is changed regularly to indicate when the ewe was served and thus when it might *lamb, and, if more than one colour is seen, to indicate that the first service was unsuccessful.

ragwort (*Senecio jacobaea*) A poisonous plant with yellow flowers in summer that grows as a *weed in pastures, verges, and waste ground. It is toxic to grazing animals, especially horses and cattle, as the alkaloids it contains can accumulate over a period of time to cause liver failure. The alkaloids are concentrated and thus more toxic when the plant is dried, so cutting and leaving the plant on the ground or incorporating it into hay exacerbates the problem. It is a biennial plant appearing as a flat rosette in the first year before flowering in the second. The seed may be spread by the wind or, in the case of road or railway verges, by the movement of vehicles. It can be controlled by herbicide, but livestock must be prevented from eating it until it is entirely decomposed. Sheep may eat the rosette with little apparent effect, which may limit its spread. The only really effective control method is to pull it up by the roots, but this is a time-consuming process. It is designated as a noxious weed in many countries, including in the UK, where it is one of the five weeds covered by the *Weeds Act.

rainwater harvesting The process of collecting rainwater (usually from the roof of a building or buildings) in a tank so that it can be used on site for applications such as flushing toilets. Systems vary significantly in complexity depending on the end-use of the water and therefore the degree to which it needs to be cleaned before it can be used. *See also* FOUL WATER.

ram A male *sheep.

Ramblers' Association The largest walking charity of Britain, created in 1935 to promote walking in the countryside. It campaigns to increase *public access and to protect and expand the *Public Rights of Way network. Following devolution of UK governments, **Ramblers Scotland** and **Ramblers Cymru** have become separate entities in Scotland and Wales respectively. Members are affiliated through local groups that organize events, campaigning, and volunteer work parties for path maintenance etc. *See also* COUNTRYSIDE AND RIGHTS OF WAY ACT; OPEN ACCESS; NATIONAL PARKS AND ACCESS TO THE COUNTRYSIDE ACT 1949; NATIONAL TRAIL.

SEE WEB LINKS
• Full details of membership information, local groups, and activities.

Ramsar sites Areas designated for the conservation of *wetlands, based on the Convention of Wetlands, an intergovernmental treaty signed in Ramsar, Iran, in 1973. Wetlands that meet the Ramsar criteria are included in the global list of Wetlands of International Importance, and their protection and international cooperation are promoted. Wetlands are very rich in *biodiversity and have a particular emphasis on waterbirds, and consequently many sites are also designated as *Special Protection Areas.

SEE WEB LINKS
• Ramsar website, detailing protection policies and listing all global sites.

ransom value Value attributed to a parcel of land (or potentially a building) because owning it would allow a neighbouring landowner to do something (commonly *development) which would increase the value of the neighbour's land. A typical example of this would be where a farmer wishes to develop a parcel of land, but does not have appropriate access to the public highway from that land because another farmer owns the land between the potential development site and the public highway (the land between would sometimes be known as a **ransom strip**). The farmer who wishes to develop the land may be willing to pay more for the ransom strip (in order to make the development on their land possible) than they would in other circumstances (say for ordinary agricultural land), and this additional value is known as ransom value.

rape *See* OILSEED RAPE.

raptor Belonging to a group also described as birds of prey that include eagles, buzzards, falcons, harriers, and owls. The term is derived from the latin verb *rapere*, to grab or seize, and raptors are characterized by their ability to detect and seize small mammals by using their adapted talons and beak. Although protected in the UK, raptors can be a controversial subject in that some species can prey on *game bird populations and can be illegally persecuted.

RASE *See* ROYAL AGRICULTURAL SOCIETY OF ENGLAND.

rat The brown rat (*Rattus norvegicus*) is an omnivorous mammal with grey-brown fur and a distinct long, scaly tail. They are now widespread throughout the world, and were introduced into the UK in the eighteenth century. They have a particular relevance as a pest of stored *cereals and other foods, a predator of the eggs of wild birds and *game birds, a cause of structural damage to buildings, and as a vector of certain diseases that can spread to humans. The **black rat** (*Rattus rattus*) is the longer-established in the UK, but has declined since the introduction of the brown rat.

rateable value *See* BUSINESS RATES.

rate of return *See* PAYBACK PERIOD.

RDP *See* RURAL DEVELOPMENT PROGRAMME.

real burden The equivalent, under Scottish law, of the *freehold *covenant in English law.

real property Under English law, a term used to describe *freehold land. Scottish land and property law is different from English law, with the terms *immoveable property and *heritable property being broadly equivalent to real property.

real rights Rights in land and *moveable property in Scotland. These rights divide into **ownership** (in the context of land this is broadly equivalent to *freehold in England and Wales) and **subordinate real rights** (i.e. rights of others in relation to the owner's property). Subordinate real rights are also known as **encumbrances**. There are several types of subordinate real rights which can exist in land under Scottish law:

• Servitude (broadly equivalent to an *easement in the rest of the *United Kingdom; for example, a right, attached to the land, allowing Person A, in order to access their land from the public highway, to cross Person B's land).
• Negative *real burden.
• Proper *liferent (giving the holder the right to use the property for their lifetime).
• Right in security (for example, a standard security, i.e. the Scottish equivalent of a *mortgage).
• *Lease of *immoveable property.

Subordinate real rights are shown on the *title sheet of land registered on the *Land Register of Scotland.

In terms of land, a real right is a right in the land itself (for example, an owner's right to own land is enforceable against everyone else). This contrasts with a **personal right** which is only enforceable by and against those who are specifically party to that right. So if Person A enters into a *contract to sell a house to Person B, at that point they each have a personal right against each other. It is only at the point when the transfer of ownership is finished and is registered at the *Land Register of Scotland that Person B now has a real right (of ownership) in that house. Person B's new right of ownership may, however, also itself be subject to subordinate real rights which remain with the land even when its ownership changes (for example, Person C may retain the right to use a pathway across the garden of Person B's newly purchased house in order to access their own property).

It is also possible, in Scotland, to own a **separate tenement** (for example, mineral rights, salmon fishing rights, or a flat). These are not treated as subordinate real right, but is instead another right of ownership in land, so in effect is a part of the land owned separately from the land itself.

For further detail on the position in England, Wales, and Northern Ireland, where there are significant differences in land and property law, *see* ESTATE IN LAND; FREEHOLD; LEASEHOLD.

Recognized European Valuer Status *See* THE EUROPEAN GROUP OF VALUERS' ASSOCIATIONS.

record of condition *See* INVENTORY.

recreation *See* OUTDOOR ACTIVITIES.

Red Book (RICS Valuation: Global Standards) A set of global and United Kingdom standards published by the *Royal Institution of Chartered Surveyors for the *valuation of real estate (land, buildings, and the related interests) and other assets. The standards, which are regularly updated, cover a number of aspects including definitions used in valuations (for example, *market value), how the valuation process should be approached (from accepting an instruction to value something, through the process of carrying out the valuation, to the writing of the valuation report), and the requirements demanded of valuers in respect of professionalism, skills, competence, knowledge, ethics, and conduct. The Red Book incorporates the International Valuation Standards issued by the International Valuation Standards Council, and the 2017 edition has a number of national supplements, including a **United Kingdom Valuation Professional Standards** document. All RICS and *Institute of Revenues Rating and Valuation members providing a valuation are required to comply with the specified Global Valuation Standards (except in certain circumstances identified in the Red Book). The additional United Kingdom Valuation Professional Standards are mandatory in the United Kingdom and supplement, expand, or change the Global Valuation Standards in order to meet national laws and *regulations. *See also* REGISTERED VALUER; EUROPEAN VALUATION STANDARDS.

redundancy Situation where an individual or group of individuals lose their jobs because their employer decides that because of changes within, or affecting, the organization, there is no longer a requirement for the particular job in which that individual (or group of individuals) has been engaged. For example, if sales within a manufacturing company fall dramatically, some or all of the workers making the goods may be made redundant because there is no longer a demand for the items that they were making, and hence for the work that they were undertaking. In the *United Kingdom there are specific legal protections for workers if their employer is considering making redundancies; for example, for large-scale redundancies the employer must consult with the employees first. Specified periods of notice must also be given to employees, who are also legally entitled to redundancy payments. The minimum redundancy payment that an employee must receive is also set out in law, with those who have worked for the organization for longer periods being entitled to higher payments than those who have joined the organization more recently. Legal disputes concerning redundancies are heard at **employment tribunals** (*see* COURTS, LAW).

reedbed A *wetland *habitat dominated by the common reed (*Phragmites australis)* and where the water table is normally at or above ground level. They often form a mosaic of areas of *grassland, open water, and ditches, and are a highly biodiverse habitat including for birds and invertebrates. Reeds are harvested to supply traditional material for thatching roofs. Artificial reedbeds can be constructed as a system of water treatment where water trickling through the reedbed is cleaned by microorganisms living on the root system. *See also* WETLAND.

referendum Vote, conducted either at a local, regional, or national level, which allows all those people on the electoral role (i.e. those who would be eligible to vote in a general or local election) to vote on a specific issue. Rather than voting for an individual to represent the electorate (as in elections), a referendum asks voters to agree or disagree with a specific question; for example, whether they agree that a proposed *neighbourhood plan should form an official part of the *planning system against which applications for *planning permission are considered.

regeneration *See* NATURAL REGENERATION.

Regional Development Agency One of the nine Regional Development Agencies, established by Act of Parliament, which operated in the regions of England from 1999 to 2012, and had responsibilities for promoting economic development, regeneration, employment, business efficiency, investment, and competitiveness, and for enhancing development and skills relevant to employment and for contributing to *sustainable development. Following the *Localism Act 2011, many of their responsibilities have been passed to *Local Enterprise Partnerships.

Regional Spatial Strategy A regional *planning policy document, now defunct, sitting in between national and local planning policy in England. The regional planning structure in England was abolished under the **Localism Act 2011**, and planning policy now exists only at a national, *local planning authority, and *neighbourhood planning level.

registered battlefields Historic battlefield sites listed on the **Register of Historic Battlefields** in England or the **Inventory of Historic Battlefields** in Scotland. The lists include sites where military engagements occurred on dates ranging from the tenth century to the eighteenth century. Registered battlefields are covered by specific *planning policy, and the existence of such a designation is a *material consideration in planning decisions. Northern Ireland and Wales do not have registers of historic battlefields, but *Cadw is currently developing such a register for Wales. *See also* HERITAGE ASSET.

registered parks and gardens Nationally important parks, gardens, and designed landscapes listed on the national **Registers of Historic Parks and Gardens** (in England and Wales), the national **Inventory of Gardens and Designed Landscapes** in Scotland, and the national **Register of Historic Parks, Gardens, and Demesnes** in Northern Ireland. In England and Wales, sites are graded I, II*, and II in the same way as *listed buildings, but no equivalent grading systems operate in Scotland or Northern Ireland (although sites in Northern Ireland are distinguished as either 'registered' or 'supplementary'). Registered parks and gardens are covered by specific *planning policy and the existence of such a designation is a *material consideration in planning decisions. *See also* HERITAGE ASSET.

Registered Valuer Valuer who is registered with the *Royal Institution of Chartered Surveyors to carry out *Red Book valuations. Since 2011, all UK practitioners carrying out Red Book valuations have been required to join the RICS Valuer Registration scheme, and Registered Valuers must comply with the Red Book valuation standards. Registered Valuers are required to demonstrate prescribed levels of knowledge and experience in relation to valuation in order to register, and are subject to monitoring by the RICS.

Registers of Scotland Department of the *Scottish Government which holds public registers relating to land and property and also legal documents. Registers of Scotland are responsible for a range of registers, including the *Land Register of Scotland and the **General Register of Sasines**. In addition, Registers of Scotland hold a number of other registers, including:

- **Books of Council and Session** comprises three registers, including a **Register of Deeds** which holds originals of *deeds (including *leases and wills). Deeds can be deposited at the Books of Council and Session (usually voluntarily). The original will then be held by Registers of Scotland for safe keeping, and they will then issue an extract (official copy) which can be used in place of the original deed (known as registering the deed for **preservation**). It is also possible to register a deed for the purposes of **preservation and execution**. This then means that, providing the deed has a clause in it where the parties have agreed to registration for preservation and execution, the extract will have a warrant added to it allowing for lawful execution, which will be equivalent to a **decree** (i.e. a formal *court order) from the *Court of Session. This, therefore, potentially allows **summary diligence** to be used in certain circumstances without first having to seek a decree from the courts. (For further information on summary didligence, *see* BREACH OF COVENANT.)
- **Scottish Landlord Register** is the official compulsory online register of private residential *landlords.
- **Scottish Letting Agent Register** is the official compulsory online register of letting agents.

Registry of Deeds *See* LAND REGISTRY.

regulated residential tenancy *See* RENT ACT 1977 TENANCY; SCOTTISH RESIDENTIAL TENANCIES; NORTHERN IRISH RESIDENTIAL TENANCIES.

regulating services The diverse benefits that people obtain through the regulatory functions of *ecosystems such as the regulation of flooding, climate, and some human diseases. *Pollination is a key factor in the production of many food crops, and is a good example of a regulating service that can be overlooked and undervalued. *See also* ECOSYSTEM SERVICES.

regulation A rule or law made by a government or official body. Regulations are generally those rules or processes which must be followed in order to comply with the law or the requirements of a particular organization. They are therefore distinct from guidance or **good practice**, which are usually encouraged rather than required. *See also* LEGISLATION.

reinstatement cost *See* REINSTATEMENT COST ASSESSMENT.

reinstatement cost assessment An assessment of the total cost (the **reinstatement cost**) of replacing a building which has been destroyed (**total loss**). The costs will include the total cost of rebuilding the building, including factors such as demolition and debris removal, and professional, *planning, and *building control fees. These kinds of assessment are commonly carried out in order to determine what sum a building should be insured for, and are therefore generally on a total loss basis; however, buildings can sometimes be insured on other bases (for example, if the current building is obsolete). It is important that the assessment is correct, because if a building is under-insured, in the event of a claim the insurers will not pay out the full cost of reinstating the building (because, by under-insuring the building, i.e. insuring it for a lower amount than it will actually cost to replace it, the insured is likely to have been paying lower *premiums than they should have been). Because of this it is important that the person undertaking the reinstatement cost assessment has a robust understanding of the relevant costs so that the overall reinstatement cost is as accurate as possible. There are regularly updated standard costings available (for example, through the *Building Cost Information Service of RICS), but these will need to be adjusted for local conditions. Non-standard buildings (for example, *listed buildings) can cause significant issues in reinstatement cost assessments as the costs will, potentially, depart very significantly from standard costings. Reinstatement cost assessments are sometimes referred to as **insurance valuations**; however, they are not estimates of *market value but instead assessments of cost, and so should be referred to as reinstatement cost assessments.

related earning capacity *See* AGRICULTURAL HOLDINGS ACT 1986.

relief for heritage assets (conditional exemption) *See* INHERITANCE TAX.

remediable breach *See* BREACH OF COVENANT.

remediation The process of cleaning up and/or restoring land that has been damaged in some way, commonly either through being contaminated by some substance or, alternatively, through processes such as mineral extraction. *See also* CONTAMINATED LAND.

remediation notice *See* CONTAMINATED LAND.

rendzina A fertile shallow dark soil over chalk or limestone, rich in organic matter. *See also* SOIL.

renewable Term used to describe a natural resource which, if used or lost, can be replaced in a relatively short timescale (for example, within the time frame of a human life). Timber from *forestry is therefore described as a renewable resource because the growing cycle is relatively short, and therefore, timber which is used (say for fuel or as a building material) can be replaced through the growth of new trees relatively quickly. This contrasts with *non-renewable materials such as *fossil fuels, where the cycle of the material is so long (in the case of fossil fuels, millions of years) that once used they cannot, in effect, be replaced. *See also* SUSTAINABLE DEVELOPMENT.

renewable energy Energy which is generated from *renewable sources, including *anaerobic digestion, *biofuels, *biomass, *geothermal energy, *heat pumps, *hydro-electricity, *solar power, and *wind turbines. As well as land-based forms of renewable energy generation, there are several other types of renewables, including **tidal and wave power** (which utilize tidal and wave movements to drive generators producing electricity) and offshore wind power. *See also* CONTRACTS FOR DIFFERENCE; FEED-IN TARIFF; RENEWABLE HEAT INCENTIVE; RENEWABLES OBLIGATION; RENEWABLE TRANSPORT FUEL OBLIGATION.

Renewable Heat Incentive (RHI) *United Kingdom support scheme for heat-producing *renewable energy installations. Installations which are registered under the scheme receive payments for between seven and twenty years which are based on the heat supplied by the system, and the scheme is therefore designed to incentivize investment in renewable heat technologies. The scheme is divided into a **Domestic Renewable Heat Incentive (RHI)** scheme (which is relevant to installations serving individual households) and a **Non-Domestic RHI** scheme (which covers businesses, charities, the public sector, and installations serving more than one domestic property), and whilst both the Domestic and Non-Domestic schemes operate on similar principles there are some differences between the two; for example, payments are received for fewer years under the Domestic scheme. The technologies which are supported by the RHI include *anaerobic digestion, *biomass boilers, *heat pumps, and *solar power (thermal). Whilst the RHI scheme initially operated in the whole of the United Kingdom (and continues to exist in England, Wales, and Scotland), it was suspended in Northern Ireland in 2016.

Renewables Obligation Certificate (ROC) *See* RENEWABLES OBLIGATION.

Renewables Obligation (RO) United Kingdom *renewable energy support system introduced in 2002 (and in 2005 in Northern Ireland) in order to incentivize the production of increased levels of renewable electricity from large-scale generators. The scheme closed to new generators in 2017 and has been replaced for new large-scale generation projects by the *Contracts for Difference system; however, existing generators remain within the Renewables Obligation scheme. (Small-scale renewable electricity projects in England, Wales, and Scotland have been supported since 2010 by the *Feed-in Tariff (FIT) scheme, which continues to operate but is expected to end in spring 2019. A separate system for small-scale generators has operated in Northern Ireland and is discussed in further detail below.)

The Renewables Obligation functions by imposing a requirement on large-scale electricity suppliers to source a given percentage of their electricity from renewable sources. The system is implemented through **Renewables Obligation Certificates (ROC)**. ROCs are issued to generators of renewable electricity who then sell them to the electricity suppliers who, in turn, present them to the government regulator to demonstrate that they have met their renewable energy obligations. If a supplier does not present the required amount of ROCs in a scheme period, then they must pay an equivalent amount of money into a **buy-out fund**. This is then redistributed to renewable electricity suppliers in proportion to the number of ROCs they have presented. The purpose of the scheme, therefore, is to incentivize generators to increase the amount of renewable electricity they generate, by making it more financially attractive.

The RO was open to generators of renewable electricity using a number of different technologies, including *wind turbines, *solar power (photovoltaic), *hydro-electricity, and *anaerobic digestion. Since its introduction, alongside the FIT, there has been a significant increase in the installation of these technologies in the United Kingdom.

The Renewables Obligation replaced the previous large-scale electricity generation renewable energy support scheme, the *Non Fossil Fuel Obligation, which operated in England, Wales, and Northern Ireland, and its equivalent in Scotland, the **Scottish Renewable Obligation**.

As noted above, the FIT scheme has not been operating in Northern Ireland, where, instead, very small-scale generators (termed **microgenerators**) have been supported through the **Micro-Northern Ireland Renewables Obligation (Micro-NIRO)** scheme which operates on the same principles as the more general Renewables Obligation system. The Micro-NIRO scheme, however, closed to new entrants in 2017.

For information on *Levy Exemption Certificates, which were formerly issued to renewable electricity generators alongside ROCs, *see* CLIMATE CHANGE LEVY.

Renewable Transport Fuel Obligation (RTFO) Requirement imposed on larger suppliers of transport fuels in the *United Kingdom that a given percentage of the fuel they supply must be from *renewable sources (for example, *bioethanol and *biodiesel). The percentage is set out in *regulations, and many suppliers meet it by blending small quantities of renewable transport fuels into their conventional fossil fuel-based products (e.g. petrol and diesel). The requirement was introduced by the UK government to promote the supply of renewable transport fuels (and, in part, to meet *European Union obligations in this area).

rennet Enzymes produced in the stomach of ruminant animals of which the key component is chymosin or rennin. Traditional animal rennet is an enzyme derived from the fourth stomach of new-born calves, lambs, or goats, where it is produced to help them digest their mother's milk. The enzyme curdles milk, separating it into the curds (solid) and whey (liquid). The curds are then used to make cheese. Rennet can also be sourced from plants or microbes.

rent Regular payments (normally of money) made by the *tenant to the *landlord for a property that is let in return for the right to *exclusive possession of that property for the *term of the *tenancy. Capital sums paid at the start of a *lease or *tenancy (or on *assignment) are not rent. Purchasers of a long *leasehold, for a residential property (say a flat), may therefore pay a capital sum at the beginning of the term, but will continue to pay **ground rent** (usually a relatively small amount) throughout the term of the lease.

Rent Act 1977 tenancy Residential *tenancy in England and Wales (also referred to as a **regulated tenancy**) which is most likely to be the type of tenancy in existence if it commenced before 15 January 1989 (which was when the new types of residential tenancy under the *Housing Act 1988 came into effect). In order to qualify as a Rent Act 1977 tenancy the tenancy must be of a dwelling house let as a separate dwelling and occupied as a residence. When the contractual period of the tenancy (at that stage known as a **protected tenancy**) terminates, the tenancy then usually becomes a **statutory tenancy**, and hence Rent Act 1977 tenants have *security of tenure.

There are some types of occupation which cannot be protected tenancies. These include:

- Dwellings with a low rent, or those with a high rateable value or rent (the measure used depends on the date the tenancy commenced). Rateable value in this instance relates to a pre-1990 system of domestic local taxation.
- Lettings with more than two acres of agricultural land.
- Lettings where the dwelling house is within an agricultural holding (under the *Agricultural Holdings Act 1986) and is occupied by the person responsible for the control of the farming (either the tenant, or an agent or servant of the tenant).
- Lettings where the dwelling house is held under a farm business tenancy (under the *Agricultural Tenancies Act 1995) and is occupied by the person responsible for the management of the farming (either the tenant, or an agent or servant of the tenant).
- Lettings where the rent includes payment for board (meals) and attendance (services such as cleaning or laundry).
- Other types of let, including student or holiday lets or dwellings with resident landlords.
- *Local authority tenancies.

Tenancies protected by the Rent Act 1977 are subject to **regulated rents**, known as **fair rents**. Either the *landlord or *tenant can apply to the *Rent Officer to register a fair rent for the property, and these are held on a public register of fair rents. Generally, a new level of fair rent can only take effect every two years. The Rent Act 1977 specifies that the Rent Officer must take into account a number of factors when assessing the rent, including disregarding **scarcity** (so there is an assumption in determining the rent that the number of properties available is not substantially less than the number of tenants looking for property to rent). The amount by which the landlord can increase the rent is also, however, subject to an inflation-linked cap, regardless of the level of fair rent potentially determined.

When the protected tenancy terminates (either by effluxion of time, the serving of a *notice to quit, or other *common law methods of termination, for example *forfeiture), then under the Rent Act 1977 a statutory tenancy comes into effect on broadly the same terms as the protected tenancy, allowing the tenant to remain in the property.

A landlord can only recover vacant possession of a protected tenancy by obtaining a *court order for *possession (and in the case of a contractual tenancy whereby the statutory tenancy has not yet arisen then landlord will also need to terminate the agreement by one of the common law methods, such as a notice to quit, or forfeiture). In order to obtain a court order for possession, the landlord will need to either demonstrate that suitable alternative accommodation is available (and that it is reasonable to make a possession order), or alternatively that one of the **mandatory** *grounds for possession included in the Rent Act 1977 has been satisfied, or that one of the **discretionary** grounds applies and that it is therefore reasonable to make a possession order (in the case of discretionary ground the court will need to be satisfied that it is reasonable to make a possession order).

There are nine discretionary grounds, including circumstances such as unpaid rent or breaches of the tenant's *obligations. Many of the ten mandatory grounds are designed to cover situations where the property had been used, or intended, for some purpose before it was let, and the tenant now wishes to use it for that purpose. However, given that the landlord must, in most cases, have given notice of these circumstances before the start of the tenancy, and that Rent Act 1977 tenancies have long since ceased to be the main type of residential tenancy for new tenants, the application of these grounds is now significantly less relevant than previously.

There is also the potential opportunity for two *successions to occur under the Rent Act 1977. There are complex rules relating to entitlement to succession; however, it is in essence available for spouses or civil partners and other family members who reside with the tenant. Depending upon the date of death of the tenant(s) and whether it is a spouse/civil partner or other family member succeeding to the tenancy, the new tenancy will either be a statutory tenancy or an **assured periodic tenancy** (under the *Housing Act 1988).

Rent (Agriculture) Act 1976 Legislation protecting qualifying agricultural workers in England and Wales where their occupation of a dwelling house began before 15 January 1989 and who would otherwise fall outside the protection of the *Rent Act 1977 because their occupation does not comply with the minimum rent threshold existing under the Rent Act 1977 (being at a very low, or no, rent). In order for the Rent (Agriculture) Act 1976 to apply, the agricultural worker (the **protected occupier**) must meet qualifying criteria in the legislation; for example, in relation to the length of time they have worked in agriculture (which has its own definition within the Act). They must also be occupying under a *tenancy or *licence (for exclusive occupation) a dwelling house owned by their employer, or by someone with whom their employer has made an agreement to provide housing for their agricultural workers.

Once the employee is a protected occupier then they retain that status even if they leave the original employment (and potentially even if they leave agriculture). However, once the employment ends, the landlord can charge *rent. The security of tenure provisions under the Rent (Agriculture) Act 1976 operate in a very similar way to those of the Rent Act 1977, i.e. that once the original contractual agreement between the parties is terminated by a *notice to quit, or a notice to increase the rent, a **statutory tenancy** then arises and the protected occupier becomes a statutory tenant. They can then remain in the property unless the landlord obtains a *court order for possession using one of the *grounds for possession included within the Act. As with the Rent Act 1977, these are divided into **discretionary** and **mandatory** grounds. The discretionary grounds include suitable alternative accommodation is available, and are generally similar to those under the Rent Act 1977; however, there are significantly fewer mandatory grounds.

Under a statutory tenancy, then either the landlord or tenant can apply to register a **fair rent** in the same way as for the Rent Act 1977. They can also, in the absence of the registration of a fair rent, agree a rent between themselves; however, the Act sets out restrictions on how it can be determined.

One succession is available under the Rent (Agriculture) Act 1976 for either a spouse or civil partner or family member residing with the deceased before their death. Depending on whether the deceased was a protected occupier or a statutory tenant, and whether the person succeeding was the spouse/civil partner or a family member, the potential successor may become a protected occupier, or a statutory tenant, or an assured tenant (under the *Housing Act 1988).

Similar tenancies and licences granted to agricultural workers after 15 January 1989 are addressed through the Housing Act 1988.

See also TIED COTTAGE; SERVICE OCCUPANCY.

Rental Valuation Officer An official of *Rent Service Scotland who determines and registers *fair rents. *See also* RENT OFFICER.

rental value The amount received as *rent for a property which is let or leased.

rent arrears Outstanding *rent which has not been paid on the due date by a *tenant to their *landlord and is therefore still owed to them. *See also* BREACH OF COVENANT.

Rent Assessment Committee *See* First-Tier Tribunal Property Chamber; Rent Assessment Panel; Residential Property Tribunal Wales.

Rent Assessment Panel Independent panel, funded by the Northern Ireland government, from which Rent Assessment Committees (usually comprising a Chair and one member) are established in order to consider if rents determined by a *Rent Officer are appropriate.

rent book Record of *rent paid by a residential *tenant. In the *United Kingdom, certain tenants are legally entitled to a rent book (for example, tenants who pay their rent weekly in England, Wales, and Scotland, and most private residential tenants in Northern Ireland). The law sets out the information that must be included in a rent book.

Renting Homes (Wales) Act 2016 *Legislation introducing new **occupation contracts** for rented residential property in Wales. At the time of writing, the legislation has not yet come into effect, but it includes provisions for the introduction of a **standard contract** for private *landlords which will replace the assured shorthold tenancy and operate on similar principles (*see* Housing Act 1988), and a **secure contract** which will be used for *local authority and housing association lettings.

Rent Officer Official of the *Valuation Office Agency in England, the *Welsh Government, or *Rent Service Scotland (where they are also sometimes referred to as *Rental Valuation Officers); the Rent Officer for Northern Ireland is funded by the Northern Ireland Department for Communities. They assess levels of *fair rent for *regulated residential tenancies in England, Wales, and Scotland, or maximum rent for rent controlled *Northern Irish residential tenancies. In addition, they also determine the levels of housing benefit to which private-sector tenants can be entitled. *See also* First-Tier Tribunal Property Chamber; First-Tier Tribunal for Scotland (Housing and Property Chamber); Rent Assessment Panel; Residential Property Tribunal Wales.

rent review The process of reassessing the level of *rent of a property which has been subject to a *lease or *tenancy for some time (and either increasing or decreasing it). The lease or tenancy agreement may include provisions for the frequency of rent reviews, and the basis on which the new rent is to be assessed; however, for many types of tenancy the law sets out provisions for rent reviews (in terms of how much notice must be given for a rent review, how frequently they can happen, and on what basis the new rent can be assessed).

Rent Service Scotland Scottish Government organization which provides valuations for *fair rents for *regulated tenancies (in Scotland generally, residential tenancies starting prior to 2 January 1989) or for certain secure tenants of housing association property. A Rent Service Scotland *Rental Valuation Officer (or *Rent Officer) will determine and register the fair rent. If either the landlord or tenant object to the fair rent which has been set, then generally the case is referred to the *First-Tier Tribunal for Scotland

(Housing and Property Chamber), which will then reconsider the fair rent which has been determined, either raising, lowering, or confirming it.

rent tender Written proposal, put forward by a prospective *tenant, explaining what level of *rent they would be prepared to pay for an *agricultural tenancy, and their proposals for the farm or *bare land they wish to rent. Commonly, a farm, or bare land, for rent is advertised and then prospective tenants are invited to submit rent tenders by a specified date, usually after a viewing day when the potential tenants can see the farm. The prospective *landlord, or their agent, will then consider the various rent tenders they have received (and usually interview the potential tenants) in order to determine who they wish to offer the tenancy to. The prospective landlord may not necessarily always offer the tenancy to the prospective tenant who has proposed the highest level of rent, because the landlord will also be interested in how they are planning on running the farm, and the likelihood that they will be able to continue to pay the rent at the level they are suggesting (i.e. how what they are planning on doing with the farm will generate sufficient money to pay the rent). Therefore, rent tenders will often include things such as the background of the prospective tenant, including their experience and qualifications, how much land (if any) they are currently farming, their general approach to farming (if they are keen on environmental projects, for example), and their financial background (including how they will finance any proposals for the farm or land, including details of any potential borrowing). A rent tender would also normally include detailed budgets and *cash flows for the enterprises that the prospective tenant is proposing undertaking on the farm (and how these will then support the level of rent they are suggesting). A similar process of tendering is also sometimes used for *contract farming agreements.

repairing covenant *See* REPAIRING OBLIGATION.

repairing obligation Obligation to keep the premises let under a *lease or *tenancy in repair. Generally, a lease or tenancy agreement will set out who has responsibility for keeping what in repair, the general principle being that the longer the tenancy the less likely that the landlord will be responsible for repairs. However, in the case of residential tenancies of less than seven years in England and Wales, the **Landlord and Tenant Act 1985** sets out specific responsibilities that the landlord has for keeping the structure and exterior of the property repaired, as well as facilities such as water supply and heating. In Scotland, private landlords of most residential properties are required to meet legal **Repairing and Tolerable Standards** which encompass the same broad requirements as in England and Wales, with some additional provisions. In Northern Ireland, similar provisions are included within a requirement for let private residential properties to meet a **Housing Fitness Standard**. *See also* COVENANT; BREACH OF COVENANT; HOUSING HEALTH AND SAFETY RATING SYSTEM.

repairing tenancy *See* SCOTTISH AGRICULTURAL TENANCIES.

reporter (Scotland) *See* LOCAL PLAN.

Reporting of Injuries, Diseases, and Dangerous Occurrences Regulations (RIDDOR) Health and safety regulations, applying in the *United Kingdom, which require employers, the self-employed, and, in some circumstances, people who have some control over work premises, to report serious workplace injuries, plus some specified diseases which are thought to be linked to workplace exposure, and also near-misses with the potential to cause serious injury or death, to the *Health and Safety Executive in England, Wales, and Scotland (or the *Health and Safety Executive Northern Ireland in Northern Ireland).

reserve *See* AUCTION.

reserved matters Those matters which are not included in an application for *outline planning permission and which therefore require subsequent approval by a *local planning authority.

Residential Property Tribunal Wales Independent tribunal which resolves disputes in Wales relating to private rented or leasehold property. It has a broadly equivalent role to the *First-Tier Tribunal Property Chamber: Residential Property, in England. *See also* COURTS, LAW.

residential tenancies For England and Wales *see* RENT ACT 1977 TENANCY; RENT (AGRICULTURE) ACT 1976; HOUSING ACT 1988 (assured tenancy, assured shorthold tenancy, and assured agricultural occupancy); RENTING HOMES (WALES) ACT 2016. For Scotland *see* SCOTTISH RESIDENTIAL TENANCIES. For Northern Ireland *see* NORTHERN IRISH RESIDENTIAL TENANCIES.

residual valuation method Valuation method used to assess the value of sites with *development potential and based upon the principle that the value of a development site is likely to be the amount left behind when all of the costs of development, plus the developer's *profit, are deducted from the value of the final development (gross development value or GDV). Residual valuation approaches also underpin many of the methods of *development appraisal. In essence, the following formula describes how the residual valuation method operates:

value of the completed development (gross development value)
- costs of development (including developer's profit)
= value of the site in its current condition (residual value)

In order to use this method, it in essence requires the valuer to determine the GDV and costs of a hypothetical project (which, therefore, usually requires the valuer to determine what is likely to be the most appropriate use of the site that is likely to be given *planning permission). The types of cost that will be included are construction costs, planning fees and related costs, taxation costs, legal and other professional fees, marketing costs, **contingencies** (i.e. an allowance for unexpected costs), and the costs of borrowing money to carry out the development (or alternatively the related *opportunity cost, i.e. the potential loss of income which could have been generated by doing something else with the money invested in the development; at its simplest, say the loss of interest had the money been left in the bank). An appropriate amount which represents the amount of profit acceptable to a developer undertaking the project will also need to be determined and included.

The residual valuation method has been the subject of many criticisms because of the number of assumptions that underpin it and therefore the potential for the residual value estimated to be significantly affected by any changes in these assumptions. For example, there is an inherent assumption in this method that the gross development value of the final completed project has been correctly estimated (and indeed that the right kind of hypothetical development has been used), and also that all the other costs of the development have been accurately determined. To try to address this it is possible to use more sophisticated, computer-based models to evaluate the impact of changing some of these factors on the calculated residual development value in order to test that figure (this is referred to as conducting a **sensitivity analysis**); however, it still does not eliminate the possibility of the 'wrong' figures being used in the calculation.

At its most simple, the approach discussed above also assumes that any borrowings are evenly spread through the life of the project and that costs are incurred right from the start (neither of which is likely to be the case). Hence, the residual valuation method can also be developed at a more sophisticated level by using *discounted cash flow approaches which try to incorporate a more realistic interpretation of cash coming into and out of the hypothetical project. There are a number of computer software packages that can be used to carry out this kind of analysis.

However, despite these potential refinements to the process, the residual valuation method is still open to significant criticism owing to the range of underpinning assumptions, and for that reason is now often used alongside the *comparative valuation method to assess the value of development sites. It is not, however, always possible to find comparables for development sites, and therefore the residual valuation method continues to be used in some circumstances.

See also DEVELOPMENT APPRAISAL.

resilience A common term applied to *farms and *farmers to describe an ability to recover from adverse impacts such as unfavourable weather or disease and maintain agricultural production. It can also commonly describe resilient *ecosystems that have an increased ability to recover to their original state after, for example, man-made or natural disasters. *See also* ECOLOGICAL NETWORKS.

restricted byway A category of *Public Rights of Way introduced in the *Countryside and Rights of Way Act 2000. It is defined as a PROW and more commonly allows usage by foot, horse, and by vehicles that are not mechanically propelled, such as a horse and cart. They are intended to replace and clarify the older category of a '**road used as a public path**' (RUPP) which was originally introduced under the *National Parks and Access to the Countryside Act 1949.

restrictive covenant *See* COVENANT.

Retail Prices Index (RPI) Measure of *inflation in the *United Kingdom, published by the government's **Office for National Statistics** and based on the costs of a basket of commonly purchased household goods and services (including owner-occupier housing costs). The Retail Prices Index, although still published, is not now used as a main reporting measure by the government, having been replaced by the *Consumer Prices Index Including Housing Costs. *See also* CONSUMER PRICES INDEX.

retrospective planning application Planning application submitted after a development has been started or completed. *See also* ENFORCEMENT, PLANNING.

Revenue Scotland *Scottish Government body responsible for collecting the taxes that the Scottish Government has the powers to collect. At the time of writing, these taxes are the **Land and Buildings Transaction Tax** (*see* STAMP DUTY LAND TAX) and the Scottish Landfill Tax (*see* LANDFILL TAX). Other national taxes paid by individuals, businesses, and organizations in Scotland continue to be collected by *Her Majesty's Revenue and Customs.

reversion The part of an interest in land which is left after the owner of that interest has granted another, lesser, interest out of it but has retained what is left of their own interest. So, if a *lease is granted by a *freehold owner, then the freehold is the reversion. *See also* INVESTMENT METHOD.

RHI *See* RENEWABLE HEAT INCENTIVE.

rhizome An underground stem from the nodes of which new shoots and roots can emerge as a form of vegetative propagation. An example of a plant that spreads in this way is the weed *couch grass.

rhododendron A woody plant of the heath family *Ericaceae*, introduced into the UK in the nineteenth century where it was planted in gardens and woodlands as an ornamental shrub, as a *shelterbelt, or as *game cover. Subsequently, rhododendron has spread extensively into the *countryside in some areas, outcompeting native *flora. Control is labour intensive and costly, and requires coordinated efforts at a landscape scale and over a period of years. *See also* NON-NATIVE SPECIES.

rice A cereal of the *Oryza* family, used primarily for human consumption; its grain is the staple diet of half the world's population, especially in Asia. It is planted in early summer, usually in flooded *paddy fields, and is harvested in autumn. *See also* CEREALS.

RICS *See* ROYAL INSTITUTION OF CHARTERED SURVEYORS.

RICS UK Commercial Real Estate Agency Professional Statement
Royal Institution of Chartered Surveyors' (RICS) publication setting out
standards, best practice guidance, and mandatory requirements, in relation to

the selling, letting, and management of commercial property in the United Kingdom. The document has the status of an RICS *professional statement.

RICS UK Residential Real Estate Agency Professional Statement (Blue Book) Royal Institution of Chartered Surveyors' (RICS) publication setting out standards, best practice guidance, and mandatory requirements, in relation to the selling, letting and management of residential property in the United Kingdom. The document has the status of an RICS *professional statement.

RICS Valuation: Global Standards *See* RED BOOK.

RIDDOR *See* REPORTING OF INJURIES, DISEASES, AND DANGEROUS OCCURRENCES REGULATIONS.

ride An unsurfaced and simple track through *woodland, used for access, extracting timber, marking out *compartments, and *game *shooting, and as an aid to deer control and pest control. Rides can have significant ecological benefits through promoting habitat diversity within woodlands and increasing *biodiversity, as well as some being of historical and cultural interest. *See also* BRIDLEWAY.

ridge The raised area of ground caused by the operations of a *plough. As the soil is turned over, it creates a peak called the *ridge, and a trough called the *furrow. Where this took place repeatedly in mediaeval times it can still be seen in the land formation today, and is known as *ridge and furrow.

ridge and furrow An archaeological pattern of mediaeval cultivation. Seen usually in pasture today, it is a relic of the open field strip system of farming common in Europe in the Middle Ages. *See also* BEETLE BANK.

Riding for the Disabled (RDA) An established UK charity that uses *horses and ponies to provide therapy and enjoyment for people who are physically disabled or have learning difficulties.

rifle A long-barrelled firearm designed to fire a single bullet. The barrel contains spiral grooves and ridges which spin and stabilize the bullet in flight. A barrel thus spiralled is said to be 'rifled'. There is detailed legislation in the UK governing the use and ownership of rifles, and also legislation that requires the use of minimum rifle calibres and specifications of the ammunition used for specific species. The bolt-action rifle is a common type of sporting rifle and is universally popular for shooting deer and other large mammals. *See also* DEER STALKING.

right to light A right, under the law in England and Wales, for a property to benefit from continued access to natural light. It is a form of *easement, and is in effect acquired after having had twenty years uninterrupted access to light across another's land by right (rather than permission). If, therefore, someone is proposing to do something (or has done something) to block that light (for example, by erecting a building on their property which blocks the natural light going into a window of their neighbour's property), and the conditions entitling the neighbour to a right to light apply, then the neighbour may be able to take the matter to the *courts (as infringing a right to light is considered to be the *tort of *nuisance). The courts may award *damages (i.e. *compensation), or alternatively issue an *injunction requiring changes to the building project or, in extreme cases, preventing the building from being erected at all.

ring-barking The removal of a ring of bark from the entire stem of the tree, either deliberately or by livestock that will cause the tree to die. Mammalian forest pest species such as *deer and hare can also be responsible for ring-barking. Ring-barking can be a useful management tool to destroy unwanted trees within woodland, and to additionally provide high-value dead-wood *habitats.

ringworm A skin disease caused by one of a number of *fungi, the main symptom of which is an itchy red or silvery circular rash on the skin. It is highly contagious and can be spread from animals such as cattle and dogs to humans. It is treated by a fungicidal cream or ointment applied to the rash, or occasionally by an oral medicine that is circulated around the body.

riparian Derived from latin *ripa*, meaning bank, it is the area of land immediately adjoining a river or other natural *watercourse. The riparian vegetation that characterizes these margins often includes hydrophilic plants, and can be particularly important for their role in *soil conservation. The riparian zones can be more tightly defined; for example, riparian *woodland or riparian *buffer zones. A person who owns land on the bank of a watercourse is a riparian owner having both rights and responsibilities described as riparian rights. These might also include allocation of water resources and fishing rights.

risk assessment Process of identifying the risks of an undertaking in a systematic way. Whilst risk assessment principles can be used in a variety of scenarios, risk assessment plays a particular role in health and safety law in the *United Kingdom, with *regulations requiring those with responsibility for health and safety to undertake risk assessments to identify potential hazards and to assess the risks that they pose (including who might be harmed and the probable severity of that harm), in order to enable steps to be taken to control that risk. *See also* HEALTH AND SAFETY AT WORK ACT 1974; HEALTH AND SAFETY AT WORK (NORTHERN IRELAND) ORDER 1978; HEALTH AND SAFETY EXECUTIVE; HEALTH AND SAFETY EXECUTIVE FOR NORTHERN IRELAND.

riverbank *See* RIPARIAN.

ROC *See* RENEWABLES OBLIGATION CERTIFICATE.

rod An old unit of measurement, used in the United Kingdom, equal to 5½ yards (approximately 5 metres). In practical terms, a surveyor would use a rod of this length to measure short distances. One rod is also equal to one pole or one perch. *See also* CHAIN.

rodenticide A poison used to kill rodents, particularly rats and mice. In the UK, the use of rodenticides is regulated according to categories, and only those who have completed an approved course can use the latest anticoagulant rodenticides. *See also* PESTICIDE.

rod licence Any angler aged over 12 is required by law to have such a licence when angling for freshwater fish. Different licences are available for *game fishing or *coarse fishing. This is in addition to permission to fish in a particular river or lake, as all fishing rights in freshwater are owned by someone. There is no requirement to have a licence for sea angling except when fishing for salmon or sea trout. For example, the *Environment Agency in England issues the rod licence for the use of up to two fishing rods. Equivalent organizations in the devolved governments of the UK, with some variations, also issue rod licences.

rogue Literally, something that is out of place, used to describe a plant that is not wanted in the specific circumstances. It might be a plant that is not genetically true in a crop grown for seed, a plant of the wrong species such as barley in wheat, or a weed such as a *wild oat in a cereal crop. The term is also used as a verb to remove the offending plant by pulling it up, preferably before it can shed seed.

rollover relief *See* CAPITAL GAINS TAX.

rood An old unit of measurement used in the United Kingdom, equal to a quarter of an *acre. *See also* CHAIN.

roost A place, more often used in reference to birds, that provides shelter from the weather and provides protection from predators. The places used to roost are often species-specific, and include *woodland, *reedbeds, cliffs, and buildings. Bats also use roosts for shelter, which can include hollow trees, roof spaces, or caves. Roost shooting is a type of *pigeon shooting which targets birds that are approaching the woodland where they are going to rest or sleep. This is normally carried out at dusk.

rooster A male chicken, especially in the North America, Australia, and New Zealand. *See also* CHICKEN.

root The part of a plant below the surface of the soil through which water and nutrients are absorbed into the plant. A tap root is a single large root that may penetrate deep into the ground, whilst there may be a less deep structure of branch roots that spread through the soil to anchor the plant. In *leguminous plants there are nodes or swellings in the roots that contain the rhizobium bacteria that fix atmospheric nitrogen. Some plants have aerial roots that are above ground, and others have *rhizomes that are stems below ground. For the highest productivity in growing crops, it is essential to have a good root structure so that the plant can take up optimum quantities of water and nutrients. Root growth may be limited by waterlogged or compacted soils. In some cases, notably grape vines, the upper parts of a plant may be grafted onto the root of another plant which is called the rootstock.

rootstock *See* ROOT.

rotation The growing of different crops in ensuing seasons. As different species have different nutritional requirements and are affected by different pests and diseases, the rotation of crops can improve the fertility of the land and prevent the build-up of pests and diseases by interrupting the life cycle or removing the habitat. In particular, the inclusion of a *legume in the rotation can increase the available nitrate content of the soil, and is frequently used for this purpose in organic farming. Grazing livestock can also aid fertility and may be included in the rotation. The concept was developed by Viscount 'Turnip' Townshend in the mid-eighteenth century as the Norfolk Four Course Rotation of wheat, clover, turnips, and barley. In livestock farming the grazing of paddocks may be in rotation known as the Morrey System.

rotavator An agricultural tillage implement that disturbs and aerates the soil by means of blades, usually L-shaped, mounted on a rotating horizontal spindle. It may be a small hand implement driven by its own small petrol engine for garden or small-scale use, or it may be towed behind or mounted on the *three-point hydraulic linkage of a tractor and driven by the *power take-off.

Rothamsted Research Founded in 1843 and previously known as the Rothamsted Experimental Station, this is the longest-running agricultural research station in the world. Situated in Hertfordshire and funded by the *Biotechnology and Biological Sciences Research Council, research is concentrated on plant science integrating biotechnology with other areas of science such as agronomy and agro-ecology, so both existing and new knowledge can be implemented through agricultural practice. It has become renowned recently as the centre for plant genetic research in the UK.

(⊕) SEE WEB LINKS

• Website of Rothamsted Research, providing details of the research work being undertaken.

rough grazing *See* UPLANDS.

rough shooting Can also be described as *walking up. It is an informal style of shooting that is not controlled and managed to the same extent as *driven shooting. It involves walking up *game or other *quarry species, in a *wood, a *field, or along a *hedgerow, which is then flushed ahead of the shooter or 'gun'. It can range from a group walking up in line with other shooting participants or 'guns', to walking up individually or in pairs.

roundwood The logs and small branches from thinning and lop and top operations, and often specifically referred to as small-diameter roundwood. The products from this low-value timber are mainly paper, panels, and fencing, with *broadleaf roundwood also supplying the turnery and firewood markets. *See also* SAWLOG.

roundworm *Endoparasites, roundworms are *nematodes that affect the intestinal tract or lungs of animals, causing loss of production either by diarrhoea and loss of appetite or respiratory problems. Pasture is infected by the faeces of grazing animals and then ingested by others, prolonging the cycle. Control is by good pasture management or by the oral administration of an appropriate medicine such as an *anthelmintic wormer.

rowan *See* TREE.

Royal Agricultural Society of England Founded in 1838 to promote the scientific development of agriculture, with the motto 'Practice with Science'. Based at Stoneleigh Park in Warwickshire, from 1839 it ran the Royal Show to demonstrate the latest advances in agricultural husbandry. After the last Royal Show was held in 2009, the Society has concentrated on its core activity of working with others to promote scientific advances in farming.

((⊕)) SEE WEB LINKS
• Website of the RASE, providing details of membership and activities.

Royal Agricultural University Formerly the Royal Agricultural College, it is the first and oldest agricultural college in the English-speaking world. It was established and granted a Royal Charter in 1845.

Royal Assent The formal agreement, by the monarch, of a new *Act of Parliament or *Act of Assembly in the United Kingdom which brings it into legal effect.

Royal Forest A former area of unenclosed countryside designated and introduced by the Norman kings for their pursuit of *hunting and generation of income. It was a legal term, and Forest Law was a distinct legal system operating within the Forests which protected the deer and wild boar. Contrary to what their name suggests, Royal Forests were not exclusively covered in trees, but could include tracts of agricultural land and even settlements. The few remaining Royal Forests were transferred to the *Forestry Commission in 1924. *See also* FOREST.

Royal Forestry Society (RFS) Founded in 1882, it is the largest and oldest forestry charity, with a particular focus on education, and support for those actively involved in *woodland and *forestry management across England, Wales, and Northern Ireland. *See also* FORESTRY COMMISSION; ROYAL SCOTTISH FORESTRY SOCIETY.

((⊕)) SEE WEB LINKS
• RFS website, with membership details and information on events and awards relating to woodland management.

Royal Institution of Chartered Surveyors (RICS) Professional body accrediting and regulating individuals and firms involved in the development and management of land, real estate, construction, and infrastructure. The history of the RICS extends back to nineteenth-century London. It holds a Royal

Charter which sets out the organization's objectives and requires it to promote the usefulness of the profession for the public advantage in the United Kingdom and in other parts of the world. Qualified members of the RICS fall into three classes: Fellows (who are senior, highly regarded professionals entitled to use the letters FRICS after their name), Members (denoted by the designatory letters MRICS), and Associates (denoted by the letters AssocRICS, who have passed the relevant assessment for Associate level and may aspire to subsequently qualify as a Members). Fellows and Members are also known as *chartered surveyors. There are a number of qualification routes, most of which involve degree-level education and a period of employment experience, followed by an RICS assessment. The RICS regulates its members and sets out specific requirements for them, in relation to being competent in their area of expertise, complying with a code of professional and ethical standards, and engaging in ongoing lifelong learning. The RICS also publishes a wide range of property-related standards and best-practice guidance (for example, the *Red Book), and seeks to influence governments and organizations throughout the world to support best practice in the property sector. *See also* BUILDING COST INFORMATION SERVICE OF ROYAL INSTITUTION OF CHARTERED SURVEYORS; CODE OF PRACTICE; GUIDANCE NOTES; INFORMATION PAPER; INTERNATIONAL STANDARD; PROFESSIONAL STATEMENT; RICS UK COMMERCIAL REAL ESTATE AGENCY PROFESSIONAL STATEMENT; RICS UK RESIDENTIAL REAL ESTATE AGENCY PROFESSIONAL STATEMENT (BLUE BOOK).

(((⊕))) SEE WEB LINKS

• RICS website, with information on the work done by chartered surveyors, a digest of news relating to the property sector, and related research.

Royal Scottish Forestry Society (RSFS) A membership charity for those involved in the *woodland and *forestry industries in Scotland, including practitioners, academics, and amateurs. *See also* FORESTRY COMMISSION; ROYAL FORESTRY SOCIETY.

(((⊕))) SEE WEB LINKS

• RSFS website, with information on membership details and information on news and events.

Royal Society for the Prevention of Cruelty to Animals (RSPCA) A charity founded in 1824 to promote animal welfare. One of the oldest animal organizations in the world, the RSPCA operates in England and Wales and has stimulated the creation of equivalent organizations worldwide, including, for example, the Scottish Society for the Prevention of Cruelty to Animals. Although primarily involved with domestic animals and livestock, it has become involved with more politically sensitive wild-animal issues including *badgers and *fox hunting.

(((⊕))) SEE WEB LINKS

• RSPCA website, providing details of membership, activities, and campaigns, and information on rehoming unwanted pets and horses.

Royal Society for the Protection of Birds (RSPB) A charitable body established at the end of the nineteenth century, it promotes the conservation and protection of birds and other wildlife. It is the UK's and Europe's largest nature conservation charity, with more than one million members. It operates by owning and managing more than 200 nature reserves, lobbying government at local, national, and international levels, and promoting education and campaigns.

(((⊕))) SEE WEB LINKS
• RSPB website, detailing all aspects of its work, including publications and events.

Royal Town Planning Institute (RTPI) United Kingdom-based professional membership organization and chartered institute that works to maintain professional standards within the *planning sector, and to advance the science and art of planning for the benefit of the public. The RTPI has several classes of membership, including chartered town planner status (which can be achieved by professionals with a degree in planning and/or a number of years experience). Members are bound by a code of professional conduct. The RTPI also runs Planning Aid England, which provides planning advice and support to communities and individuals.

(((⊕))) SEE WEB LINKS
• RTPI website, with information on the work of town and country planners, and a digest of news relating to planning.
• Website of the Royal Town Planning Institute's Planning Aid England, including guidance on the planning system in England, and web links to similar guidance for Scotland and Wales.

RPA *See* RURAL PAYMENTS AGENCY.

RPI *See* RETAIL PRICES INDEX.

RSPB *See* ROYAL SOCIETY FOR THE PROTECTION OF BIRDS.

RSPCA *See* ROYAL SOCIETY FOR THE PREVENTION OF CRUELTY TO ANIMALS.

RTPI *See* ROYAL TOWN PLANNING INSTITUTE.

ruddle *See* RADDLE.

rumen The first of the four stomachs of a *ruminant animal, the others being the reticulum, omasum, and abomasum.

ruminant An herbivorous animal that has adapted to digest *herbage such as *grass. Cellulose is extremely difficult to digest, so the ruminant has four stomachs, the first and largest of which is the *rumen, which contains bacteria and other microorganisms that help with digestion. Ruminant animals chew the cud, which means that partly digested material is regurgitated from the rumen to the mouth for further chewing. The main ruminant animals in farming are cattle, sheep, goats, and deer.

runoff, surface Term used to describe the movement of *water across the surface of the ground and into *watercourses (as opposed to soaking down into the *soil). The movement of water in this way can result in the *erosion of soil and *diffuse pollution, as the water can carry soil particles and chemicals (such as *phosphates, *nitrates, and *ammonia) from the surface of the soil into watercourses, where they can contribute to issues such as *eutrophication and loss of *biodiversity. Excessive runoff of water from land can be indicative of poor soil conditions, and issues such as *compaction which can prevent water from soaking into the soil. Excessive livestock stocking densities and over-application of *fertilizers, *manure, and *slurry to agricultural land (i.e. in excess of what is required for plant growth) can increase the risk of diffuse pollution being caused by runoff. Consequently, governments internationally are increasingly requiring farmers to undertake measures to maintain and improve soil structure and to limit over-stocking and the excessive application of fertilizers, slurry, manure, and chemicals, as well as preventing the poor storage of materials such as *silage and agricultural fuel oils. For example, in the *United Kingdom, controls include the *Nitrate Vulnerable Zone regulations, *cross-compliance rules, and the **silage, slurry, and agricultural fuel oil regulations**. In 2018, new **Farming Rules for Water** were introduced in England which require farmers to take further actions to protect water quality through reducing soil erosion and the improved planning and monitoring of the application of manure and fertilizers. Improving water quality is also one of the core objectives of the UK's *agri-environment schemes, as well as other mechanisms such as the *Catchment Sensitive Farming scheme.

rural Derived from the latin term *ruralis*, 'of the countryside', it is a common description of areas outside of urban developments such as cities and towns. Farmland and *forest areas would normally be described as rural, and the term encompasses not just the geographic areas, but also a general description of, for example, populations, housing, development, and crime. Different countries have differing definitions of what constitutes rural, with the areas usually having lower densities of population and smaller settlements such as villages. For example, the *Department for Environment, Food and Rural Affairs (DEFRA) defines rural areas in England as those outside of settlements with more than 10,000 residents.

Rural Development Programme (RDP) The instrument that details the various rural development policy measures founded in the EU obligations under the second pillar funding stream of the *Common Agricultural Policy. RDPs are part financed by the **European Agricultural Fund for Rural Development** (EAFRD), and are supported by specific EU Articles and Regulations. Member states and the devolved governments of the UK have some flexibility in adopting and adapting from a European 'menu of measures'. For example, the **Rural Development Programme for England (RDPE)** allows the *Department for Environment, Food and Rural Affairs (DEFRA) to fund schemes such as the *Countryside Stewardship Scheme, woodland grant schemes (*see* FORESTRY GRANTS), and rural development projects such as *LEADER.

rural economy The management of resources in rural areas, including both
*farm and non-farm industries. It also encompasses a range of issues focused on
rural areas, including land use, housing, populations, and government policies.
There can be a tension between stimulating rural economies and protecting the
*natural resources of the countryside.

rural enterprise dwelling (Wales) *See* RURAL WORKER'S DWELLING.

rural exception site A site where *planning permission has been granted for
predominantly *affordable housing on a site where planning permission for
open-market housing would not normally be given (such as within a small village
with very limited facilities). Rural exception sites are provided for within national
*planning policy in England, Scotland, and Wales (where the sites are known as
affordable housing exception sites). The policy is designed to facilitate the
provision of affordable housing, where there is an identified need, even in small
rural settlements where housing development would not normally be permitted.

Rural Payments Agency (RPA) As an executive body sponsored by the
*Department for Environment, Food and Rural Affairs (DEFRA), the RPA acts as
the paying agency for *Common Agricultural Policy payment schemes in
England, including the Basic Payment Scheme, *agri-environment and woodland
schemes, and various other *Rural Development Programme schemes. Its
additional responsibilities are varied, and include managing the Rural Land
Register that details farm holdings and land ownership. Its officers also visit and
inspect farms to ensure compliance with scheme rules. Bodies equivalent to the
RPA are the **Rural Payments and Inspections Division** in Scotland, the **Rural
Payment Wales in Wales**, and the **Department of Agriculture and Rural
Development** in Northern Ireland. *See also* NATURAL ENGLAND.

rural practice chartered surveyor A *rural surveyor who is qualified as
either a Member or Fellow of the *Royal Institution of Chartered Surveyors.

rural proofing A term used to describe considerations given to government
policies to include or address specific *rural issues that might be generated
because of, for example, distances, population sparsity, or demographics in
rural areas.

Rural Stewardship Scheme *See* AGRI-ENVIRONMENT.

rural surveyor A person, often a *chartered surveyor, who either manages a
rural estate on behalf of its owner or owners, or provides professional advice in
relation to the management of land and property in rural areas to a number of
clients. The term 'rural surveyor' is often used interchangeably with that of *land
agent. Rural surveyors undertake a broad range of work, acting for *landlords,
*tenants, and *owner-occupiers, and deal with the full range of property found in
rural areas, from farms, commercial offices, and residential properties, through
to *renewable energy projects and *telecommunications equipment. They have
expertise in a variety of areas, including negotiating the sale and letting of rural

property, reviewing rents and *leases, planning and farm *diversification matters, farm *subsidies, auctioneering, *valuation, *compulsory purchase, environmental management, and woodland management. They often work as consultants, advising farmers and landowners on the long-term strategic management of their properties, taking into account factors such as maximizing the value of their assets, development opportunities, and taxation. They will often work closely with other professionals such as solicitors and accountants.

rural worker's dwelling A dwelling which has been given *planning permission specifically on the grounds that there is an essential need for a rural worker to live at a particular site where there is no existing accommodation available, and where, otherwise, planning permission would not have been granted for a new dwelling. *Planning policy in the United Kingdom seeks to protect open countryside from uncontrolled development (and particularly so in areas where there may be some form of planning designation). Consequently, planning policy sets out specific justifications that are normally required in order to develop an isolated new dwelling in open countryside. These include the need for a rural worker to live at a particular site, perhaps, for example, on the grounds of animal welfare. Historically, these types of permission have tended to be limited only to workers in agriculture or forestry, but they have now been broadened to include rural workers more generally, in recognition of the wide range of business activities which occur in the countryside. However, the essential need to live at a particular site still must be established for such permission to be granted. Rural worker's dwellings are often subject to planning *conditions restricting their occupation to specific types of rural worker, and may also be subject to an *agricultural tie set out in a legal agreement.

rust One of a number of fungal diseases that infect the aerial parts of plants, particularly the leaves. The spore pustules vary in colour according to the species of fungus, giving rise to the name of some of the diseases. Examples are yellow rust on cereals, brown rust on wheat, and crown rust on oats and ryegrass. Control is by *fungicide application.

rustle To round up and steal animals, usually cattle or sheep, from a farm or other site. Rustling is carried out by a rustler, originally by driving the animals on foot but frequently on horseback. In the modern age it is more probable that the animals will be loaded onto a lorry to be driven away, either to another farm, to common land, or to an abattoir. As with any theft, it is a criminal offence, punishable by a fine or a prison sentence.

ryegrass A *grass of the *Lolium* family, the most widely grown in the world. It is fast growing, has a long growing season, is high yielding, and is palatable as a grazing sward if conserved in hay or silage. There are two main species: perennial ryegrass (*Lolium perenne*) and annual or Italian ryegrass (*Lolium multiflorum*). Ryegrasses are also widely used in recreational swards such as lawns and sports grounds.

s106 agreement **(England and Wales)** *See* SECTION 106 PLANNING OBLIGATION AGREEMENT (ENGLAND AND WALES).

s75 agreement **(Scotland)** *See* SECTION 75 PLANNING OBLIGATION AGREEMENT (SCOTLAND).

SAC *See* SPECIAL AREA OF CONSERVATION.

sainfoin A leguminous perennial herb of the *Onobrychis* family. It provides excellent grazing for livestock, and the voluntary intake by cattle and sheep is said to be 20% higher than grass. It is drought-resistant, grows well on dry alkaline soils, and is often used for a cut of *hay or *silage with subsequent *grazing. It can be grown on its own or in a mixture, and, as a legume, it needs no nitrogen fertilizer.

salinization The build-up of salts in the soil, eventually to toxic levels that make the growing of crops difficult or impossible. Salts reduce the osmotic potential of soils, thus preventing plants from absorbing water. Whilst common salt, sodium chloride, may be the most frequently occurring salt in soils, other salts may also cause similar problems. The cause is inundation by sea water and the evaporation of water leaving the salts behind. Treatment is by water management and by growing salt-tolerant plants.

salmon Several species of fish which belong to the family *Salmonidae*; for example, the Atlantic salmon (*Salmo salar*). They are migratory fish, and the adult returns from the sea to freshwater to spawn. They are of significant economic importance for commercial fishing, *game fishing, and *aquaculture. The latter includes the farming of salmon in coastal waters, with particular concerns raised on its ecological impacts and disease effects on wild salmon.

salmonella A bacterium that causes food poisoning, diarrhoea, fever, and abdominal cramps in humans. It can be present on foodstuffs, especially meat and eggs, and contamination can occur through poor hygiene. The bacterium is killed by cooking at high temperature, so cooked food should be kept separate from raw food, and hands should be thoroughly washed before handling foods.

saltmarsh Coastal wetlands that are subject to tidal flooding and draining, with a substrate which is commonly deep mud and *peat. They are an important *habitat, and an integral part of the economy and culture of local fisheries. Saltmarshes occur worldwide, especially on protected coastlines, and are a common habitat in estuaries. They are also an important feature protecting shorelines from erosion, and they promote water quality. Managed retreat is an approach to managing erosion by allowing areas such as saltmarsh to continue to flood and so to protect other identified and important areas.

SAM *See* SCHEDULED MONUMENT.

sapling A young and immature tree that can normally range in height between 0.6 and 3.0 metres, and normally having characteristics that differ from the mature tree, including smooth bark and a more flexible trunk or stem. It would not normally produce fruit or flowers. A *whip is a younger sapling of about 0.6 to 0.9 metres in height, with only a central stem and more limited side-branching.

saprophyte An organism that lives and feeds upon dead organic matter. Such organisms commonly include certain fungi and bacteria. They play a vital part in the recycling of organic material in the soil, and deriving simpler compounds that can then be assimilated by other organisms. Saprophytes are more simply known as 'decomposers'.

sash window *See* APPENDIX 2 (ILLUSTRATED BUILDING TERMS).

saved policy Policy within a *local plan which is still consistent with national *planning policy, even if the rest of the local plan is (usually due to its age) out of date. Local plans are subject to regular review, however, plans which are at the end of their plan period may no longer be consistent with national planning policy if this has been subject to substantial change. Under the **Planning and Compulsory Purchase Act 2004**, local planning authorities in England were allowed to apply to central government to 'save' policies which were still consistent with national policy (thereby extending their life until the adoption of a new local plan).

sawfly (*Symphyta*) Insects of the *Hymenoptera* order, together with bees, wasps, and ants. There are at least 400 different species, many with specific host plants, such as the apple sawfly or turnip sawfly. The adult insect lives for around two weeks before laying its eggs in cuts in plant leaves. The larvae resemble moth caterpillars and feed on the leaves and fruit of plants. The damage to plant tissue may be severe but is usually localized. The larvae live for around four to five weeks before pupating in the soil. Sawfly may be controlled by an insecticide such as **pyrethrum**. Sawfly larvae are a high-protein source of food for the chicks of farmland birds such as the grey partridge.

sawlog Higher-value *hardwood and *softwood timber of at least 14 cm diameter, cut to various lengths, usually in the *forest and prior to transport to a *sawmill. Mainly destined to be sawn into planks or boards for joinery and the building and construction markets.

sawmill A factory dedicated to cutting up timber or more specifically *sawlogs, into planks and boards. It can be a relatively small unit serving a single rural estate or large farm, or a larger industrialized unit processing timber from a regional area. *See also* WOOD; WOOD CHIP; WOODFUEL; WOOD SHAVINGS.

scarcity value An element of additional value that can be attributed to the scarcity of a particular type of property in a market; essentially, that where there are more potential purchasers or tenants looking for a type of property within an area than there are available properties, then the value of that property may be higher than would otherwise be expected. The law provides that in *rent reviews of properties which are subject to some types of tenancy, scarcity value should be discounted if any *comparables are being assessed.

scarecrow A device designed to scare birds away from crops. Traditionally, a scarecrow was made of straw in the shape of a man dressed in human clothes and fixed to the ground by a stake. It was of limited value, as it did not move and the birds soon became accustomed to it. More recently, many different types of birdscarer have been developed, from kites to gas-powered bangers.

scarify 1. To break up or loosen the soil to a shallow depth prior to the planting of a crop. **2.** To remove dead or decaying material from the soil surface, as in removing dead grass from a lawn. **3.** To scratch the outer surface of seeds to aid the absorption of water and hasten germination.

scheduled monument Nationally important monument or archaeological site included on the national lists of scheduled monuments (also sometimes known as **scheduled ancient monuments**) in England, Wales, Scotland, and Northern Ireland. To be designated as a scheduled monument, a site must generally have been created, or heavily influenced, by people (as opposed to being entirely naturally occurring). They range from prehistoric tombs to earthworks, castles, and Second World War defences. Whilst there is some commonality with *listed buildings, most scheduled monuments are in some way ruined or partially ruined, whereas listed buildings tend to be more complete and have potential economic uses. Works that may potentially disturb the site of a scheduled monument require *scheduled monument consent (a consent, separate from *planning permission, that can be required for works to repair the scheduled monument, as well as operations such as metal-detecting in its vicinity). Where scheduled monument consent is required, it is an offence to carry out works without obtaining it. Certain works, such as some agricultural operations (though not those which may cause significant damage, such as *subsoiling), or some works urgently required for health and safety, are covered by **class consents**, which have the effect of granting consent without the need for individual applications. Scheduled monument consent applications are dealt with by *Historic England, *Cadw in Wales, *Historic Environment Scotland, and the **Northern Ireland Department for Communities Historic Environment Division**. *See also* HERITAGE ASSET; AREA OF SIGNIFICANT ARCHAEOLOGICAL INTEREST.

scheduled monument consent *See* SCHEDULED MONUMENT.

schedule of condition *See* INVENTORY.

scope 1, 2, and 3 emissions International **Greenhouse Gas Protocol** for measuring the amount of carbon produced by an organization. Direct and indirect emissions are divided into three different scopes:

- Scope 1 emissions are those which are directly produced by the organization itself (for example, from boilers owned or controlled by the organization).
- Scope 2 emissions are those from electricity which has been used by the organization, having been purchased from other producers.
- Scope 3 emissions are those produced as a consequence of the organization's activities, but not directly by them (for example, those produced in the process of disposing of waste produced by the organization, or by employees commuting to and from work in their own vehicles).

The purpose of using such a protocol is that it allows standardized comparisons of carbon outputs from different organizations (and hence the ability to operate systems such as *carbon trading). The scope 1, 2, and 3 emissions protocol is often used by organizations to monitor their carbon emissions within their *environmental management system. However, because scope 3 emissions are not directly in the control of the organization itself they are more difficult to measure (for example, an employee travelling to work in the organization may be sharing a lift with the employee of another organization). Consequently, organizations who may measure their scope 1 and 2 emissions will not necessarily monitor scope 3 emissions.

See also CARBON ACCOUNTING; CARBON CALCULATOR; CARBON CYCLE; CLIMATE CHANGE.

scoping *See* ENVIRONMENTAL IMPACT ASSESSMENT.

Scottish agricultural tenancies Tenancies of agricultural land for the purposes of trade or business in Scotland. There are several different pieces of *legislation governing agricultural tenancies in Scotland, with the three main ones being the **Agricultural Holdings (Scotland) Act 1991 (AH(S)A 1991)**, the **Agricultural Holdings (Scotland) Act 2003 (AH(S)A 2003)**, and the **Land Reform (Scotland) Act 2016 (LR(S)A 2016)**. Under these pieces of legislation there are several different types of tenancy:

- **1991 Act tenancies.** A *lease under this legislation is known as an **agricultural holding**. In the main, 1991 Act tenancies will have commenced before the AH(S)A 2003 came into effect, although it has been possible to enter into a 1991 tenancy by agreement between the parties since 2003. *Rent review provisions are set down in *statute, and if negotiations between the parties do not reach a satisfactory agreement, rents can be determined by the *Scottish Land Court according to criteria set down in the legislation and subject, usually, to a minimum period between reviews of three years. Under the 2003 Act, a statutory right was introduced for 1991 Act tenants to assign (*see* ASSIGNATION) their lease to someone who would be entitled to succeed to the tenant's estate under the Scottish laws of *intestacy. This has been further changed by the LR(S)A 2016, which introduces an opportunity for the tenant to offer to give up their tenancy to the landlord in return for a capital sum. If the landlord does not accept this, then the tenant will be entitled to assign the tenancy to a third party (although the landlord can, as with other assignations, contest this if there are reasonable grounds to do so; for example, that the proposed new tenant would not have the knowledge or skills to manage and maintain the land properly). Similarly, a 1991 Act tenancy can, in some circumstances, be inherited by **succession**, either through being left in the tenant's will, or in the case of *intestacy, by the *executors transferring it to a successor. The 2003 Act gave 1991 Act tenants the right to diversify by using the land for a non-agricultural purpose. There are also *security of tenure provisions within the 1991 Act. A 1991 Act tenancy will automatically extend by *tacit relocation, and the tenancy will continue unless the landlord is able to demonstrate one of the *statutory **grounds of removal**. These are similar (though not exactly the same) to those under the *Agricultural Holdings Act 1986 in England and Wales, and the tenant can, in some circumstances, challenge them through the *Scottish Land Court. *Compensation is available to an outgoing tenant for certain *tenant's improvements specified in the legislation, and also for disturbance (the provisions for tenant's improvements and disturbance are also similar to those under the Agricultural Holdings Act 1986). The landlord may also be entitled to compensation from the tenant if the tenant has been responsible for a deterioration of the holding. The 2003 Act gave 1991 Act tenants the right to buy the land from the owner of it (although it is a **right of pre-emption** only, so it only applies if the owner intends to sell it). In order to do this, the 2003 Act required that the tenant must have previously registered their interest in acquiring the land in the **Register of Community Interests in Land** (however, the requirement to register in advance is due to be removed as a consequence of the LR(S)A 2016).

- **Short Limited Duration Tenancy (SLDT)**. These were introduced under the AH(S)A 2003. They cannot exceed five years in length. An SLDT which runs over the five-year limit will automatically become a Modern Limited Duration Tenancy (see below). Compensation for tenant's improvements is available. There are also limited rights of succession under an SLDT.
- **Limited Duration Tenancy (LDT)**. Also introduced under the AH(S)A 2003, these cannot be less than ten years in length (though this was fifteen years when they first came into effect). As a consequence of the 2016 legislation it is, however, no longer possible to constitute a new tenancy as an LDT, and they have been replaced by the Modern Limited Duration Tenancy discussed below. Under an LDT, at the end of the fixed term, if the landlord wishes to end the tenancy then they must follow a procedure which requires the serving of two notices. If they do not do this (or the tenant does not serve notice) then the tenancy continues in what is effectively a statutory form of tacit relocation (though it does not operate in exactly the same way); the mechanism for an LDT instead operates by a **cycle of continuations**, with the length of each continuation (of which there are three before the cycle starts again) being set out in law. The landlord can then only end

the tenancy at the end of each continuation (subject to the correct serving of notices). Rent review provisions are often set out in the tenancy agreement, but there are default provisions within the legislation with the right to have the rent set by the Scottish Land Court. Tenants are able to assign an LDT with the landlord's consent, although the party to whom the tenancy would be assigned does not have to be one who would be entitled to inherit the tenant's estate, and, as with 1991 Act tenancies, the landlord can contest it on certain grounds. It is also possible to inherit an LDT, on a similar basis to a 1991 Act tenancy. As with 1991 Act tenancies, the tenant has the right to diversify, and compensation for tenant's improvements is available.

- **Modern Limited Duration Tenancy (MLDT).** This was introduced by the Land Reform (Scotland) Act 2016, and it must not be less than ten years in length, although if the tenant is a 'new entrant' it may be terminated after five years in certain circumstances. It replaces the Limited Duration Tenancy (and is fairly similar), and it is possible to convert, by agreement, a 1991 Act tenancy, or an LDT, into an MLDT. A tenant may assign their MLDT, subject to some limited rights of objection by the landlord. Like an LDT, an MLDT can continue beyond its initial ten-year length through a statutory form of tacit relocation (however, the length of time for which it continues is different from an LDT). Termination at the end of the tenancy by the landlord will require a double-notice notification process (with a single notice for tenants), i.e. similar requirements to the LDT. An MLDT can, potentially, be *sub-let if the lease allows it.

- **Repairing Tenancy.** The LR(S)A 2016 includes provisions to introduce repairing tenancies which allow for an initial period of five years (or more by agreement) when the tenant can improve the land, rather than farm it, and therefore will not have to comply with the *good husbandry rules. It is anticipated that the tenancy itself will have to be for at least thirty-five years, with the other rules in relation to continuation, termination, and extension, being the same as for an MLDT.

- **Lease for Grazing or Mowing.** These are leases which are only for the purposes of grazing or mowing at a specified part of the year. The maximum period allowed is 364 days, and at least one day must elapse after the expiry of the lease before it can be re-let to the same tenant for the same reason. If such a lease is allowed to run on after its termination date with landlord's consent it will automatically default to a SLDT.

- **Small Landholders.** There is specific legislation governing small landholdings which stretches back to the nineteenth and early twentieth centuries. In order to qualify, the tenant must cultivate the land themselves. They (or a predecessor in the same family) must also have provided, or paid for, all, or the greater part, of the buildings and permanent improvements on the holding themselves (if they have not then they are instead known as a **statutory small tenant** and have fewer rights). The *Scottish Land Court has jurisdiction under the small landholders legislation, and the tenant generally has a right to succession, security of tenure, compensation for improvements, and the right to apply to the Scottish Land Court for a rent (referred to as a **fair rent** for small landholders and **equitable rent** for statutory small tenants) to be fixed according to principles set out in statute in the absence of agreement between the parties. However, there are now very few small landholdings in existence. There is, in addition, a body of law relating to *crofts. *See also* GRAZING.

Scottish Assessor Local government official in Scotland responsible for the valuation of domestic and non-domestic properties for local taxation purposes within their valuation area. They carry out a similar role to District Valuers in England, Wales, and Northern Ireland, but differ in that they are appointed by local authorities rather than a central body (although they cooperate through the **Scottish Assessors Association**). They are responsible for compiling and maintaining lists of *rateable values of non-domestic properties for *business rates and *Council Tax bandings of domestic properties. Appeals against the valuations made by Scottish Assessors are heard by the *Scottish Valuation Appeal Committee.

Scottish Environment Protection Agency (SEPA) A non-departmental body of the Scottish government that is responsible in Scotland for the protection and improvement of the environment. SEPA acts as a regulator and as an adviser to businesses and industry to ensure awareness and compliance with environmental legislation. It has some similar functions to the *Environment Agency in England. *See also* NATURAL RESOURCES WALES; NORTHERN IRELAND ENVIRONMENT AGENCY.

Scottish Government *Devolved government for Scotland with responsibilities for health, education, justice, rural affairs, housing, and the environment. The Government comprises the First Minister, Cabinet Secretaries and other Ministers and two Law Officers (Lord Advocate and Solicitor General for Scotland). The Cabinet is supported by the Cabinet Secretariat. The First Minister is the leader of the Scottish Government and is appointed by the monarch following nomination by the *Scottish Parliament. The First Minister, in turn, appoints the Scottish Ministers making up the Cabinet (with the approval of the monarch). The Scottish Parliament is the elected body that makes laws for Scotland in devolved areas, and scrutinizes the work of the Scottish Government. Elected members of the Scottish Parliament are identified by the letters MSP after their name (i.e. Member of the Scottish Parliament). For those areas where the Scottish Parliament does not have the right to make laws, the United Kingdom Parliament and Government is the main national law-making body.

Scottish Government Planning and Environmental Appeals Division *See* APPEAL, PLANNING; LOCAL PLANNING AUTHORITY.

Scottish Housing Regulator *See* HOMES ENGLAND.

Scottish Income Tax *See* INCOME TAX.

Scottish Land and Estates A membership organization that represents the interests of Scottish landowners and rural businesses. It has evolved and been created mainly from the former Scottish Landowners' Federation and subsequently the Scottish Rural Property and Business Association. In addition to its advisory work, it also has a key role as a lobbying voice to government. *See also* COUNTRY LAND AND BUSINESS ASSOCIATION.

Scottish Land Court Court of law in Scotland which has the authority to resolve disputes in relation to agricultural *landlord and *tenant issues, including some matters relating to *crofts. The Scottish Land Court has very specific areas of jurisdiction and does not deal with all land-related disputes in Scotland, as it operates alongside the *Lands Tribunal for Scotland which has responsibility for many other potential areas of dispute which relate to land and property. Matters relating to ownership of land are also dealt with by the *Sheriff Court and *Court of Session. *See also* CROFTING COMMISSION.

Scottish Landfill Tax *See* LANDFILL TAX.

Scottish Landowners' Federation *See* SCOTTISH LAND AND ESTATES.

Scottish Natural Heritage A statutory body under the responsibility of the *Scottish Government, with roles that include regulatory and advisory functions on a wide range of *biodiversity, *landscape, and *public access issues. It was formed by the amalgamation of the **Nature Conservancy Council for Scotland** and the **Countryside Commission for Scotland**, and more latterly also took over the functions of the **Deer Commission for Scotland**. *See also* NATURAL ENGLAND; NATURAL RESOURCES WALES; SCOTTISH ENVIRONMENT PROTECTION AGENCY.

SEE WEB LINKS
• Provides details of individual roles and operational functions.

Scottish Parliament Elected Parliament in Scotland which is authorized by the United Kingdom Parliament in Westminster, under *devolution powers, to make laws relating to Scotland. *See also* SCOTTISH GOVERNMENT.

Scottish Planning Policy *See* PLANNING POLICIES.

Scottish Renewable Obligation *See* NON FOSSIL FUEL OBLIGATION.

Scottish residential tenancies Forms of *tenancy for occupation of rented residential property in Scotland. There are currently four main types of residential tenancy in Scotland:
• **Regulated tenancies**. Tenants under these regulated tenancies (which were mainly entered into prior to 1989) enjoy *security of tenure, with the *landlord only being able to obtain vacant possession by successfully demonstrating one of the *grounds for possession to the *First-Tier Tribunal for Scotland (Housing and Property Chamber). Regulated tenancies in Scotland share significant similarities with the English and Welsh *Rent Act 1977 tenancies, and the terminology of a **protected** and **statutory** tenancy is also used. As with Rent Act 1977 tenancies the landlord or tenant may apply to the *Rent Officer

to register a **fair rent**. There is also the potential for *succession under a regulated tenancy.
- **Assured tenancy**. These were introduced by the **Housing (Scotland) Act 1988** and operate on similar principles to assured tenancies in England and Wales (*see* Housing Act 1988). The tenancy starts as a **contractual assured tenancy** and will continue by *tacit relocation until the landlord serves a *notice to quit. At that point it will become a **statutory assured tenancy** and the tenant will benefit from *security of tenure, with the landlord only being able to end the tenancy through successfully demonstrating one of the grounds for possession under the Act to the First-Tier Tribunal. A *market rent can be charged for an assured tenancy.
- **Short assured tenancy**. These were also introduced by the 1988 *legislation, and are, in effect, a form of assured tenancy. They also operate on similar principles to the assured shorthold tenancy in England and Wales. Short assured tenancies must have a minimum length of six months, but there is no security of tenure beyond the *ish.
- **Private residential tenancies**. As a consequence of the **Private Housing (Tenancies) (Scotland) Act 2016**, with effect from 1 December 2017 most new private residential tenancies in Scotland must be private residential tenancies, as it is no longer possible to create new assured, or short assured, tenancies. A tenant occupying a property under a private residential tenancy will have *security of tenure, with the landlord only able to recover possession by demonstrating one of the grounds for eviction to the First-Tier Tribunal. There are also some controls on rent, with a tenant able to refer rent increases to a *Rent Officer, and the Scottish Ministers being able to cap the level of rent increases in certain localities.

As with England and Wales, agricultural workers living in dwellings provided by their employer have additional protections under Scottish law.

Scottish Rural Property and Business Association *See* Scottish Land and Estates.

Scottish Valuation Appeal Committee Committee hearing appeals against property valuations made by *Scottish Assessors in relation to *business rates and *Council Tax. Members of the Committee are drawn from an Appeal Panel and are independent of the Assessor or local authority. *See also* Lands Tribunal for Scotland.

Scott Report 1942 The report of the Scott Committee on Land Utilisation in Rural Areas. In an attempt to integrate *agriculture and *land use planning, a committee was appointed by government to consider issues surrounding construction and development in the *countryside. One of the consequences of the report was the Town and Country Planning Act 1947.

scraper An implement for moving animal slurry through a livestock building or across a concrete yard. It may be hand held or mounted on the *three-point hydraulic linkage on the back of a *tractor. The base in contact with the concrete is usually rubber to ensure and maintain good contact.

scrapie A fatal degenerative disease of the central nervous system of sheep and goats. The cause of the disease has yet to be identified, but is thought to be a microscopic virus-type organism. One strain is highly contagious and is thought to be spread largely from mother to offspring through the placenta and colostrum. A spongiform encephalopathy, it is related to *bovine spongiform encephalopathy, or Mad Cow Disease. It is a *notifiable disease in the UK, which means that it must be reported to the appropriate authorities, which will impose movement restrictions and ensure that the premises are thoroughly disinfected.

screefing The operation by which vegetation is scraped away from the immediate area with a spade prior to digging a hole or slot into which a *sapling or *whip is planted. This both aids the digging and reduces the impact of the competing vegetation. Spot spraying using a *herbicide may be used as a preferred method prior to planting. The surface of small areas of land can also be broken up by hand tools or a rotavator to aid *natural regeneration to promote *seedling growth. *See also* MAST; SCARIFY.

screening *See* ENVIRONMENTAL IMPACT ASSESSMENT.

scrub An area of poor and stunted *trees or *woody shrubs, often associated with poorer soils and generally of low timber value. Frequently described as within a complex mosaic of scrub or **scrub woodland**. May also be described as a single scrub species, e.g. *hawthorn scrub or *birch scrub. Can be of high *conservation value.

SDA (Severely Disadvantaged Area) *See* LESS FAVOURED AREA.

SDLT *See* STAMP DUTY LAND TAX.

SEA *See* STRATEGIC ENVIRONMENTAL ASSESSMENT.

season One of the four divisions of the year: namely, spring, summer, autumn, and winter. Of particular relevance and common usage in *agriculture in relation to growing and *harvesting periods and *livestock reproductive cycles. The **wet** or **rainy season** when it is particularly and consistently wet, or **dry season** when there is little or no rain. *See also* CLOSE SEASON; OPEN SEASON.

seawall A structure built to protect coastal shores from wave damage and erosion. Commonly built of concrete, or can also be constructed from steel or wooden structures. Boulders or rubble can be used for further protection to these structures or as stand-alone barriers. *See also* COASTAL CHANGE MANAGEMENT AREA.

section 106 planning obligation agreement (England and Wales) Private legal agreement (provided for under section 106 of the Town and Country Planning Act 1990, as amended), between a local planning authority and a developer which attaches to the grant of planning permission in order to make the proposed development acceptable in planning terms. The type and nature of the measures which may be included in a section 106 (s106) legal agreement varies between sites; however, the underlying principle is that they should only be required if, without them, permission would not be grantable. The agreements relate to the land rather than the applicant, and therefore their effects can last beyond the initial construction phase.

s106 agreements are used for a range of reasons: examples may include requiring a developer to provide *affordable housing within a housing development or to fund an infrastructure improvement, or they may require that land is used in a specified way. Whilst they may potentially involve the payment of a financial contribution towards infrastructure, they are distinct from the *Community Infrastructure Levy (CIL) in that they are determined on a site-specific basis and relate to a particular development, rather than being set out in a published charging schedule which is designed to fund infrastructure more generally in the local area. s106 agreements cannot be used in relation to those things that are intended to be funded through the CIL.

Whilst the Town and Country Planning Act 1990 applies only in England and Wales, equivalent planning obligation systems exist in Scotland (*section 75 planning obligation agreement) and Northern Ireland (section 76 planning agreement), although not the Community Infrastructure Levy which only applies in England and Wales. *See also* CONDITIONS, PLANNING.

section 75 planning obligation agreement (Scotland) Planning obligation (or developer contribution) provided for under section 75 of the Town and Country Planning (Scotland) Act 1997 as amended. *See also* SECTION 106 PLANNING OBLIGATION AGREEMENT.

security of tenure The *statutory right of a *tenant to remain in a property and restrict the right of a *landlord to regain *possession, and, under some statutory codes, to lawfully continue in occupation of a property after the expiry date of their current *tenancy or *lease. Security of tenure provisions differ according to the statutory code (law) under which the tenancy is operating.

sedge Flowering plants of the family *Cyperaceae* resembling grasslike plants or rushes. More commonly found in wet and poorly drained areas. Of little agricultural and nutritional value, sedge is an important element of *wetland *habitats. It can also provide thatching material, especially for ridging. *See also* REEDBED.

seed The means of reproduction of a flowering plant. Once the flower has been pollinated, the seed is produced. This can then germinate and produce another plant. In an agricultural or horticultural context, the seed is harvested and planted in soil to produce the next crop. *Grain is the seed of *cereals and other species. A seed crop is one that is specifically grown to produce seed to be planted for the next crop, and may be subject to inspection to ensure that it is genetically pure.

seed bank A store of seed of many different species to preserve genetic purity and thus, effectively, a gene bank. For example, the viable seeds of crops grown in ancient Egypt found in tombs might be stored in a seed bank, or seeds of plants from all over the world. These seeds can be grown if required or used for *crossbreeding to create new varieties.

seedbed Soil that has been prepared for sowing seed. It is important that the structure of the soil provides a suitable medium for the germination and establishment of the seed. That preparation may involve a number of tillage operations from none at all, known as *direct drilling, to *ploughing, *cultivating, *harrowing, and rolling. Such tillage may result in a fine tilth, but may also lose moisture in a dry soil.

seeding The process of sowing or planting *seeds such as *cereal *crops, including *wheat and *barley. In reference to *trees or *shrubs, seeds are planted in an area where initial care can be provided, such as in a nursery, before plants are of a suitable size and robustness to be transferred to a permanent site. *See also* DRILL; DIRECT DRILLING.

seedling A *seed that has germinated and emerged from the soil before it grows into a mature plant.

segmentation Method used in marketing where people are categorized into different groups, or market segments, in order to allow the design and marketing of a product, service, or brand to be targeted specifically at a particular segment or segments. If the process of market segmentation has been carried out effectively, all the potential customers in a segment will share similar characteristics and needs. The market can be segmented by using a number of different variables, and arguably operates most effectively when several are used in combination. These variables can include geographic segmentation (where potential customers live), psychographic segmentation (based around personality types), behavioural segmentation (for example, whether a product or service is going to be used for business or social purposes), or demographic segmentation (for example, age or economic factors). Whilst an organization selling a product or service which is likely to be relevant to a broad range of people may not adopt a segmentation approach, with other products or services the same organization may consider that they can use their marketing resources more efficiently by targeting a small number of segments of the market, or possibly an individual segment sometimes known as a *niche market. *See also* DIFFERENTIATION.

selective felling *See* FELLING.

semi-natural A description frequently used to qualify a *habitat or feature such as semi-natural *woodland or semi-natural *grassland, where these have been modified by human impact but retain a degree of naturalness. The vegetation is not cultivated or intensively managed for *agriculture, and would be composed of species that are native to the area. There are very limited areas of the UK that can be described as natural; for example, in more remote upland areas this description would apply. *See also* Ancient Semi-Natural Woodland; landscape.

semi-natural woodland *See* Ancient Semi-Natural Woodland.

sensitivity analysis *See* development appraisal.

SEPA *See* Scottish Environment Protection Agency.

separate tenement (Scotland) *See* tenement.

septic tank drainage Tank and *soakaway-based system for treating *foul water in situations where a property is not connected to the mains drainage system. Sewage (for example, from a bathroom) is drained into a tank, which is located underground and which normally has two sections. In the first section, the sewage settles and is partially broken down by bacteria, with the solid sludge sinking to the bottom. The liquid from the first section of the tank then flows into the second section, where the process is repeated. The remaining liquid from the second section then drains away into the surrounding soil through a system known as a soakaway (a series of perforated drainage pipes often laid in shingle). The solid sludge remains in the tank and is periodically emptied by a lorry tanker. The system is usually passive and relies on gravity; however, because of the importance of the soakaway, septic tanks are not suitable for use in some soils (for example, poorly draining clay), nor in areas which are environmentally sensitive. *See also* cesspit; reedbed; sewage treatment plant.

sequential approach, flood risk Provision within *planning policy in England that sets out the principle that *development should only be located where it is at a very low, or no, risk of flooding, and that where this is not reasonably possible, planning decisions should take into account, in a sequential way, both the level of flood risk of the site and also the degree of vulnerability to flooding of the proposed use of the site. For example, the most potentially vulnerable uses should only be sited in the areas of lowest, or no, flood risk unless there is some exceptional requirement for them to be in an area of higher risk (known as the application of the **Exception Test**), in which case, specific flood risk mitigation measures will be required to ensure that the development is safe for its life. Although the specific terminology is only used in England, the same approach to considering flood risk within development control is also used within Wales, Scotland, and Northern Ireland.

service charge A charge levied by a *landlord on their *leaseholders in order to cover maintenance and insurance of the property, and any services that may be provided by the landlord (for example, heating, lighting, cleaning of common areas, and so on). Most service charges are variable (so are based on the costs of the services in any one particular year); however, a landlord is only allowed to make a service charge if the right to do so is included within the *lease, and the lease will also set out how the service charge can be made (for example, its frequency). *See also* sinking fund.

service occupancy Form of occupation of a residential property where an employer requires an employee to live in accommodation provided by them as part of their job for the better performance of that job (perhaps because someone is needed on the employer's premises at all times). In this case, because they are only in the property as a condition of their job in order to do it effectively, *case law indicates that their occupation is not usually considered to be a *tenancy, although agricultural workers may still have significant protection; for example, under the *Rent (Agriculture) Act 1976 or the *Housing Act 1988 (assured agricultural occupancy) in England and Wales, and similar legislation in Scotland and Northern Ireland. *See also* EXCLUSIVE POSSESSION; SERVICE TENANCY.

service tenancy A *tenancy granted to a tenant holding a specific office, appointment, or employment from the *landlord, and which only continues for as long as the tenant holds that position. Agricultural workers may, though, still have significant protection, for example, under the *Rent (Agriculture) Act 1976 or the *Housing Act 1988 (assured agricultural occupancy) and similar legislation in Scotland and Northern Ireland. *See also* EXCLUSIVE POSSESSION.

servient tenement *See* TENEMENT.

servitude Legal right in Scotland which is broadly equivalent to an *easement.

set-aside The policy measure of taking land out of production to reduce *crop surpluses. Introduced by the EU in 1988, in 1992 it became compulsory for *arable farmers claiming *Common Agricultural Policy support payments. The area of land to be set aside was set at 15% of arable land, with an exemption for small and *organic farmers. The rate was subsequently varied, and was effectively abolished in 2009. By this time, a secondary objective of the scheme which had gained greater awareness was to benefit the environment; for example, by growing grass mixtures or cover crops on set-aside land to benefit wildlife and soils. The USA has a longer history of using set-aside as a means to reduce surpluses that predates the EU policy. *See also* GREENING MEASURES.

setting The area around a *listed building, or other building or structure with a heritage *planning designation, which contributes to the way in which it is experienced, and which may be significantly larger in size than the *curtilage of the designated building or structure, and may be in different ownerships from it. Applications for *planning permission for *development within the setting of a designated heritage building or structure will be considered with regard to their impacts on the setting of that building, as a *material consideration in the decision-making process. The setting of a heritage building may include a number of other existing buildings or structures (for example, other non-designated dwellings situated close to a house which is listed), and may change over time (for example, if an existing building is demolished it may open up new vistas to, and from, a designated heritage building or structure, and therefore alter its setting).

settled land *See* LIFE TENANT.

settlement of sale *See* CONVEYANCING.

settlor Person (known as a **truster** in Scotland) who transfers property into a *trust.

7 Ps *See* MARKETING MIX.

severance *See* COMPULSORY PURCHASE.

Severely Disadvantaged Area (SDA) *See* LESS FAVOURED AREA.

sewage sludge A by-product of the treatment of **sewage** and waste water from sewage farms, also known as *biosolids. It can be spread on agricultural land as a soil enhancer and fertilizer, raising the organic matter content and providing plant nutrients. However, sewage sludge can contain heavy metals, and its sale to farmers and spreading on the land is heavily regulated under the Sludge (Use in Agriculture) Regulations.

sewage treatment plant Tank-based system for treating *foul water in situations where a property is not connected to the mains drainage system. There is a variety of different types of sewage treatment plants (some of which use electricity, and some of which do not). Whilst *septic tank systems are, arguably, a type of sewage treatment plant, they have largely been replaced by more modern types of systems (sometimes known as **package treatment plants**) which clean sewage water to a higher standard, and therefore do not always require the installation of a *soakaway. *See also* CESSPIT; REEDBED.

sewelling A string, cable, or similar draped with strips of coloured material that hang down at intervals, laid out across a row of forked sticks or similar. It can be jerked occasionally by a hidden *beater, and if correctly constructed it will form an effective barrier to encourage *game birds, particularly *pheasants, to rise from a flushing point towards the line of waiting guns. *See also* BEATING; DRIVEN SHOOTING; SHOOTING.

shale gas *See* FRACKING.

share farming Farm business arrangement where two separate farmers jointly operate a farming enterprise on the same land, sharing resulting returns in agreed proportions relative to their contributions to the enterprise. Unlike a *partnership (where all of the individuals involved are linked in the same business and are therefore liable for each other's debts), in share farming each of the individuals involved operates an independent business with their own individual liabilities. One party contributes the land, with financial capital, equipment, livestock, seeds, fertilizers, sprays, and responsibility for management and labour being split between both parties by agreement. The parties agree to split the returns (in practice, receipts from the sales of, for example, crops, animals, or milk) amongst themselves relative to their respective contributions to the arrangement. *See also* CONTRACT FARMING.

shareholder An individual or organization that holds shares in a business. Each **share** effectively represents part of the ownership of that business and normally entitles the owner to a **dividend** (a share of the *profit of that business). In a public limited company (denoted by the abbreviation plc after its name), shares can be bought and sold publicly. However, in a private limited company (Ltd) they cannot.

shares *See* SHAREHOLDER.

sheaf (*pl.* **sheaves**) A bundle of *cereal stalks including the straw and the ears tied together lengthways, having been cut by a scythe or a binder. The sheaves are then stood up with others in a *stook in the field to dry out before being carted and stacked in a barn or rick in a rickyard at the farmstead to await *threshing.

shear Literally, to break away or remove one part from another. In livestock farming, to remove the fleece of a *sheep or other animal kept for its fibre by cutting or clipping. It may be carried out using hand shears, but more commonly with electrically driven shears. Sheep are sheared or shorn in summer so that the animals do not overheat, and for the fleece of *wool that is used to make clothing or carpets or as a padding or insulating material. *See also* SHEARING.

shear bolts Bolts that hold two components of an implement together that are designed to shear or break if the tension becomes too great and thus prevent damage to any other part of the machine.

shearing The process of removing the fleece or coat of sheep or other animals kept for their fibre. It is a highly skilled operation often carried out by a group of professional shearers who may travel from continent to continent throughout most of the year. *See also* SHEAR.

sheep A ruminant animal, *Ovis aries*, kept primarily for meat but also for its fleece of wool and for milk. The male is a *ram or *tup, the female a *ewe, and the offspring *lambs. They tend to be more hardy and graze a broader range of herbage than cattle, and are thus kept on moors, rough grazing, and permanent pasture of the uplands. The hill breeds are then *crossed with lowland breeds to produce the ewes used to produce prime lamb in the lowlands. In mediaeval times, the wool was very valuable for clothing and the trade was an important component of the economy in England and parts of Europe. Today, when clothing tends to be made from cotton and synthetic fibres, the value of the wool may not even cover the cost of *shearing. Reaching sexual maturity typically at six to eight months, most breeds of sheep are seasonal breeders. Ewes are put to the *tup in autumn and, after a gestation period of around 152 days, give birth in spring, in late March and April in the northern hemisphere. They are capable of multiple births, but the ewe has only two *teats and so twins is ideal. Depending upon breed and nutrition, the lamb reaches slaughter weight at around three months off grass in the summer, but may be kept round to be finished in winter either in a shed or on turnips or other forage. When a lamb reaches a year old it is

Breeds of sheep in the UK

Beulah Speckled Face	Hampshire Down	Romney
Blue Faced Leicester	Hebridean	Scottish Blackface
Border Leicester	Herdwick	Shetland
Charollais	Jacob	South Down
Cheviot	Kerry Hill	Swaledale
Clun Forest	Lleyn	Suffolk
Cotswold	Lonk	Texel
Dorset Down	Masham	Welsh Mountain
Dorset Horn	Oxford Down	Wensleydale

Traditional and not uncommon are crossbred sheep that have been reared in the uplands. For example: Scotch half-bred, Border Leicester ram on Cheviot ewe, Scotch Mule, Blue Faced Leicester ram on Blackface ewe. Some new breeds have been developed from selective crossbreeding, such as Easycare and Meatlinc.

known as a *hogget. The price of lamb is highest in spring, so some producers induce early lambing in mid-winter to catch the market at its peak. There are numerous vernacular terms for sheep of various ages; for example, a ewe of eighteen months old as it is put to the tup may be described as a two-tooth.

sheep-dip A chemical compound in which the sheep is immersed to control *ectoparasites such as *ticks, *blowflies, *mites, and lice. The active ingredient of most dips is an organophosphate chemical which is highly toxic, and thus use and disposal is strictly regulated. There has been controversy surrounding the illness of shepherds and other workers exposed to sheep-dip.

shelterbelt *See* WOODLAND.

shelterwood A system of *silviculture that promotes the establishment through *natural regeneration of a young crop of *trees which is protected by the overhead or side shelter of an old crop. It is not so commonly practised in Britain, but can be suitable for good seeding species and *woodland that is not prone to windblow. *See also* CONTINOUS COVER FORESTRY.

shepherd Someone who keeps or tends sheep. In the past and still in some cultures this involves a nomadic lifestyle looking after sheep on pastures or other forage with open access. Today, most sheep are kept within enclosed pastures and may be housed in winter.

Sheriff Appeal Court *See* SHERIFF COURTS.

Sheriff Courts Scottish civil (and criminal) courts located in cities and towns throughout Scotland. Civil cases are heard by a **Sheriff**, and Sheriff Courts have the authority to determine cases in a range of civil matters. Whilst the Sheriff Courts deal with the majority of cases in Scotland, many civil matters can be dealt with by either the *Court of Session or Sheriff Courts. There are some matters, however, which can only be dealt with by the Court of Session (including *judicial reviews). Sheriff Courts would normally deal with small claims where one person wishes to claim money back from another person or business (i.e. for non-payment of a debt). Civil appeals from the Sheriff Courts are heard in the **Sheriff Appeal Court** and, in some circumstances, can be further appealed to the Court of Session or, in turn, the **UK Supreme Court**. *See also* COUNTY COURT; HIGH COURT; SCOTTISH LAND COURT.

shire *See* HORSE.

shoot 1. A general term to describe the engagement of hunting of *game birds, normally with *shotguns. Hence the common use of terms such as shoot day, shoot lunch, or shooting party, for example. **2.** The first parts of a plant, normally the stems but can include leaves and flowering appendages, to show above the surface of the *soil. *See also* DRIVEN SHOOTING.

shooting *See* CLAY SHOOTING; DEER STALKING; DRIVEN SHOOTING; ROUGH SHOOTING.

shooting syndicate A group of shooters, usually termed 'guns', who buy shooting, normally from landowners, or who manage their own shooting operation. *See also* SHOOTING; SPORTING RIGHTS; GAMEKEEPER.

Shoreline Management Plan (SMP) Policies and management plan of an identified stretch of shoreline that address issues relating to flooding and erosion. Developed by Coastal Groups which include key personnel from local government and the *Environment Agency in England or equivalent devolved organizations. *See also* SALTMARSH.

short assured tenancy *See* SCOTTISH RESIDENTIAL TENANCIES.

short limited-duration tenancy (SLDT) *See* SCOTTISH AGRICULTURAL TENANCIES.

short-rotation coppice Trees that are normally grown on a cycle of less than ten years, and commonly include *willow and *poplar. The production of *biomass from this system is now most normally used for generating energy. Planting and harvesting of short-rotation coppice is heavily mechanized, and there has been some interest, aided by grant schemes for growing this crop for non-agricultural use of farmland. *See also* COPPICE.

shotgun A smooth-bored gun that come in a variety of sizes but with a barrel usually not less than 24 inches in length. They are primarily used for shooting *game, smaller *quarry species, and vermin. Shotguns fire *lead shot that spreads after leaving the barrel, allowing faster and smaller targets to be hit. They must be legally held on a shotgun certificate.

The certificate holder must ensure that the gun and cartridges are stored securely. *See also* GAME.

showjumping A competitive equestrian sport in which the aim is to jump a round of fences clear within an enclosed or marked-off arena. Penalties are incurred if fences are knocked down or refused, and the time of the round is also taken into account. Showjumping can be ridden as a single competition or as one of the elements of *eventing. *See also* EVENTING.

shrubs *See* WOODY SHRUBS.

silage A fermented *herbage material, usually *grass, used as a feedstuff for *cattle and *sheep. It is one of a number of methods that allows the high level of plant growth in the spring to be captured and utilized during the less productive period in winter. The silage system uses anaerobic fermentation to conserve the grass in an edible condition. As with *hay, the quality of the silage is determined by the ratio of plant leaf to stem (leaf is more nutritious than stem) and the silage-making process. *Ryegrass species are best suited to silage production. The process of silage-making involves *mowing the grass, wilting, harvesting, and storage. Silage is stored in *clamps, sheeted with plastic and weighted down, or *bales, which are bagged or wrapped in plastic to create the anaerobic conditions. *See also* BALE; BALER; HAYLAGE; STORE.

silage, slurry, and agricultural fuel oil regulations *See* SSAFO REGULATIONS.

silo A bunker or storage container for bulk materials, especially *grain or *silage. Silage may be stored in a *clamp, which is a type of silo. Many silos are cylindrical structures in which grain or other bulk feedstuffs are stored.

silt Medium-sized *soil particles between 0.002 and 0.05 mm, particularly those that have been carried by water and settled out elsewhere, such as a watercourse or *reedbed.

silvicultural systems *See* SILVICULTURE.

silviculture The care and cultivation of *forest *trees with a primary objective being the production of timber products. A silvicultural system is the process of this care and cultivation, and may include *planting, *brashing, *pruning, thinning, and various forms of *felling, including *coppicing. The selection of different silvicultural systems will reflect the different objectives and priorities set for the forest management. *See also* FOREST RESEARCH.

simplified development zone *See* SIMPLIFIED PLANNING ZONE.

simplified planning zone Area designated by a *local planning authority in the *United Kingdom where *planning permission for a specific *development (or class of development) is granted (i.e. without the need for individual applications for planning permission to be submitted). The permission can be granted with or without *conditions or exceptions. It has, in practical terms, a similar effect to a localized form of *permitted development, and can sometimes be used, for example, where a local authority wishes to encourage businesses to move to a particular area, or wishes to promote the construction of new housing, as it has the effect of reducing the requirements for individual planning applications. The Scottish Government is currently proposing to rebrand simplified planning zones as **simplified development zones**.

sinking fund A fund set up in order to ensure that the owner of a *wasting asset is able to maintain the capital (money) that they have invested in that asset. For example, the purchaser of a *leasehold investment will find that that leasehold will reduce (or 'sink') in value as the lease grows nearer to the end of its term. At the end of the lease it will no longer have value as it will no longer give permission to occupy the property, so no one else would be interested in purchasing it. Therefore, in order to allow the owner of the leasehold to have funds at the end of the lease so they can purchase another leasehold, or other investment that will continue to give them an income, they may set up a sinking fund. To do this, during the life of the lease they will invest a proportion of the rent that they (as leaseholder) will receive from the *tenant into a sinking fund. That amount, plus the *compound interest they will receive on the money invested in the sinking fund, should mean that by the end of the lease they will have a sufficient sum of money in order to put them in the same position as they would have been before they purchased the leasehold in the first place.

Sinking funds are also sometimes used to save money in order to pay off a long-term debt. They are also used by landlords to save funds in order to pay for the routine maintenance (for example, exterior painting) of a property where the landlord may place a proportion of the *service charge they receive from their leaseholders into a sinking fund to pay for routine work on the property.

Sites of Special Scientific Interest (SSSI) A range of statutorily protected areas of the best examples of the Britain's *biodiversity, geological, and physiological features. The majority of sites are designated for biodiversity and also underpin international designations including *Special Protection Areas, *Special Areas of Conservation, and *Ramsar sites. Many sites are managed as woodland or are farmed, but usually not intensively and are subject to legal restrictions. They are frequently managed as part of an *agri-environment or woodland scheme. A large proportion are privately owned, or owned by non-government organizations such as the *Royal Society for the Protection of Birds or wildlife trusts. SSSIs were originally notified under the *National Parks and Access to the Countryside Act 1949, and subsequently under the *Wildlife and Countryside Act 1981 and the *Countryside and Rights of Way Act (and parallel legislation in Scotland). Northern Ireland has equivalent protection afforded by **Areas of Special Scientific Interest (ASSI)**.

((⊕)) SEE WEB LINKS

• Provides detailed information relating to the management of sites in England.

sitting tenant *See* VACANT POSSESSION.

size class Timber species are commonly categorized by certain size classes recognized by distinct ranges of measurements, and these can equate to stages of development; for example small *sapling, *standard, and *veteran trees. *See also* AGE-CLASS; STANDING TIMBER VOLUME; TREE.

skewbald *See* HORSE.

skidder A tractor which is used to pull felled timber. It lifts the butt end with the thinner crown end dragging along the ground.

skylark plot Small patches in winter *cereal fields that are not *drilled, leaving a patch of shorter vegetation that benefits ground-nesting birds such as the skylark. They normally do not nest in the skylark plot itself, but it enables easier foraging. These have been included as an *agri-environment option in some schemes, including the *Countryside Stewardship Scheme, which stipulates a minimum area of 16 square metres and a minimum density of two plots per hectare.

SLU *See* STANDARD LIVESTOCK UNIT.

slumping *See* SOIL CAPPING.

slurry Liquid or semi-liquid manure or animal waste. Animal waste may be mixed with bedding such as straw, when it becomes *farmyard manure, or scraped from the concrete floor or yard as a semi-liquid into a slurry store, tank, or lagoon from which it is pumped and spread on the land.

small claims court *See* COUNTY COURT.

small-diameter roundwood *See* ROUNDWOOD.

small landholders *See* SCOTTISH AGRICULTURAL TENANCIES.

SMD *See* STANDARD MAN DAYS.

SMR *See* STATUTORY MANAGEMENT REQUIREMENT.

SNH *See* SCOTTISH NATURAL HERITAGE.

soakaway Perforated drainage pipes normally laid in shingle which form the final stage of treatment of sewage within a *septic tank system (or, if installed as a way of dispersing rainwater collected in *surface water drains, a soakaway can be a simpler pit filled with rubble).

When forming part of a septic tank drainage system, the sewage has been through a two-stage settlement and bacterial digestion process (which separates out the liquid and solid elements of the sewage) before the liquid part enters the soakaway, where further bacterial digestion of the liquid occurs as it gradually filters into the surrounding soil.

Soakaways can also sometimes be used with more modern *sewage treatment plant systems (although this is not always the case, as modern sewage treatment plants themselves clean sewage water to a higher standard than septic tank-based systems).

social exclusion The exclusion of some individuals, or groups, within society from the ability to access services, facilities, and opportunities (such as employment) which the majority of people are able to access. Social exclusion is often regarded as being caused by financial issues (for example, not being able to afford to run a car in an area where there is little public transport); however, it can also be caused or exacerbated by other factors such as age, disability, or poor levels of education and/or skills. These factors may often exist in combination with each other. Social exclusion caused by a lack of public transport is a particular concern in rural areas. The availability of public transport is therefore an important consideration in determining where to locate *affordable housing in rural localities.

social housing *See* AFFORDABLE HOUSING.

sod A clod or lump of earth usually with grass growing on it, a **turf**.

sodbuster A *farmer who tills the land, or the implement such as a *plough or heavy *cultivator that is used to break up the soil. It sometimes referred to someone who broke open virgin land; for example, in the taming of the American West.

softwood General description of *wood products derived from *conifers or of conifers themselves, in direct contrast to *hardwoods derived from *broadleaves. The description normally reflects the relative softness of this type of wood, generally linked with its faster growth and reduced density and hence it is easier to cut.

soil The upper layer of earth made up of minerals, water, air, and organic matter in which plants grow. It is formed by the erosion of rock by wind and water over long periods of time; it is estimated that 3 cm of soil takes 1,000 years. Soil particles vary greatly in size and are categorized into three: clay are the smallest, less than 0.002 mm, silt is between 0.002 mm and 0.05 mm, and sand from 0.05 mm to 2.0 mm. Any particles larger than 2 mm are called gravel or stones. Soils also contain *organic matter, growing or decaying plant material, and microorganisms which break down the decaying plant material into humus. These microorganisms are vital to soil health. There are gaps between the soil particles that are filled with air or water. Nutrients may be adsorbed onto the surface of soil particles, especially clay. All soils are made up of these particles, with the soil type determined by the precise composition and the underlying rock. The main soil types are: clay, silt, sand, peat, chalk, and loam, with names for particular soils, such as *podzol and *rendzina. The profile of the soil consists of the *topsoil over the *subsoil over the underlying rock.

Soil structure is important to the fertility and ability of plants to grow well. An open soil structure with good organic matter content has pores for air and water and thus allows rain to percolate through the profile. It also allows the roots of plants to grow. Where the soil has aggregated, often by the pressure of heavy machinery, there is *compaction or a pan of solid material that limits the percolation of rainfall through the soil profile and the growth of plant roots.

Main soil types and descriptions

Soil type	Description
Chalky or lime rich	Soils with high levels of calcium carbonate over chalk or limestone. They may be light or heavy; outbreaks of clay cap are common, and are very alkaline.
Clay	Soils with more than 25% clay particles. They are heavy and fertile, high in nutrients, but may be wet and cold in winter and bake dry in summer.
Loams	Soils with a mixture of clay, sand, and silt particles that avoid the extremes of each type.
Peat	Soils with very high organic matter content and thus moisture-retentive. They also contain high levels of carbon and are thus best for sequestration.
Sandy	Soils with mostly larger sand particles with little clay. They are light and warm but may be vulnerable to drought, low in nutrients, and often acidic.
Silt	Soils with predominantly medium-sized particles. They are fertile, light, and moisture-retentive, but may easily become compacted.

When this occurs it is important to break the pan by deep cultivation by a subsoiler. If a soil is compacted and rainfall cannot percolate through the soil profile, heavy rainfall can lead to soil *erosion and *diffuse pollution. To allow this to happen may be a breach of *cross-compliance. With intensive arable cropping, soils have become less fertile with poor structure, and in particular, low organic matter content. Erosion by wind and rainwater has led to the loss of millions of tonnes of topsoil.

Soil Association The largest *organic certification organization in the UK, certifying the largest proportion of UK organic farms and farming systems. It is a membership organization and a charity that campaigns nationally and internationally, and raises awareness of the benefits of organic farming and food.

((⊕)) SEE WEB LINKS
• Provides details of activities, events, and certification standards.

soil capping Condition of *soil where a thin crust of soil forms at the surface as a consequence of raindrops battering the soil and causing the soil particles to bind together. The cap then prevents *water from soaking down into the soil, and makes it difficult for plant shoots to emerge through the surface. It manifests itself as a fine dense layer of soil at the surface, potentially with areas where the *topsoil has broken up and collapsed (slumped). This is known as **capping** and

slumping. Soil capping can increase the risk of the excessive surface *runoff of water (and potentially soil *erosion and *diffuse pollution of *watercourses), as well as poor *crop *yields. Fine sandy and *silty soils tend to be more prone to it. Soil capping is often indicative of low levels of organic material in the soil. Organic material can be increased by measures such as applying *farmyard manure, incorporating *cover crops and *grass *leys into arable *rotations, and returning crop residues to the soil.

soil wash A means of removing pollutants from contaminated *soil by use of water or other aqueous solution. Where the contaminants are soluble or able to be held in suspension, they can be removed directly by the liquid. Where they are adsorbed onto soil particles such as clay, the process separates these finer contaminated soil particles from others. The contaminants and any contaminated soil particles must then be disposed of safely.

solar photovoltaics (solar PV) *See* SOLAR POWER.

solar power A type of *renewable energy which uses energy directly from the sun. The sun's energy can be utilized in a number of ways.

Passive solar design principles can be used within a building in order to utilize the energy naturally coming into it from the sun in the most efficient manner possible in order to reduce the need for other forms of heating. This is done through approaches such as using energy-efficient building materials and incorporating the most effective orientations and layouts of glazing.

Solar thermal panels can be utilized to provide hot water. Where used, these are often mounted on the roof of a building. The panels contain a liquid (often a water and anti-freeze mix) which is heated up by the sun's energy and is then circulated around to a coil in the hot water cylinder, thereby transferring the heat from the liquid which has flowed through the solar panel into the building's hot water system.

In contrast, **solar photovoltaic** systems (also known as **solar PV**) convert the sun's energy into electricity. The most commonly used types of solar PV system in the *United Kingdom are currently solar PV panels; however, other types of system such as solar PV tiles are also available. Solar PV panels can either be mounted on buildings, or on the ground (where large collections of them are sometimes referred to as **solar farms**). In solar PV systems the sun's energy is captured by **solar cells** which convert the energy into direct current. This can then be converted to alternating current by an **inverter** and then used either directly in the immediate building(s) or exported to the national electricity grid. Because solar PV systems tend to generate most electricity when the sunlight is at its strongest (which is also often the point at which electricity demand is at its lowest), they are increasingly being combined with battery storage systems which allow electricity generated by the solar PV system to be stored and then made available when demand is higher.

sole trader A business owned by one person. Sole traders are taxed in the same way as individuals in the *United Kingdom; so, for example, they will pay *Income Tax on the profits of their business, rather than *Corporation Tax.

They will also have unlimited personal liability for any debts of the business. Small tradespeople, for example, decorators, carpenters, plumbers, and so on, often operate as sole traders.

solum *See* TENEMENT FLAT.

sow A female *pig.

soya (soya bean, soybean) A *legume, *Glycine max* originating in East Asia and widely grown in warmer climates around the world. It is sown in spring and harvested in summer, with a growing season of less than three months. The seed, a bean produced in pods similar to peas, is high in protein and is a common ingredient of animal feeds. But the seed also has a high oil content and the plant is categorized as an *oilseed rather than a pulse by the United Nations Food and Agriculture Organization.

SPA *See* SPECIAL PROTECTION AREAS.

Special Areas of Conservation (SAC) A designation to protect rare and threatened *habitats and *species, under the EC Habitats Directive. Together with *Special Protection Areas, these sites are called **Natura 2000** sites, and are an important and internationally recognized network of European sites of particular ecological value. *See also* SITES OF SPECIAL SCIENTIFIC INTEREST.

(⊕) SEE WEB LINKS
• Includes selection and regulations, and a full list of UK sites.

Special Countryside Area *Planning designation used in Northern Ireland for an area of the countryside where the quality of the landscape or its amenity value is considered to be such that development should only be allowed in exceptional circumstances. The areas are designated in *local plans which will include *planning policies to protect them. *See also* LOCAL LANDSCAPE POLICY AREA.

Special Landscape Area (SLA) Local *planning designation used in England, Wales, and Scotland for an area of *landscape that the *local planning authority considers is worthy of protection, but which may not be designated as *National Park, an *Area of Outstanding Natural Beauty (England and Wales), or a *National Scenic Area (Scotland). In Scotland they are also sometimes known as Areas of Great Landscape Value. *See also* SPECIAL COUNTRYSIDE AREA.

Special Protection Area (SPA) A designation to protect the habitats of rare and vulnerable migratory birds, under the EC Directive on the Conservation of Wild Birds. Together with *Special Areas of Conservation, these sites are called **Natura 2000** sites, and are an important and internationally recognized network of European sites of particular ecological value. *See also* SITES OF SPECIAL SCIENTIFIC INTEREST.

(⊕) SEE WEB LINKS
• Includes selection and regulations, and a full list of UK sites.

special waste Term used in Scotland for *hazardous waste.

species A grouping of organisms of similarity and that cannot be further subdivided into two or more species. There is some flexibility in defining a species group but a common divider is their ability to breed amongst the group and fertile offspring. A **genus** is a higher grouping of similar species, and some individual species can be further subdivided into **subspecies**. *See also* BIODIVERSITY; NON-NATIVE SPECIES.

Species Action Plan *See* BIODIVERSITY ACTION PLAN.

spinney *See* COPSE; WOODLAND.

spoil Waste material, for example earth, mud, stones, and so on, removed during excavation, dredging, or ditching.

sporting rights These are essentially the right to access and to shoot *game and other *quarry species over specified land. Importantly, any such rights also include the right to take away wild animals which would, on their death, be in the ownership of the landowner. Sporting rights are normally 'in hand', i.e. owned as part of the land, or they may be in a separate ownership. The landowner may sell or let out the land but retain the sporting rights, or permit an individual or group the rights for a specified and limited period only. *See also* SHOOTING SYNDICATE.

spot price The price of a product or commodity at the time it is offered for sale rather than a contract or forward price. *See also* GRAIN MARKETING.

SPPS Strategic Planning Policy Statement for Northern Ireland (*see* PLANNING POLICIES).

spray drift Liquid blown by the wind during an application of a *pesticide or liquid *fertilizer through a *sprayer. This can be significant if the chemical used is toxic and there have been claims of human ill health caused by spray drift as well as contamination of watercourses and damage to property or gardens. To avoid spray drift, spraying should only be carried out when there is little or no breeze. A *buffer strip around a field may be used to avoid contamination by spray drift, especially when it adjoins a watercourse.

sprayer An agricultural machine used to apply pesticides or liquid fertilizers to land or a growing crop. It may be self-propelled, towed behind a tractor, or mounted on the *three-point hydraulic linkage of a *tractor or *quad bike. It consists of a tank to hold liquid, a boom with nozzles that folds in on itself for transport, and a pump to force the liquid through the nozzles. The boom can be up to 50 metres wide, but a standard size might be 24 metres. For more localized use such as spraying specific small areas of target weeds such as docks or thistles in a grass sward, there are hand sprayers where the pressure is maintained using a handle or lever and the nozzle is on the end of a lance.

The tank may be carried on the operator's back, in which case it is known as a knapsack sprayer.

spruce (Norway, Sitka) *See* TREE.

squatting Unlawfully occupying land without the owner's permission. *See also* ADVERSE POSSESSION; PRESCRIPTION; TRESPASS.

squirrel (*Sciurus*) The grey squirrel (*Sciurus carolinensis*) is a woodland and parkland-living rodent with a distinct bushy tail. It was introduced into the UK from North America in the nineteenth century, and is now much more common than the native **red squirrel (Sciurus vulgaris)**, which has been severely impacted by the grey squirrel. Grey squirrels compete for habitat, and carry and transmit the squirrel pox virus which fatally affects the red squirrel. They have a particular relevance because of the potential and substantial damage caused to woodland by bark-stripping, and are the focus of wide-ranging initiatives to control their numbers. There are increasing concerns that they also have an adverse impact on woodland birds by predating the eggs and young in the nests. *See also* NON-NATIVE SPECIES.

SSAFO Regulations *Regulations applying in the *United Kingdom which set out requirements relating to how, and where, *silage, *slurry, and agricultural fuel oil can be stored, with the objective of minimizing the risk of any of them causing *water pollution. Although England, Wales, Scotland, and Northern Ireland each have their own SSAFO Regulations, the principles underpinning them are the same. These include rules concerning where, and how, structures for storing slurry are constructed and also their maintenance and capacity, requirements that tanks for storing agricultural fuel oil are **bunded** (i.e. they have a double wall to prevent leakage in case one wall fails), restrictions on how close to water silage can be made and stored (so as to avoid polluting liquids from the silage entering water), and also how any store is constructed. Land located in *Nitrate Vulnerable Zones is also subject to additional controls to prevent pollution of water bodies by *nitrates.

SSSI *See* SITES OF SPECIAL SCIENTIFIC INTEREST.

stable A building normally associated with housing *horses and ponies that is usually divided into separate stalls or *looseboxes. There are a large variety of different types and sizes of stable, ranging from accommodation for a single animal up to the American barn stable, which describes a building that houses everything under one roof, including the horses, *tack room, and feed storage.

stack Sometimes known as a rick in past times, a collection of sheaves, hay, or straw for storage in the open. In modern times the material is in *bales stored on top of each other, perhaps with a waterproof top, such as a tarpaulin, to keep the hay or straw dry.

stag hunting Although this can refer more generally to the *stalking or shooting of *deer, more specifically within England it refers to the *hunting of red deer using a pack of **staghounds**. The stag or male deer is separated from the herd and pursued until brought to bay where it will be shot at close quarters. The hind or female deer is also separately hunted in this way. Stag hunting continues in operation within the exceptions allowed within the Hunting Act 2004. *See also* FOX HUNTING; POINT TO POINT.

stakeholder An individual or organization with an interest in a particular issue or activity. Stakeholders can be members of the public, private and public companies, charities, or government bodies.

stakeholder pension Private pension schemes that meet specific requirements set out by the *United Kingdom government. These requirements include aspects such as caps on the level of charges that can be made by the pension provider, and the ability to make flexible contributions. Stakeholder pensions were introduced in 2001 and are offered by a number of pension providers. They are funded through individuals and their employers making contributions to the respective pension scheme. They are therefore different to the **State Pension** (which is funded by the government, and which individuals build up an entitlement to through making *National Insurance contributions during their working life). Therefore, on retirement, people contributing to a stakeholder pension scheme (or other private pension scheme) will receive a pension funded from their (and their employer's) payments into the stakeholder (or other private) pension, as well as any State Pension to which they are entitled. All employers in the UK are now required, by law, to automatically enrol eligible employees into a private pension, and the government sets out minimum levels of contribution that must be made by the employer and the employee. The government also effectively contributes to private pensions by providing tax relief on contributions (i.e. the *Income Tax paid by the employee is calculated after their pension contribution has been taken out of their salary, so they do not pay Income Tax on the element of their salary which is paid into their pension).

stalking The more traditional pursuit of hill stalking commonly involves extensive climbing or walking over open mountainside to locate and approach a *deer, in order to get into a position in which to shoot and kill it cleanly. This activity may include the services of a professional stalker who accompanies the guest or paying client. Woodland stalking targets lowland deer species in a *forest, *woodland, or farmland environment. It is generally undertaken very early or late in the day, and may involve stalking on foot, concealed, or in a *high seat. *See also* FIREARMS; OPEN SEASON.

Stamp Duty Land Tax (SDLT) Tax, applying in England and Northern Ireland, which is paid on the purchase of a property, or the signing of a *lease where the rent which will be paid over the life of the lease exceeds a specified amount. The purchaser, or *tenant, is responsible for paying the tax. The amount payable is based on a percentage of the amount the property was bought for (or on the total amount of rent which will be paid), with properties purchased for higher prices paying a higher percentage. There are different rates for residential and non-residential property. Whilst the requirement to pay SDLT is relatively clear for the purchase of properties, it can be more complex for leases. For example, where a lease does not have a definite end, or where there are successive leases relating to the same property between the same *landlord and tenant, it is possible that the requirement to pay SDLT will only be triggered some years after the initial lease was signed when the total value of rent paid exceeds the relevant threshold.

If a residential property is being purchased and the purchaser already owns residential property, then normally a higher rate of SDLT will be paid (although there are some exceptions; for example, if the new property is replacing the purchase's existing main residence). Most buyers buying their first home, however, are exempt from SDLT.

Until 2018, SDLT also applied in Wales, but it has now been replaced by a **Land Transaction Tax**. It operates on a similar basis to SDLT, with second residential property purchasers also potentially paying a higher rate. Leases are also subject to the tax, but unlike with the SDLT, most residential leases are exempt from it.

Scotland has operated a separate system (the **Land and Buildings Transaction Tax**) since 2015. It again operates on a similar basis to SDLT, with second residential property purchasers also potentially paying a higher rate. Leases are again also subject to the tax; however, like the Land Transaction Tax in Wales, most residential leases are exempt from it.

standard An individual tree within *coppice woodland allowed to grow to fuller maturity, and to produce timber suitable for construction and so on. Can also be used to describe a large nursery-grown tree normally produced for planting in streets, parks, or gardens. They are normally 2–3 metres in height and 10–18 cm in girth. They do not have extensive root systems, so require a high degree of immediate aftercare post planting. *See also* SAPLING.

Standard Livestock Unit A unit that can be used to calculate *stocking rate or density on a standardized basis. Thus a mature cow or bull is 1 livestock unit, young cattle of 6–24 months 0.6, all sheep and goats 0.15. Thus, if a pasture can support a stocking density of 2 SLU per hectare, the actual number of animals to be turned out can be calculated.

standard man days System which can be used in planning the labour that is required for a farm, based on standard assumptions about how many days labour will be available from each worker per year.

A standard man days (SMD) calculation is made based on the number of hours a year a typical worker will be available (taking into account holidays, illnesses, and any overtime). All of the activities (or enterprises) being undertaken on the farm will require a certain number of SMDs per unit of production (say per animal for livestock, or per hectare for cropping). The SMD requirement can therefore be calculated by multiplying the SMD requirement per unit of production by the overall size of the enterprise (say number of hectares). The SMD requirements for all of the enterprises on the farm are then added together, and an allowance for the maintenance and management of the farm is then made, to give a total SMD requirement for the farm. This is then divided by the number of SMDs the average worker will be available for each year to give a total number of full-time employees needed.

This can be a relatively effective system if the labour requirements are relatively consistent throughout the year (for example, in a livestock unit); however, it is less effective if the requirement is more variable throughout the year (for example, in arable cropping). It also does not take into account factors which will affect how much fieldwork can be undertaken (for example, weather and soil types and daylight hours). In order to address this issue, **field work days available** figures can be used (these are standard averages of the likely number of days per month that it will be possible to work in fields, attempting to allow for factors such as weather and soil conditions) in combination with *field work rates.

standard security *See* MORTGAGE.

standing timber volume A measurement of the volume of timber within standing trees, normally expressed as cubic metres overbark standing (m^3 obs). Commonly separated into **coniferous standing timber volumes** and **broadleaved standing timber volumes**, and further categorized into regional and the public or private sectors. A **tariff system** can be used to assess the standing timber volume of more uniform **stands** of trees by measuring a sample.

statement of financial position *See* BALANCE SHEET.

statute Law established through the formal approval of a **legislature** (i.e. an elected assembly or parliament). *Acts of Parliament or *Acts of Assembly are examples of statute law in the United Kingdom. *See also* CASE LAW; LEGISLATION.

statute law Body of law established through the formal approval of a **legislature** (i.e. an elected parliament or assembly). In the United Kingdom this includes laws made by the *Houses of Parliament, *National Assembly for Wales, *Scottish Parliament, or *Northern Ireland Assembly. Statute law is contrasted from common law, which is established through custom and the decisions of the courts (*case law). In the United Kingdom, the law comprises both statute law and case law. *See also* LEGISLATION.

statutory Something that is established by law, i.e. by *statute. For example, a rule applying to a business which is described as a 'statutory requirement' is a legal obligation which must be complied with.

statutory consultee Organization with whom the *local planning authority is, by law, required to consult in relation to planning applications. The consultees will vary depending on the nature of the proposed development, and regulations set out who must be given the opportunity to comment on each type of proposed development. The list of potential statutory consultees is extensive and includes adjoining landowners, *highways authorities, water and sewerage undertakers, *Historic England, the *Environment Agency, the *Health and Safety

Executive, and *Natural England (and their equivalents in Wales, Scotland, and Northern Ireland). Local planning authorities are also required to advertise applications so that members of the public are able to comment on them, and are also able to consult other non-statutory bodies that they think are relevant to particular applications. *See also* PLANNING OFFICER.

statutory declaration A written legal statement made by someone about a particular factual matter, and signed in the presence of one of a number of people permitted by the law to witness such statements (for example, a solicitor). In the *United Kingdom, within the statutory declaration must be a specific form of words stating that the person signing it believes it to be factually correct to the best of their knowledge. They are used when someone needs to make an official statement about some matter where there is no other evidence available (for example, someone has changed their name and wishes to apply for a passport).

statutory instrument (SI) *See* DELEGATED LEGISLATION.

Statutory Management Requirement (SMR) Certain and specific legal standards that *farmers and *landowners are required to be compliant with in order to receive Common Agricultural Policy payments. They include, for example, legislation for the protection of habitats and species, and the protection of soil and water resources. Non-compliance may result in a reduction in payments and even prosecution. The SMRs together with the Good Agricultural and Environmental Conditions make up the *cross-compliance requirements. *See also* GREENING MEASURES.

statutory nuisance Provision within the law in the *United Kingdom which allows *local authorities to require an activity to be stopped or reduced if it is causing problems for others (in relation to a number of defined issues including premises that are kept in a poor state, noise, smoke, smells/dust, artificial light, insect infestations, animals, or other accumulations such as rubbish). The **Environmental Protection Act 1990, as amended** (in England, Wales, and Scotland, and equivalent *legislation in Northern Ireland) requires local authorities to investigate complaints made to them about issues which could be a statutory nuisance. For something to be considered to be such a nuisance it must fall into one of the above categories, and must be either causing, or being likely to cause, health issues, or be substantially and unreasonably preventing others from using or enjoying their home or some other premises.

If the local authority considers that the issue is a statutory nuisance then they can serve an **abatement notice** on the person, or persons, responsible for it (or if they cannot be found, then on the owner or occupier of the property). This notice can require the activity to be stopped or reduced (perhaps limited to certain times of the day). Non-compliance with an abatement notice can result in a fine, and in some circumstances the local authority may take direct actions to stop the activity (for example, confiscating equipment) and make the person given the notice pay the related costs. Alternatively, the local authority may apply to the courts for an *injunction (or the equivalent *interdict in Scotland) to stop the activity. *See also* PRIVATE NUISANCE.

statutory tenancy *See* RENT ACT 1977 TENANCY; RENT (AGRICULTURE) ACT 1976; SCOTTISH RESIDENTIAL TENANCIES; NORTHERN IRISH RESIDENTIAL TENANCIES.

statutory undertaker Body, either public, or privately owned, providing some form of public utility or service under *statutory powers (such as water, sewerage, gas, electricity, or railway companies). These bodies have various powers, under legislation, to carry out activities in connection with the provision of their respective service. Depending on the nature of the service, this may include aspects such as additional *planning rights, or the ability to acquire land under *compulsory purchase powers.

steer A castrated male of the cattle family, also known as a *bullock. Unless reared for breeding or for slaughter at around one year old, the male *calf is usually castrated to prevent any unwanted sexual activity or aggressive behaviour, and to avoid any taint in the meat associated with entire bulls. *See also* CATTLE.

STEP analysis *See* PESTLE ANALYSIS.

steppe A large area of temperate grassland, semi-arid with hot dry summers and cold winters, without tree cover except those that may occur by rivers or bodies of water. The word comes from the Russian and, whilst these grasslands appear in all continents except Australasia and Antarctica, also known as savanna, the Great or Eurasian Steppe stretches from the plains of Hungary and Ukraine through Asia to Manchuria.

stewardship Relates and refers to careful management and responsibility of a property, institution, business, or natural resources. More commonly used within the *farming and land sectors to relate to, for example, stewardship of the land, *soils, or the *environment. Also favoured by representative organizations, such as the *Country Land and Business Association and *Scottish Land and Estates, to promote a more responsible attitude to land ownership. *See also* COUNTRYSIDE STEWARDSHIP SCHEME.

stile A construction, usually of wood or stone, that allows for the passage of a person on foot over a wall or fence line. At the same time, a stile will act as a sufficient barrier to retain livestock. It would be commonly found along the route of a *footpath. *See also* PUBLIC RIGHTS OF WAY.

stirrup cup A traditional parting drink given to departing guests when they had their feet in the stirrups. Now used to refer to the drinks served to the mounted participants or **field** at a *fox hunting or *stag hunting meet.

stock 1. A shortened form of *livestock such as *cattle or *sheep, but also used in other contexts. In valuation, deadstock refers to inanimate assets such as farm

machinery. A *stockman is someone who looks after livestock. Stockproof means
that livestock cannot pass, as in stockproof fence or hedge. **2.** In forestry and
horticulture, stock refers to plants ready for planting or used in propagation, as in
bare root stock for trees or *rootstock for **grafting**.

stock fencing A barrier to contain livestock. *See also* FENCE.

stocking The keeping of livestock, often used in terms of grazing animals. Thus
stocking rate is the number of animals kept in relation to the area in which they
are kept. It has a wide range of uses; for example, it might be applied to the
number of fish released into a river or lake. Stocking density is similar but may
refer to animals kept in confined spaces; for example, poultry or pigs in a shed.

stockman Someone who looks after *livestock. *See also* STOCK.

stocktaking valuation A valuation of the stock of a farming business for use
in that business's end-of-year *accounts in the *United Kingdom. The purpose of
a stocktaking valuation is to account for costs that have been borne by the
farming business at the date of valuation but which have not yet produced any
income (for example, crops that have been drilled but not yet harvested).
Because stocktaking valuations are used in accounts that are, in turn, used to
determine the tax payable by farmers, *Her Majesty's Revenue and Customs
(HMRC) produces guidance on acceptable ways to approach stocktaking
valuations. The *Central Association of Agricultural Valuers also produces
models to assist in determining some of the costs which are accounted for in a
stocktaking valuation.

A stocktaking valuation will typically include some, or all, of the following:

- Direct costs which are related to the buying, production, and growing of
  livestock and crops. These are likely to differ, slightly depending on the
  enterprises undertaken on the farm, and can include items such as:
  ○ Livestock: feed costs, veterinarian's fees and employee/contract labour costs.
  ○ Growing and harvested crops: sprays, seeds, fertilizers, and employee/
    contract labour and machinery costs for carrying out *cultivation or
    harvesting.
- Indirect costs; which may include general employee/contract labour and
  machinery costs, and the maintenance of farm buildings.
- Consumables; such as *fertilizers, feedstuffs, and diesel which have been
  purchased but not yet used.
- Livestock; in some cases it is difficult to determine a value for livestock, as it
  may not be clear exactly how much they have cost to 'produce' to that date, as
  they may have been bred on the farm or acquired young for maturing on the
  farm. In this instance, HMRC provides guidance on an alternative **deemed
  cost** approach where a percentage deduction can be made from open market
  value (HMRC also defines what they mean by this).
- Deadstock (most commonly crops in store which have already been
  harvested). As with livestock, it is also possible to value these using a

deemed cost approach with HMRC producing standard percentage deductions from open market value.

Some farming businesses who keep production livestock may use a **herd basis** in their accounts. This is suitable for farms where livestock are kept not to trade but to produce another product; for example, hens kept for eggs, or sheep to produce lambs. By electing with HMRC to use a herd basis, these production animals are treated more like capital *assets rather than as trading stock.

Stoneleigh A village in central England, formerly the home of the National Agricultural Centre and the site of the now elapsed **Royal Show**, before ceasing one of the UK's largest national agricultural events. The site known as **Stoneleigh Park** is a base for a range of agricultural and rural-based businesses and organizations, including the *Agriculture and Horticulture Development Board, the *National Farmers Union, and the *Royal Agricultural Society of England.

stook Several *sheaves stood up together on their cut ends in the field to allow them to dry out.

stool The roots and stumps of felled trees, also commonly referred to in respect of *coppice stools, i.e. the remaining stump from which new growth sprouts after felling. The existence of large coppice stools can indicate but not confirm that the *woodland is ancient.

stop notice *See* ENFORCEMENT, PLANNING.

store 1. A building for storage, as in *grain store, feed store, or *fertilizer store. 2. An animal that has been reared but not finished for slaughter. Thus, instead of rearing beef cattle or lambs through to slaughter, a livestock farmer might sell them for another to finish. As this point they are called stores. An arable farmer who has some pasture on his farm but who has no breeding animals might buy store cattle or lambs to graze the pasture until ready for slaughter. Some store cattle or lambs may be finished in straw yards or on other forage such as *stubble turnips.

strategic development plan (Scotland) *See* LOCAL PLAN.

strategic environmental assessment (SEA) Assessment, required and in some circumstances prescribed by law, of the possible environmental impacts of a plan, programme, or set of policies (usually drawn up by a public body) which may result in, or control, changes to the physical or natural environment. Circumstances in which an SEA can be required include when a new *local plan or *neighbourhood plan is being developed, as they contain policies which control *development. SEA can be regarded as a strategic version of environmental impact assessment (EIA), with EIA considering the potential environmental impact of individual projects, whilst SEA aims to address the potential cumulative impact of the series of individual projects which may be permitted under, say, the local plan for an area.

strategic housing land availability assessment (SHLAA) Procedure, carried out as part of the process of developing a new *local plan, by which a *local planning authority identifies suitable sites for future housing development in order to meet the housing needs they have determined through their *strategic housing market assessment. The SHLAA involves the local planning authority identifying sites with the potential for development (i.e. which comply with *planning policy requirements), and which are likely to come forward during the period of the plan (i.e. there are no factors limiting the likelihood of this happening; for example, issues with access or infrastructure). The terminology SHLAA is most commonly used in England; however, parallel procedures exist in Wales, Scotland, and Northern Ireland.

strategic housing market assessment (SHMA) Assessment of current and future housing needs carried out by a local planning authority in England in the process of preparing their *local plan. SHMAs are designed to identify the scale and mix of housing types that are likely to be required over the period of the local plan, taking into account factors such as household and population projections. The local authority then uses this information as a basis for identifying the number of housing sites that they need to consider in their *strategic housing land availability assessment. The terminology SHMA is most commonly used in England; however, parallel procedures exist in Wales, Scotland, and Northern Ireland.

Strategic Planning Policy Statement for Northern Ireland (SPPS) *See* PLANNING POLICIES.

Strategy for Sustainable Food and Farming *See* CURRY REPORT.

straw The stalks or stem of a *cereal crop once it has been harvested. It may be chopped and incorporated into the soil to improve the *organic matter content, or it may be *baled and removed for animal feed or *bedding. Wheat straw is normally used for bedding, whilst good-quality barley straw may also have value as roughage fodder.

strip grazing A system of *grazing that limits the area in which the animals can graze, usually by means of an electric fence, so that fresh grazing can be allocated on a regular basis, perhaps daily. It prevents the animals from damaging the uneaten fodder by trampling or contamination by faeces. It is commonly used for dairy cows at grass, and for sheep on winter fodder such as stubble turnips.

stripper header A type of header for a *combine harvester developed in the 1980s but not in common use. Whereas a conventional header has a reciprocation knife that cuts the crops and a horizontal auger that takes the whole crop into a *threshing drum, a stripper header has fingers on a rotating spindle that largely threshes the crop in the header. The grain and some other material is then taken into the body of the combine, whilst the straw is not. The principle is that this improves the efficiency of the machine, as considerably less plant matter goes through the machine, which is a particular advantage when the straw is not entirely ripe or the crop is weedy.

stubble The residue left after the harvest of a crop, particularly a *cereal, or the field itself. There may be straw, usually chopped, or just the cut stalks of the crop sticking out of the ground. At one time in the UK it was common for stubbles, with or without straw, to be burnt to kill weeds or pests and effectively sterilize the surface of the soil. However, this was thought to be harmful to the environment and the practice was banned in 1993. Over-wintered stubbles are thought to be a good habitat and source of food for farmland birds, but such value has been reduced by the efficiency of *herbicides in killing weeds and *combine harvesters in not shedding small or cracked grains. The value of stubble as a habitat is improved by sowing or *broadcasting a seed such as *mustard.

stubble turnip A type of fast-growing turnip that is planted in *stubble, usually of winter *barley which is relatively early to harvest, or broadcast into the crop before harvest, to produce fodder for ruminant animals in autumn and winter. Both the leaves and the bulbs provide nutrients for the animals, which are usually *strip grazed across the field.

stumpage A measure of the price or value of *standing timber, and originally based on numbers of trees harvested, or 'per stump'. It can indicate the right to fell timber on the land of a third party, or indicate a tax charged on each felled tree. Stumpage price reports can provide a guide to the market value of standing timber.

stumping back The practice of cutting back top growth of broadleaves at, or a few years after, planting to stimulate a vigorous new shoot. There are therefore some similarities to *coppice, but it occurs much earlier in the life cycle of the tree. It can be used as a traditional method to improve poorly formed *oak and *ash, and to promote clean and straight timber.

subject plan (Northern Ireland) *See* LOCAL PLAN.

sub-letting The letting of a property by a *lessee. So, in a situation where Person A is the *landlord and Person B is their *tenant, Person B may then **sub-let** to Person C who is then entitled to enter into occupation. So Person C will then pay *rent to Person B, who will, in turn, pay rent to Person A. This is different from a situation where Person B may **assign** their lease (*see* ASSIGNMENT; ASSIGNATION (SCOTLAND)). If Person B has assigned (transferred) their lease to Person C then Person B will have no further interest in the land so Person C will pay rent to Person A. Some leases will have an **alienation clause** which will either expressly allow sub-letting or assignment (either with or without the landlord's specific permission) or forbid it. *See also* ALIENATION; TERMS USED WHEN PROPERTY IS SUBJECT TO LEASE OR TENANCY.

subordinate real rights (Scotland) *See* REAL RIGHTS.

subsidence Situation where a building, or part of a building, sinks into the ground because either the *foundations of the building are not substantial enough, or the ground itself is disturbed (possibly because of the action of external factors such as water, tree roots, or mining). Subsidence causes significant cracking in buildings as they move. Whilst small cracks (known as **settlement cracks**) are often found in newly constructed buildings as a consequence of the initial compression of sub-soil due to the weight of the building, and are considered normal, subsidence is a more significant problem as it can affect the stability of the whole structure. Buildings which have significant issues with subsidence often need **underpinning** (that is, the addition of extra foundations or the strengthening of the existing foundations).

subsidy *See* COMMON AGRICULTURAL POLICY.

subsoil The layer of *soil beneath the *topsoil. To subsoil is to break up compaction by deep cultivation using a subsoiler.

subsoiler A strong metal-framed piece of *farm machinery with a single or several *tines attached. It is a *tillage tool that is used to break up compacted *subsoil, a process called **subsoiling**. This allows for, and increases, the flow of air and water which improves *crop growth.

sub-tenancy *See* TERMS USED WHEN PROPERTY IS SUBJECT TO LEASE OR TENANCY.

succession The procedures and law which govern what happens to someone's *assets and *liabilities (including their property) on their death. This includes *intestacy (where someone does not leave a will), as well as how assets are passed by a will. The word 'succession' is also used to describe the mechanisms, under some *statutory codes, covering how, in some circumstances, the spouse/civil partner or close relative of a tenant can **succeed** to (in effect, take over) a *tenancy on the death (or, in the case of some *agricultural tenancies, the retirement) of a tenant.

suckler An animal that suckles its young. It particularly applies to *cows when instead of *weaning the *calf so that the cow can be milked for human consumption, the cow is left to suckle the calf. Thus it is often used to describe a beef rather than dairy cow, and a group may be described as a suckler herd or even as a single-suckle herd, as the cow usually has only one calf.

SUDS *See* SUSTAINABLE DRAINAGE SYSTEMS.

sui generis **uses** *See* USE CLASSES ORDERS.

sulphate A chemical compound formed of sulphur, oxygen, and another compound, an acid or **ester** of sulphuric acid. Sulphur is an essential nutrient for plant growth, but sulphate is soluble and thus may be leached readily. Where there is a shortage of sulphur, it may be applied as *ammonium sulphate, which also contains nitrogen as a nutrient, as magnesium sulphate or as calcium sulphate, also known as gypsum, which can improve soil structure and moisture absorption and retention. Sulphate may be a constituent of a *compound fertilizer. Copper sulphate mixed with slaked lime (calcium oxide) or **Bordeaux mixture** is a traditional fungicide, used in vineyards and fruit orchards to control powdery *mildew and other fungal infestations.

summary diligence *See* BREACH OF COVENANT.

supplementary feeding To provide extra nutrients when the basic diet is not adequate for optimum production. It usually applies to grazing animals but can apply to other animals such as fish in rivers or lakes in winter. When grazing animals are out on grass or other forage crops, such as turnips, particularly at times of adverse weather or in winter, it may be necessary to supplement the diet. *Hay or *straw can be given to allow extra *fodder, *concentrated feeds containing cereals and protein may be put out in troughs or scattered on the ground whilst extra vitamins and minerals may be provided, usually in the form of a block or lick.

supplementary planning document Document (also known as supplementary planning guidance) which sits underneath a *local plan, and which provides further detail in relation to the interpretation of a specific aspect of the *local plan (for example, building design).

supplementary planning guidance *See* LOCAL PLAN.

supporting services Underpinning services that maintain the conditions for life and are necessary for the production of all other *ecosystem services. Some examples include production of atmospheric oxygen, soil formation and retention, nutrient cycling, water cycling, and provisioning of habitat. *See also* NATIONAL ECOSYSTEM ASSESSMENT.

surface water Water which is found on the surface of the ground (including in rivers, streams, lakes, ponds, and springs). Surface water contrasts with groundwater, which is water found underground between the pores of soil and within rocks. In building construction, the term 'surface water' includes rainwater which is collected from the paved areas and roofs of a building. It is therefore managed differently from *foul water, which is contaminated with waste and requires cleaning up before being released into the environment (e.g. that from bathrooms and kitchens). *See also* ABSTRACTION LICENCE; IMPOUNDMENT LICENCE.

surrender Situation where a *lessee surrenders (gives up) their *interest to their immediate *landlord. This can, potentially, be done either by agreement (in which case it must usually be by a *deed) or through actions accepted by the landlord and tenant as ending the *lease; for example, by the *tenant physically giving up *possession (by abandoning the property) and the landlord reoccupying it.

sustainability *See* SUSTAINABLE DEVELOPMENT.

sustainable agriculture In principle, sustainable agriculture means farming systems that do not reduce reserves of natural resources such as energy, water, and soil fertility. In practice, it is almost impossible to farm without any reduction, so it has come to mean farming techniques which have, at their heart, principles of protection of the *environment, animal health and welfare, public health, and communities. One example may be *organic farming, another the generation and use of *renewable energy to minimize fossil fuel consumption.

sustainable development Term, widely adopted by governments worldwide, since being defined in the 1987 United Nations' Brundtland Commission's report, 'Our Common Future', and used to describe a way of carrying out human activities which allows current needs to be met in such a way that the ability of future generations to meet their own needs is not affected (by not, for example, using up, or damaging, resources so much that they are not available to future generations). Whilst identifying how sustainable development can apply in practice is open to a wide range of interpretations, a common principle is that sustainable development should balance economic, environmental, and social aspects. In common with many governments internationally, delivering sustainable development is a key objective of United Kingdom government policy and underpins a broad range of government activities. The most direct link is, however, to *planning policy, where policy in England, Wales, Scotland, and Northern Ireland requires all applicants to demonstrate that their proposals will deliver sustainable development before they can be granted *planning permission. Hence, United Kingdom planning policy sets out what government considers sustainable development looks like at an individual application level.

sustainable drainage systems (SUDS) Term used to describe mechanisms of dealing with surface water in ways that do not increase the risks of localized flooding (for example, by overwhelming existing drainage systems). Current planning polices and legislation place significant emphasis on SUDS; for example, new developments are generally expected to incorporate SUDS surface water management. This follows concerns stemming from an increasing use of hard impermeable surfaces around buildings (for example, replacing front gardens with hard-standing for parking), which then pushes increased amounts of rainwater into drains which were not designed to cope with it, rather than it draining through to the soil where it falls. At a practical level, SUDS can range from using permeable gravel for car parking (allowing water to drain through) through to green roofs, or larger schemes which may incorporate ponds to manage larger volumes of water.

swamp *See* WETLAND.

sward The top layer of soil when covered with grass or other forage. The composition of the forage is, for example, *grasses, and *legumes such as *clover, and in permanent pasture, perhaps herbs, wild flowers, and other plants. The nutritional quality of the sward will depend upon what species are present.

swath Strictly, the space taken when a crop is cut, originally by one stroke of a scythe. However, it is usually taken to mean the row of grass or other material cut as it lies on the ground; for example, when grass is cut and left to dry in *hay or *silage-making. The word is also a verb, to swath, meaning to cut a crop in rows. For example, a crop of *oilseed rape might be swathed to allow it to dry out for a period or for the green material to senesce.

sweeper bull A *bull that is run with dairy cows to *cover any cows that are not pregnant. Most dairy cows are served by means of *artificial insemination, often known as AI, but this is not always successful. By having a sweeper bull in the herd, any cows that have not been made pregnant by AI can be identified and served by the bull instead.

sweet corn Maize eaten as a vegetable, also referred to as corn on the cob. *See also* CORN.

SWOT analysis Tool used in business management and marketing which involves undertaking a strategic analysis of the strengths, weaknesses, opportunities, and threats of an organization, project, or business venture. Strengths and weaknesses both relate to internal factors and can include particular skills, characteristics, or advantages the organization has which offer a competitive advantage over others, versus potential areas of weakness, or limitations, in relation to others. Opportunities and threats relate to external aspects such as areas or aspects of the business environment that a company can exploit, versus potential external threats. SWOT analysis is used to inform strategic development and marketing within organizations, often in combination with *PESTLE analysis.

sycamore *See* TREE.

synergistic value *See* MARRIAGE VALUE.

S

tacit relocation Principle within Scottish *landlord and *tenant law (although not applying to all types of *lease) whereby if neither landlord nor tenant have served a *notice to quit to end a lease when it reaches its *ish then the lease automatically extends for a further period. Unless there are any *statutory provisions providing otherwise, then leases of less than a year will normally extend for the same duration as the original lease, with leases of a year or longer being extended by a year. If no notice to quit is served to take effect at the end of the period of extension then it will extend again, and so on. A lease on tacit relocation usually continues on the same terms as the original lease contract. It is possible to *contract out of tacit relocation for some types of lease.

tack Saddlery is often referred to as 'tack', and includes all the basic requirements needed to equip a *horse or pony for work. These are the saddle and all its fittings including the girth, stirrup-leathers, and stirrup-irons, and the bridle and its bits.

TAN (Technical Advice Note (Wales)) *See* PLANNING POLICIES.

tapeworm A parasitic flatworm of the *Cestoda* class, the most significant being *Taenia solium*, the pork tapeworm, or *Taenia saginata*, the beef tapeworm. The latter are the only tapeworms found in UK cattle, as adult worms live in the small intestine with the head attached to the intestine wall. Segments are shed in the faeces thirty days after infection in quite long chains. Immunity develops rapidly, and, although diarrhoea and unthriftiness can be associated with tapeworm, clinical and economic losses are generally minimal. Treatment is by an *anthelmintic wormer which is administered orally, by injection, or by pouring onto the skin.

targeted inputs A system of husbandry management that seeks to ensure that inputs are targeted to gain optimum returns. In *arable farming, for example, this applies particularly to applications of nitrate *fertilizer and *pesticides at a rate, time, and place as required by the crop. *Precision farming is an example of this, where rates may be varied across a field according to the potential to respond. This method of farming is advocated by conservation groups, notably *Linking Environment and Farming (LEAF).

target species Species that are the focus of any activity. For example, when using a selective *pesticide, the *active ingredient controls the particular species whether that be a weed, insect, or other pest. In the opposite sense, it is also used to designate species that are the focus of conservation; for example, specific management to support the target species.

teat The nipple of the mammary gland of a female mammal, from which the milk is suckled by the young, or something resembling a teat or nipple; for example, a perforated plastic bulb by which an infant or young animal can suck milk from a bottle.

Technical Advice Note (Wales) *See* PLANNING POLICIES.

technical details consent Application made to a *local planning authority in England which provides the additional details of a proposed *development which are required in order to convert a *planning permission in principle into *full planning permission for a development.

tedder A machine using in hay-making. After the *grass or *sward has been cut, it is spread by means of a tedder, usually comprising rotating tines, to allow the sun and air to dry the hay prior to *baling.

TEGoVA *See* THE EUROPEAN GROUP OF VALUERS' ASSOCIATIONS.

telecommunications equipment Equipment used to transmit and receive telecommunications, including voice, data, and video communications. The equipment required varies depending on the technology, and includes that required for fixed-line and mobile networks, as well as for broadcasting. The advent of mobile 'phone technology resulted in new requirements for mobile 'phone equipment to be located in rural areas. In the United Kingdom this is commonly done through operators leasing sites from landowners on which to locate such equipment.

Mobile 'phones operate via radio waves, and mobile 'phone networks comprise a series of **radio base stations**. These can either be mounted on buildings or other structures, or on free-standing **mobile 'phone masts**. Their locations are organized through a structure in which the country is divided up geographically into areas called **cells**, each of which has a base station. There are different types of cell, each of which has a finite capacity; therefore, where there are a greater number of users, more cells and base stations are needed:

- **Macrocells**. These provide the main infrastructure for the mobile 'phone networks, and antennas for macrocells can either be on masts or mounted on roofs or other structures. They are located in positions where the radio waves can cover large areas without interruption, and therefore may be relatively visible.
- **Microcells**. The antennas for microcells are smaller and cover shorter distances, providing supplementary capacity in areas where the network is busy with many users. They are therefore often found in urban areas where they may be mounted on buildings or on street furniture such as lampposts.
- **Picocells**. These provide coverage over even smaller distances and hence are often found inside buildings, with high numbers of users (such as transport terminals) or where coverage is poor.

- **Femtocell**. These provide the most localized coverage, and can be either indoor (perhaps in an individual house) or outdoor, where several of them may form a mini-network, say in a rural village, where traditional mobile networks cannot reach. They connect into broadband services to provide a mobile signal.

temporary stop notice *See* ENFORCEMENT, PLANNING.

tenancy 1. Word used as an alternative to the term *lease. There is little practical distinction between a lease and a tenancy (in that both grant the right to *exclusive possession of a property), apart from the requirement for a lease to normally be made by *deed (which a tenancy does not always have to be). **2.** More broadly, the interest of someone who has a right or *title to hold land. This includes situations of *tenancy in common and *joint tenancy.

tenancy agreement Written agreement between a *landlord and *tenant establishing the *terms of a *tenancy. *See also* LEASE.

tenancy at sufferance Legal position, under English law, if a *tenant continues to occupy a property at the end of their *tenancy without the *landlord's specific permission. If a landlord then refuses permission for the tenant to remain in the property, then the tenant is, in effect, *trespassing, and in this position their only legal protection to remain in the property would be that under the *Protection from Eviction Act 1977. If the landlord, however, accepts rent, then the courts would be likely to regard the tenant as having a *periodic tenancy. *See also* TENANCY AT WILL.

tenancy at will Arrangement, under English law, which can be immediately ended by either party, where a *landlord allows a *tenant to occupy, or continue to occupy, a premises (usually for a short period of time whilst negotiating a *lease). Tenancies at will are relatively rare because they confer little in the way of certainty or protection to either a tenant or landlord, and therefore have some of the features of a *licence. They also tend to be fairly short-lived, because if *rent is regularly accepted from the tenant then the *courts may regard the tenant as having a *periodic tenancy (although this is not always the case if other circumstances indicate otherwise). A tenancy at will can be established with or without a written agreement. *See also* TENANCY AT SUFFERANCE; TENANCY-AT-WILL (SCOTLAND).

tenancy-at-will (Scotland) Rare, informal, type of land occupation agreement, without a defined *ish, existing in some parts of Scotland, established through custom and usage as a consequence of a person (the occupier) renting land from a landowner in order to build a house on and paying *rent to that landowner for the ground only, but not the buildings (i.e. a **ground rent**). Because a tenancy-at-will is an informal arrangement, the occupier has no formal *title, and there are some variations in practice between different localities. Since the 1970s, however, under the law, the occupier (or someone who has subsequently acquired the house) has had the right to purchase their *landlord's interest. *See also* TENANCY AT WILL.

tenancy in common Under English property law, ownership of an *estate in land by multiple owners each of whom are each able to leave their share of the property in their will to whomever they choose. This contrasts with the position of a joint tenant in a *joint tenancy where in the event of the death of one of the joint tenants the interest of that joint tenant automatically **vests** (goes into the possession of) the other joint tenants. Under Scottish law, the broad equivalent to a tenancy in common is *title held in **pro indivisio**. *See also* TRUST.

tenant Someone who occupies a property with *exclusive possession under the terms of a *tenancy or *lease (in the case of a lease they may also be referred to as a *lessee). *See also* LANDLORD; TERMS USED WHEN PROPERTY IS SUBJECT TO LEASE OR TENANCY.

Tenant Farmers Association (TFA) A membership organization founded in 1981 by a group of farmers who felt that their interests were not being forcefully represented by existing bodies. Whilst the interests of tenant farmers may often coincide with those of owner/occupiers, there are situations where they do not, and the founders thought that at such times their particular concerns were not adequately represented by other existing farmers' organizations. The TFA lobbies at all levels of Government and gives professional advice to its members. It is open to tenant farmers in England and Wales, and there is an equivalent organization in Scotland.

(⊕) SEE WEB LINKS
• Website of the Tenant Farmers Association, providing advice and details of events.

tenant-right Right of claim by the outgoing *tenant of an *agricultural tenancy for *compensation for works of *good husbandry which will leave a benefit behind once the tenant has left (for example, the growing of crops, laying down of pasture, or sowing of seeds). The principle behind this is that the outgoing tenant will have incurred costs in, for example, preparing the soil and sowing crops, but they will not receive the benefit of the resulting income from the crops once they have grown, as the date of harvest may be after the tenant has quit the farm.

tenant's improvement An improvement to a property which is subject to a *lease or *tenancy and is made at, or partly at, the tenant's expense (and is over and above any *obligations the *tenant may have for repair and maintenance). In some areas of *landlord and tenant law, the landlord is required to give *compensation to the tenant at the end of the tenancy for some improvements undertaken by the tenant; however, this is typically subject to conditions. *See also* LANDLORD'S IMPROVEMENT.

tenement 1. Term used to identify the land that either benefits from (the **dominant tenement**) or is burdened by (the **servient tenement**) either an *easement (England, Wales, and Northern Ireland) or its Scottish equivalent (a **servitude**); or a *freehold *covenant (England, Wales, and Northern Ireland) or its Scottish equivalent (a **real burden**). **2.** Under Scottish law, a right of ownership in land which is separate from the land itself (known more specifically as a **separate tenement**); for example, mineral rights, salmon fishing rights, or a *tenement flat. *See also* REAL RIGHTS. **3.** Under Scottish law, a building, or part of a building, comprising *tenement flats.

tenement flat Under Scottish law, part of a building (itself known as a **tenement**) which comprises at least two related flats which are designed to be in separate ownership and are arranged in horizontal layers (so not a terrace, although they can involve non-residential property, so say two flats over a shop). Each tenement flat is in separate ownership. The land that the building stands on in a tenement is known as the **solum**. In Scotland, *legislation governs the ownership of the different parts of the tenement. Under the legislation the tenement is divided into **sectors**. An individual sector can be either a flat, a **close** (i.e. common stairs, landings, and passageways), a lift, or some other three-dimensional space (for example, a cellar). The law sets out specific rules as to which flats own what parts of the building; for example, which flat owns the roof, which the solum, which any garden, and how the ownership of the walls is split between the flats. The close, plus other parts of the building serving more than one flat (for example, pipes, chimney stacks, and fire escapes) are **pertinents** held in common ownership of all the flats that use that part of the building (a pertinent being some part of property which is outside the immediate boundaries of the main part). Under legislation there is also a default set of management rules (the **Tenement Management Scheme**) for the '**scheme property**' in tenements; the definition of scheme property is set out in law and includes the ground on which the tenement is built, foundations, external walls, and roof. The Tenement Management Scheme puts rules in place regarding how decisions are made by the flat owners in respect of the scheme property (predominantly those decisions relating to maintenance and insurance), including how costs are apportioned and paid. It is possible, in some instances, for the Tenement Management Scheme to be overridden if the *titles to the flats specify specific maintenance liabilities on the flats for different parts of the property (i.e. *real burdens). However, in the absence of such real burdens the Tenement Management Scheme normally applies. In some larger developments (potentially comprising more than one tenement building, and possibly also including houses) the law in Scotland also provides for a **Development Management Scheme** to exist instead of a Tenement Management Scheme. Under a Development Management Scheme, the development is run by a manager appointed by an owners' association (which is, in itself, a corporate body). There are then rules in place governing how budgets, service charges, meetings, and so on, operate under a Development Management Scheme.

The law concerning the ownership and management of buildings with multiple ownerships (for example, residential flats) in England, Wales, and Northern Ireland is different from that in Scotland, as the individual flats would normally be held as *leasehold property. *See also* COMMONHOLD; LEASEHOLD REFORM.

term 1. A period of time that a *lease is granted for; so, 'the lease had a term of five years'. *See also* PERIODIC TENANCY. **2.** A specific condition or obligation within a *contract; for example, within a lease, 'one of the terms of the lease was that the lessee must only use the building as a shop'. *See also* COVENANT; INVESTMENT METHOD.

term and reversion *See* INVESTMENT METHOD.

term of years absolute *See* LEASEHOLD.

terms used when property is subject to lease or tenancy It is possible for several different parties to all have an interest in the same property (i.e. land or buildings) at the same time, as shown in the Figure. The terms 'lease' and *'tenancy' are often used interchangeably, and practically there is essentially no difference between them, as both give the right to *exclusive possession of a property for a particular period of time. However, unlike a tenancy, a lease must generally be made by a formal *deed (itself also known as a lease).

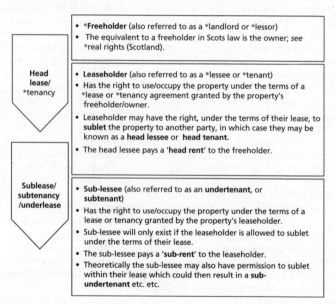

Head lease/ *tenancy

- ***Freeholder** (also referred to as a *landlord or *lessor)
- The equivalent to a freeholder in Scots law is the owner; *see* *real rights (Scotland).

- **Leaseholder** (also referred to as a *lessee or *tenant)
- Has the right to use/occupy the property under the terms of a *lease or *tenancy agreement granted by the property's freeholder/owner.
- Leaseholder may have the right, under the terms of their lease, to **sublet** the property to another party, in which case they may be known as a **head lessee** or **head tenant**.
- The head lessee pays a '**head rent**' to the freeholder.

Sublease/ subtenancy /underlease

- **Sub-lessee** (also referred to as an **undertenant**, or **subtenant**)
- Has the right to use/occupy the property under the terms of a lease or tenancy granted by the property's leaseholder.
- Sub-lessee will only exist if the leaseholder is allowed to sublet under the terms of their lease.
- The sub-lessee pays a '**sub-rent**' to the leaseholder.
- Theoretically the sub-lessee may also have permission to sublet within their lease which could then result in a **sub-undertenant** etc. etc.

terracing A means of farming sloping land by cutting a series of flat areas into the slope with vertical walls between them. This enables the growing of *crops, including vegetables. Often there may be irrigation channels integrated into the terracing to ensure adequate moisture for the crops without problems of rainfall cascading down the steps of the terrace and causing soil erosion. It is common in South and East Asia where the crop grown may be *rice, and in parts of Africa for growing tea. Terracing is also common in South America, particularly in the mountainous areas of the Andes. In Europe, vines may be grown on terraces on steep hillsides, especially south-facing to gain the maximum radiation from the sun.

tetanus A disease caused by the *bacterium *Clostridium tetani*, which can survive for long periods without a host and is commonly found in in the faeces of animals. It is relatively rare in cattle, but can cause severe losses. Horses and lambs appear the most vulnerable. The disease can be transmitted to humans, and if the bacteria enter the body through a wound they can multiply rapidly and release a toxin that affects the nerves, causing symptoms such as muscle stiffness and spasms. Immunization by vaccination is common in humans.

tether A means of restricting the movement of an animal by attaching it to a fixed point by a rope or chain. Often used on dogs or goats to prevent their straying. A tether is the rope or chain, and to tether is to attach it to the animal.

TFA *See* Tenant Farmers Association.

The European Group of Valuers' Associations (TEGoVA) Pan-European organization comprising valuers' associations, with the objective of harmonizing *European Valuation Standards and promoting the profession of the valuer through educational, ethical, and quality standards. TEGoVA's work focuses on four main areas: representing the views of the valuation profession to the institutions of the European Union; promoting minimum education requirements for membership of European professional valuers' associations; writing and promoting European Valuation Standards; and operating the Recognized European Valuer Scheme. Holding Recognized European Valuer status allows valuers to use the designatory letters REV after their name, and demonstrates that the valuer has met prescribed levels of education and valuation experience and is familiar with the European Valuation Standards. Recognized European Valuer status is obtained through application to one of the TEGoVA Member Associations who have the authorization to award the status.

(((⊕))) SEE WEB LINKS
- TEGoVA website: includes information on the work done by the Group, details of its member associations, and publications relating to valuation, including the European Valuation Standards.

thicket A densely growing area of *trees or *woody shrubs, characterized by interlacing branches and a closed canopy. It can be a single or mixed species that also provides valuable and protective habitats. **Bramble** in particular can grow to produce dense and almost impenetrable thickets.

thinning The cutting and extraction of selected trees to give more space and to promote better growth in the remaining stock. *See also* FELLING; SILVICULTURE.

thoroughbred (TB) Used to describe *horses registered in the **General Stud Book**. All thoroughbreds in Britain can be traced back to a limited pool of stallions and mares imported in the late seventeenth and early eighteenth centuries. The breed is now associated with successful horse racing throughout the world.

three-point hydraulic linkage (three-point linkage) A means of attaching agricultural implements to a *tractor invented by **Harry Ferguson** in 1926. The tractor has two arms projecting, usually from the back, which can be lifted and lowered by the tractor's hydraulic system. A fixed top link then allows the implement to be lifted when not in use. For example, a plough may be mounted on the tractor by means of the three-point hydraulic linkage. When turning on the headland of the field, it is lifted clear of the ground, to be lowered for use. The depth of the *mouldboard in the soil can be controlled by the hydraulic linkage.

thresh After harvest, grain is removed from the straw and chaff, and the inedible parts of the seed head, by means of threshing. This was achieved originally by using a flail to separate the grain from the chaff on a threshing floor in a barn. This was then replaced by a threshing machine driven by steam or by the *power take-off of a tractor. Now the process is achieved in a modern *combine harvester using a drum, wind, and sieves.

Thuja (Western red cedar) *See* TREE.

tick A small blood-sucking *ectoparasitic arthropod that feeds on the blood of its hosts. There are numerous species, of which the most common in the UK is the sheep tick, *Ixodes ricinus*. Ticks spread infection when puncturing the skin, including several diseases in lambs, and louping ill, a severe disease of grouse. Domestic pets, notably dogs, and deer are also common hosts. Tick bites may also spread *Lyme disease in humans. Control is by insecticide, which may be administered by dipping, drenching, injection, or pour-on.

tied cottage Residential property provided by an employer to their employee (typically a farm worker) for the duration of their employment. *See also* SERVICE OCCUPANCY.

tillage The movement of soil to prepare a seedbed by mechanical means such as the plough, discs, harrow, or cultivator. A soil may be tilled to encourage the growth of weed seeds, which may then be killed before the crop is sown. This is known as a false *seedbed.

tiller A side stem or auxiliary shoot produced by plants of the *Gramineae* family, known as *grasses. The yield of cereal crops depends to a significant extent on the number of tillers produced by each plant. The more tillers, the greater the number of ears and thus the lower the seed rate required, but production is weather-dependent. Tillers are produced in good growing conditions but may abort if conditions turn unfavourable. Too many tillers may result in small ears and low specific weight of grain.

tine 1. The prong on a tool such as a fork or the metal spikes attached to a *front loader for handling *bales or *silage. **2.** The pointed spikes on *farm machinery such as a *harrow or *drill. **3.** Part of a *deer's antlers.

Tir Cynnal *See* AGRI-ENVIRONMENT.

Tir Gofal *See* AGRI-ENVIRONMENT.

Tir Mynydd *See* AGRI-ENVIRONMENT.

title The right of ownership of property. Ownership of title to land has historically been evidenced through **title deeds** which have been exchanged between the old and new owners at the point of the sale or transfer of the land. Title *deeds can include *conveyances, contracts for sale, wills, *mortgages, and *leases. However, in the *United Kingdom, most changes in title and rights to land must, in order to take effect, now be registered at the respective *Land Registry in England, Wales, and Northern Ireland (and the *Land Register of Scotland, the successor to the older **Scottish General Register of Sasines**) which then guarantee title, so the owner of a property registered in such a way may no longer hold some of the historic title *deeds to their property, but can instead confirm their ownership through the respective land register.

In England and Wales, seven classes of title can be registered at the Land Registry:

- **Absolute** *freehold. This guarantees that the land is owned by the freehold owner, subject only to any other legal interests or entries entered into the Register. This is the most common form of freehold title registered; however, it will not be used if there is any degree of uncertainty about the title (and therefore any possibility that someone else may come forward in the future with legitimate rights in connection with the land).
- **Absolute** *leasehold. This guarantees that the leasehold is owned by the leaseholder, and also that the freehold of the land to which it relates is guaranteed to be owned by the registered freeholder.
- **Qualified freehold**. Used where the title of the applicant has some limitations or is weakly evidenced, so whilst the Registrar is content that the title should belong to the freeholder, there may, for example, be gaps in the evidence presented in the title deeds (which may be for a good reason), so the title cannot be fully guaranteed.
- **Qualified leasehold**. Used for the same reasons as with qualified freehold.
- **Possessory freehold**. Used, for example, when the title is based on an *adverse possession claim, or alternatively, when title deeds demonstrating ownership have been lost. Whilst, in practical terms, the owner of a possessory freehold can treat it in the same way as an absolute freehold, this is subject to the possibility that someone who is able to prove an interest or right over the land which predates the date of registration will still be able to exercise it.
- **Possessory leasehold**. Used in the same situation as possessory freehold.
- **Good leasehold**. This is used where, although there may be a *lease in existence, there is not sufficient evidence in order to be able to register the freehold to which it relates.

In Northern Ireland, because of the greater number of *estates in land, there is a larger number of classes of title that can be registered at the Northern Ireland Land Registry. As in England and Wales, these divide into absolute, qualified, possessory, and good titles, for freehold and leasehold, with freehold being divided into **fee simple** and **fee farm grant** (for an explanation of the difference, *see* FREEHOLD).

In Scotland, the Land Register of Scotland does not operate on the basis of classes of registration; however, the **Keeper of the Land Register** will give a

warranty through registering a title that the title sheet is accurate at the time of registration, but if it happens not to be, the Keeper may have to pay *compensation. However, in some relatively exceptional cases the Keeper can register a property but limit, or exclude, the warranty.

title deeds *See* DEED; TITLE.

top height The mean of the heights of the largest-diameter trees in a stand. This would exclude edge trees which may not be normal-shaped. *See also* CLINOMETER; HYPSOMETER.

topography The way the natural and physical features on the land are arranged. Hence a map showing topography will show not only a representation of where, for example, settlements are, but will also include aspects such as **contour lines** which show the steepness of hills and valleys.

topper An agricultural machine for cutting grass or other plant material to remove unwanted growth. It is used on pasture to tidy up uneven grazing or on margins such as verges. The mechanism varies and may comprise blades or flails. The same word is used to describe the removal of the green material of root crops such as carrots or sugar beet.

topsoil The uppermost level of *soil.

tort Under English law, a wrongful action or omission for which the wronged party can obtain *damages (*compensation) from the party responsible through the civil *courts. Torts include *negligence (where one party has a sustained reasonably foreseeable injury or loss, or property damage, as a result of an act or omission by another party), *trespass (where the act itself is a tort without the need for there to be any loss or damage, because in this instance the tort is concerned with protecting legal rights), and *nuisance. Under the law of tort, there is no need for the two parties to have had any kind of contractual relationship for an action in tort to be undertaken, although there may be some situations where an event can be both an example of negligence in tort and a breach of *contract. In this type of situation the injured party may choose to take action through the civil courts for breach of contract or as an action in tort. The equivalent of tort in Scottish law is **delict**.

towbar A means of attaching a towed machine or vehicle to a tractor or towing vehicle. It is usually a metal bar that protrudes from the back of the towing vehicle to which the towed vehicle is attached. This may be by means of a jaw with a pin or by means of a claw that fits over a ball *hitch. Also known as a drawbar.

town and country planning *See* PLANNING.

town council Elected body of councillors, normally serving an individual town, and forming part of the first level of local government in England and Wales, where they have an equivalent status to *parish councils and *community councils respectively. *See also* LOCALISM; LOCAL AUTHORITY.

TPO *See* TREE PRESERVATION ORDER.

tractor The main motive force used on farms. Tractors are used to tow implements or trailers, and are designed to have high torque at relatively low speeds. Implements may be mounted on the *three-point hydraulic linkage, such as a plough or cultivator. Implements may also be driven by a *power take-off, a rotating shaft protruding from the rear, or sometimes the front, of the tractor. Implements may be both mounted and driven; for example, a mower or power harrow. Tractors usually have four wheels, which may be of equal size, or the rear wheels may be larger than the front wheels. Steerage is by the front wheels. A crawler tractor has tracks rather than wheels. Tractors are manufactured in a wide range of sizes and power designed for different purposes, and some larger tractors may even be articulated in the middle.

trading account A financial record, used in a business which is trading goods, of the amount of money obtained from sales of those goods, less the cost of purchasing the goods (and taking into account goods that may be in store at the beginning and end of the trading period covered by the account); this gives a gross *profit (before any other costs of running the business are taken into account). A trading account is essentially the top part of a *profit and loss account.

trailer An agricultural implement for the transport of goods, usually towed behind a tractor or farm vehicle. It may have one or more axles, depending upon the nature and weight of the material to be transported. The bed of the trailer may be flat for transporting bales of hay or straw or material in sacks such as seed or fertilizer. There may be sides for the transport of bulk material such as grain; such trailers usually have a hydraulic mechanism for the trailer to tip up from the front to empty the contents at the back. A trailer may have a roof as well as sides; for example, for transporting farm animals or horses.

trail hunting A type of hunting that has as its primary aim the simulation of traditional *fox hunting or other forms of *hunting as practised before the ban on hunting came into operation in the UK. This is achieved by laying an artificial scent or trail which the hounds can follow. In many respects, trail hunting is very similar to *drag hunting but aims to simulate more closely the route a fox might take when crossing the countryside. The trail can also be occasionally lifted for a short distance, allowing the hounds to have to **cast**, i.e. fan out and search for the scent akin to hunting live *quarry. *See also* DRAG HUNTING; HUNTING.

tramline A pair of parallel lines across a field the width apart of a tractor axle. Tramlines are put into a crop either by not sowing the crop by blocking the appropriate seed spouts or by spraying out the plants once emerged. The tramlines are then used to ensure efficient application of fertilizers and pesticides with no misses or overlaps. The frequency of the tramlines is thus determined by the width of the sprayer or spreader, usually in multiples of 6 metres, e.g. 12, 24, or 48 metres. There is concern that having tramlines with no crop growing and a depression in the soil caused by the weight of the machinery contributes towards *soil erosion. Some farmers now have permanent tramlines used by all machinery to reduce soil compaction so that vehicles drive only on the tramlines rather than elsewhere in the field. This is known as *controlled traffic farming, but depends upon all machines having the same axle width.

transplant After one or two years in a nursery seedbed, the **tree seedling** is transplanted or lined out. This helps to stimulate root growth, as does **undercutting**, where the taproot is severed to stimulate fibrous root growth. These methods produce the *whips and *saplings which are normally used for *forest planting, and have an increased survival and growth potential when finally planted.

tree A *perennial plant usually characterized by having a single woody stem, and attaining significant height and bearing lateral branches. They have an important role in recycling energy, gases, and *nutrients. They are important in *carbon sequestration, and have a role in influencing *climate and reducing erosion, and are a significant *habitat for large numbers of wildlife *species. Trees have a commercial use and value as *firewood and as structural products or timber. Although normally cleared from agricultural *fields, they are found, for example, in *hedgerows or in groups ranging in size from a small *copse through to a *woodland or *forest. The early successional stages of trees, along with *woody shrubs, can create the understorey of a woodland.

The main woody stem lays down a layer of growth each year, termed a **tree ring**, that can be measured and correlated with the age of the tree. Trees are commonly divided into two main groups: the *deciduous or *broadleaf species, and the *conifer species. The commonest UK species of both groups are detailed in the table below. *See also* ROYAL FORESTRY SOCIETY; TREE PRESERVATION ORDER.

tree line The height above sea level where trees cannot grow. This varies, and is linked with changes in latitude. In Britain, a number of tree species will not grow above 300 metres, and the economic 'tree line' is about 450 metres. The limiting factor is the exposure to winds, and there are also practical aspects of establishing and *harvesting *forests associated with the difficulties of accessibility to these higher areas.

Tree Preservation Order (TPO) A power that local authorities have to protect an individual tree or group of trees in the interest of the general public's enjoyment. It is contained within the Town and Country Planning Act 1990 and in subsequent regulations. The TPO is legally enforceable and prohibits the tree or trees from being cut down, pruned, or damaged in any way, unless permission has been granted by the respective planning authority.

trespass Directly interfering with an owner's rights, or taking possession of someone else's property; for example, taking away someone's goods, or entering their land without the owner's permission or some other right of entry set out in

Common native and naturalized trees in the UK		
Deciduous trees		Conifer trees
Ash	Birch	Corsican pine
Beech	Cedar	Douglas fir
Horse chestnut	Hornbeam	Larch
Holly	Field maple	Norway spruce
Common lime	Poplar	Sitka spruce
Oak: English, Sessile	Sycamore	Scots pine
Rowan	Whitebeam	Western red cedar
Sweet chestnut	Wild service tree	
Wild cherry	Willow	
Wych elm	Yew	
Aspen		

the law. The act of trespass is a *tort (*delict in Scotland). *See also* ADVERSE POSSESSION; SQUATTING.

tribunal *See* COURT, LAW.

triticale A *cereal cross between *wheat and rye developed in laboratories in Scotland and Sweden in the late nineteenth century. It was thought to have the yield of wheat but the hardiness of rye, needing fewer inputs of *fertilizer and *pesticide. As such it was grown on poor or marginal land, but recent advances have made it competitive with wheat, though it is grown more widely in other countries, notably Scandinavia. It is used primarily for animal feed but also for human consumption, and is mostly found in health shops. It is also grown widely in *cover crop mixtures.

trough A long, narrow, open container used to contain water or feed for farm animals. It may be fixed in place with a water supply and ballcock for animals to drink from, known as a water trough, or for feed usually in or at the edge of a building. It may also be portable, to be taken to wherever the animals are; for example, for *supplementary feeding in a field.

truffle The fruiting body of an Aescomycete fungus. It is black and warty with a strong smell, and is buried beneath the ground in all stages of its growth. It can be found growing around the roots of *broadleaf trees such as *beech and *oak, and is commonly found by dogs and pigs that have been trained to search for it. It is among the most expensive foods in the world, and its high price and constant demand, particularly in France, can provide a lucrative trade.

trust Form of ownership of property (often of land) where the **trustees** legally own the property **on trust** for a **beneficiary** (or beneficiaries), who may be entitled to either use, or have the income from, the property held in the trust (sometimes referred to as the **trust fund**), but not, for example, sell it, as it is the trustees that would be able to do this (but *see* LIFE TENANT). Trusts are often used when the beneficiary is not able to hold the property; for example, because they are a child.

The party who puts the property into the trust is known as the **settlor** (**truster** in Scotland). Under English law, trusts have evolved from the law of *equity (for further discussion of equity, *see* COMMON LAW). The trustees therefore have **legal ownership**, whilst the beneficiaries have **equitable ownership** and enjoy the benefits of the property. Although the principles of equity do not exist in Scottish law, trusts in Scotland, in practice, operate in a similar way to the rest of the *United Kingdom, although the underpinning legal principles are different, and accordingly there are some differences in detail.

Trusts can be set up deliberately by a settlor (an **express trust**), either whilst alive or through their will on their death, and the establishment of an express trust must generally be completed in writing. The duties and responsibilities of the trustee of an express trust are set out in law, and usually in a **trust deed**.

Alternatively, trusts can arise by implication of the law (where a deliberate act in writing is not required). This is usually the case where it would be unfair for there not to be a trust. For example, if two people buy a house, with both contributing financially towards it, even if only one of them is registered as the legal owner, an **implied trust** may have arisen because both made a financial contribution towards it (although this would not be the case in Scotland). Trustees and beneficiaries can be the same people; for example, under English law, when property is co-owned, a trust is automatically imposed in law.

The main types of trust that exist in the UK are:

- **Bare trust**. The trustees hold the property on trust for a particular beneficiary who, when an adult, has an immediate and absolute right to all of the *capital and income from the trust. This type of trust is often used to hold assets for young people until they are old enough to control them in their own right.
- **Discretionary trust**. In this type of trust the trustees have the discretion to decide (within the criteria set out within the trust deed) how to use the income (and potentially capital) within the trust; for example, which beneficiary should receive payments.
- **Interest in possession trust**. The beneficiary has the right to the income from the trust (and potentially to occupy a property held within the trust), but not normally to the capital within the trust (*see* LIFE TENANT; LIFERENT, PROPER).

The taxation treatment of each of the different types of trust varies (for example, who has responsibility for any *Income Tax). Transfers of property into and out of some types of trust can also be subject to *capital taxation in some circumstances, and *Inheritance Tax can also become due after property has been held in some types of trust for several years, even if it is not being moved out. The main reason for this is that otherwise it would theoretically be possible to avoid taxes such as Inheritance Tax completely by leaving property in trust for several generations. However, there are several taxation reliefs which are available in relation to property in trust (as there are for property which is owned directly by individuals; *see* INHERITANCE TAX).

trustee Person (either an individual or corporate body) who holds property on *trust for a *beneficiary or beneficiaries.

truster (Scotland) *See* TRUST; SETTLOR.

tup Male *sheep, older than 18 months, capable of siring offspring. As a verb it describes a *ram covering a ewe or having sexual intercourse.

turkey A large *poultry bird, *Meleagris gallopavo*, domesticated from the wild turkey in America at least 2,000 years ago. It is reared for its meat and is associated with festive meals such as Christmas, or in the USA, Thanksgiving.

turning out Within a farming context, this describes letting out *livestock, normally onto *pasture after a period of being housed. Thus *dairy cows that have been kept in buildings for the winter are turned out to *grass in the spring, or *ewes that have lambed in a building may be turned out with their *lambs when they are strong enough. The release of livestock into open space can be a spectacle, and some farmers invite an audience to watch or film the event.

tussock A clump or mound of grass. Tussocky *grasses, such as **cocksfoot** and **timothy**, naturally form thick mounds or clumps as opposed to spreading out along the ground. They provide good habitat for ground-nesting birds, and can slow down and filter the passage of water. They are thus often included in *field margins and *buffer strips.

twitch A device designed to provide severe restraint of a *horse or pony. There are two main types, and the commonest is a simple length of stick with a short length of cord threaded through one end, creating a loop. This is put over the upper lip and twisted tight, which has the effect of producing **endorphins** that calm the animal, and they become quiet and amenable. It is a useful tool to aid a difficult horse being *clipped or receiving veterinary care.

U

udder The mammary gland of female *ungulate animals, including cattle, sheep, goats, and horses. A bag-like structure hanging below the animal in front of the hind legs, it has *teats or nipples arranged in pairs from which milk may be extracted. A cow has four teats, and sheep, goats, and deer have two, but there may be more, such as in pigs.

UKROFS *See* ADVISORY COMMITTEE ON ORGANIC STANDARDS.

Ulster Farmers' Union A membership organization that represents farmers and growers in Northern Ireland. Founded in 1918, its central objective is to promote their interests both at home and abroad through professional lobbying. It works closely with the Farmers' Unions of England, Wales, and Scotland, collectively supporting the Bureau de l'Agriculture Brittanique, giving local farmers a permanent staff presence in Brussels.

ultra vires Situation where a body has made a decision which goes beyond its legal powers. Decisions which are considered to be *ultra vires* can be grounds for *judicial review.

UN *See* UNITED NATIONS.

underplanting *See* PLANTING.

undersow Practice by which a second *crop is planted at the same time or soon after the main *crop with the intention that the undersown *crop experiences its main growth period and subsequent *harvest after the first. It has benefits in terms of minimizing *soil *erosion, minimizing the risk from *weed establishment, and reducing the amount of impaction of *soil through vehicles. This must be balanced against the demands of the two crops on *soil *nutrients, *water, and light, the risk of poor germination of the undersown *crop, and the reduced *yield of the main *crop through lower density of plants. Most commonly used to establish a *ley following a *cereal *crop using *grass and/or *clover mixes. *See also* INTERCROP.

UNESCO (United Nations Educational, Scientific and Cultural Organization) Agency of the *United Nations created in 1945 to support peace by establishing networks between nations based around moral and intellectual understanding and cooperation. It works in a number of areas, including supporting education, heritage protection, cultural diversity, and scientific cooperation, as well as protecting freedom of expression. UNESCO also created the *World Heritage Site designation.

ungulate Any animal with hooves, such as *cattle, *sheep, *goats, *deer, and *horses. A hoof originates from a toenail, and some animals have an odd number of hooves, such as the horse with one, whilst others have an even number, such as cows and sheep with two, and these are often described as cloven-hoofed. Many of these are *ruminants with four stomachs to digest cellulose. The precise meaning of the term 'ungulate' has changed as modern DNA technology has brought changes to taxonomy.

unique selling proposition (USP) The aspect, or aspects, of a product, service, or brand that distinguish it from the rest of the market and encourage a customer to buy it or aspire to buy it. Conveying USPs through marketing is a key component of product *differentiation.

unitary authority *See* LOCAL AUTHORITY.

United Kingdom Term describing England, Wales, Scotland, and Northern Ireland. It is distinct from the name *Great Britain, which is the collective name of England, Wales, and Scotland and their associated islands.

United Kingdom Forestry Standard A *Forestry Commission publication that is the reference standard for sustainable forest management in the UK. It provides a regulatory basis upon which guidelines on requirements are set out, and applies to all woodland. These cover matters including *biodiversity, *landscape, historic environment, and resource protection.

(⊕) SEE WEB LINKS
• Website with links to the details supplied in the full publication.

United Nations International organization founded in 1945 to support international peace, develop friendly relations between nations, and aid them in working together to address issues of poverty, hunger, disease, and illiteracy, and to promote respect for rights and freedoms. In 2018 the number of member states stood at 193. The UN undertakes a wide range of activities, from peacekeeping and humanitarian assistance through to work in the areas of human rights, international health, expanding food production, and addressing *sustainable development and *climate change issues. *See also* CONVENTION, UNITED NATIONS; KYOTO PROTOCOL; PROTOCOL, UNITED NATIONS; UNITED NATIONS FRAMEWORK CONVENTION ON CLIMATE CHANGE.

United Nations Framework Convention on Climate Change (UNFCCC) A 1992 *United Nations convention (international legal agreement) in which countries agreed to cooperate to consider how they could limit average global temperature increases and resulting *climate change, and adapt to consequential impacts. The UNFCCC is the parent treaty of the *Kyoto Protocol, which legally binds countries to greenhouse gas emission reduction targets. UNFCCC negotiations also resulted in the subsequent 2015 **Paris Agreement**, by which countries agreed to aim to limit global temperature rises in the twenty-first century.

unliquidated damages *See* DAMAGES.

upland farm

[3]3[3][3][3]33

upland farm *See* HILL FARMING.

uplands The main uplands in the United Kingdom are designated as *Less Favoured Areas, reflecting their reduced farming potential. They are not truly wild spaces, but do include significant areas of protected sites and important wildlife. Uplands include agricultural *landscapes of hill, mountain, *moor, *rough grazing, and *forestry, and they have an increasingly valued role in water cycling, including water purification and flooding control. *See also* HILL FARMER; NATIONAL PARK.

Upper Tribunal (Lands Chamber) *See* COURTS, LAW.

urban fringe The area of land at the edge of a town or city where it meets the countryside. Urban–rural fringe land typically has a range of uses, including a mixture of semi-urban uses (such as golf courses, garden centres, and airports) and some farming. Expansion of urban settlements often relies on building on urban fringe land. *See also* GREEN BELT.

urea An organic compound, found in urine, that contains nitrogen. As such it is manufactured and used as a nitrogenous *fertilizer. It is cheaper to produce than *ammonium nitrate, and contains more nitrogen; 46% rather than 34%. It may be in granular or prilled form, but has the reputation of spreading less evenly than ammonium nitrate.

use class *See* USE CLASSES ORDERS.

Use Classes Orders Orders used in the United Kingdom to classify land uses into different **use classes** for planning purposes. England, Wales, Scotland, and Northern Ireland each have a use classes order which divides up the most common types of land use into a number of different classes; examples include business uses, dwelling houses, and shops. Whilst as a general principle the *change of use of a building or area of land will require *planning permission, where that change of use is within the same use class (and where it does not involve any physical alterations), planning permission is not generally required. In addition, some specific changes in use between use classes may also be classified as *permitted development (and therefore also do not require a planning application). These are generally changes in use where the resulting impact on the surrounding area is likely to be minimal. In addition to the main use classes, the use classes orders also list a number of sui generis (unique) uses which do not fit into any of the main use classes. For example, in the **Town and Country Planning (Use Classes) Order 1987, as amended**, as it applies in England, this includes petrol stations. In Wales there are currently proposals to consolidate their current Use Classes Order and *General Permitted Development Order.

USP *See* UNIQUE SELLING PROPOSITION.

vacant possession Situation where a property (often on sale or transfer to a new owner) is not subject to any tenancy or other occupation agreement; therefore, in the event of a sale or transfer the new owner has the opportunity to choose who occupies the property. It is possible, in the *United Kingdom and other parts of the world, to sell a property to another party whilst the existing tenant retains the right to remain in that property (in this situation the tenant is sometimes referred to as a **sitting tenant**), and in this case the property would not be being sold with vacant possession.

validation, planning Process whereby a *planning application is initially assessed by the *local planning authority in order to ensure that sufficient information has been submitted for the application to be determined (decided). In the United Kingdom a list of minimum requirements is set out within planning law. These vary depending on the type of application, but normally include the completion of a standard application form, as well as the submission of plans and drawings (usually at specified scales) and, in most cases, payment of the relevant fee. Applications which do not meet the validation requirements are normally returned to the applicant without having been considered.

valuation An opinion of the value of an *asset or a *liability at a given date and on a specified basis. *See also* RED BOOK; ROYAL INSTITUTION OF CHARTERED SURVEYORS; REGISTERED VALUER; VALUATION METHODS.

valuation information paper *Information paper specifically relating to valuation issues.

valuation methods The methods used by valuers to produce *valuations. There are a number of methods, the most commonly used in the *United Kingdom being the *comparative valuation method, the *investment method, the *residual valuation method, the *profits method, and the cost-based *contractor's method and *depreciated replacement cost method. *See also* RED BOOK; ROYAL INSTITUTION OF CHARTERED SURVEYORS; REGISTERED VALUER.

Valuation Office Agency (VOA) Agency of Her Majesty's Revenue and Customs (HMRC) providing the United Kingdom government with *valuation services and property advice in relation to taxation and benefits. The VOA undertakes a range of work which includes, in England and Wales, compiling and maintaining lists of, *rateable values of non-domestic properties for *business rates, and *Council Tax bandings of domestic properties. They also register *fair rents and determine local housing allowances (the allowance paid to housing benefit claimants) in England, in addition to providing valuations and related advice to HMRC in relation to *Capital Gains Tax and *Inheritance Tax, and other taxation matters in England, Wales, and Scotland. *See also* DISTRICT VALUER; LAND AND PROPERTY SERVICES; RENT OFFICER; VALUATION TRIBUNAL.

Valuation Tribunal Independent body hearing appeals from ratepayers or taxpayers in England, Wales, and Northern Ireland in relation to *valuations for Council Tax/domestic rates and, in England and Wales, for non-domestic (business) rates, primarily where the appellant disagrees with a decision of the *Valuation Office Agency (England and Wales), the *Land and Property Services/ Commissioner of Valuation for Northern Ireland, or the local authority. There are separate Tribunals for England, Wales, and Northern Ireland. *See also* LANDS TRIBUNAL FOR SCOTLAND; LANDS TRIBUNAL FOR NORTHERN IRELAND.

Value Added Tax (VAT) Tax charged in the *United Kingdom when a **taxable person** supplies goods or services during the course of operating or furthering a business. VAT *legislation defines what is meant by a taxable person, and a business falling into this definition must register for VAT. Generally, any business which sells goods or services with a total value over the course of a year which exceeds the **VAT registration threshold** must register; however, it is also possible to register voluntarily. Once registered, the business will be issued with a VAT registration certificate and **VAT Number**, and must charge VAT on the goods and services it supplies and then pay the tax due to *Her Majesty's Revenue and Customs (HMRC). In addition, the business must also keep VAT records and a VAT account, and submit returns to HMRC. Although the business, once registered, must charge VAT on goods and services they supply (**output tax**), they can themselves **reclaim VAT** on goods and services they purchase for use in their business (**input tax**). For most businesses, the amount of VAT that that business pays, or claims back from HMRC, is usually the difference between the VAT they have charged on goods and services supplied by them, and the VAT they reclaim. However, some small businesses (including agricultural and estate agency/property management businesses) can choose to join the **Flat Rate Scheme**. Under this scheme the business pays a fixed rate of VAT to HMRC (although still charges the normal rates of VAT to customers). The business then keeps the difference between the rate they have charged their customers and the fixed rate; however, they are not allowed to reclaim VAT on items that they have bought. The administration requirements for the business are, however, generally more straightforward than for businesses in the standard VAT system.

Not all goods and supplies have VAT charged on them, as some goods and services (for example, insurance and postage stamps) are **VAT exempt**. Others are **zero rated**, or at a **reduced rate** of VAT. For zero-rated items (including, for example, most food items), although no VAT is charged, because they are still VAT-taxable they still have to be included on a business's VAT returns to HMRC. Other goods and supplies are charged at a reduced rate (for example, domestic fuel or power which is currently charged at 5%). If none of these situations applies, then the standard VAT rate is currently 20%.

Some businesses also have a **partial exemption**. If a business supplies a significant amount of goods and services that are exempt from VAT, but the business itself buys items on which it has to pay VAT in order to provide those exempt goods and services, then in some circumstances they can be considered to be a partly exempt business. Businesses classified as such will be able to claim back a specified amount of the VAT they have paid in providing the exempt goods and services. Any amount beyond that specified amount cannot be reclaimed.

Selling or letting commercial land or property is normally exempt from VAT (which then affects the ability of the business doing the selling or letting to reclaim any VAT they have paid on purchases). Therefore, in some circumstances, a business may **opt to tax land and buildings** (so they will then charge VAT on these supplies). This will then, however, allow them to reclaim VAT which they have paid themselves.

value chain analysis Tool used in business management and marketing which involves undertaking a strategic assessment of all the activities that an organization undertakes in order to identify areas where greater competitive advantage can be gained by adding further value for customers; for example, because the organization has a particular strength which can be used for *differentiation of its products or services, or because costs can be reduced. The value chain can be divided into a number of different activities; for example, manufacturing operations, logistics, marketing, and service. Analysis of the chain can also extend to the activities of suppliers and customers and the linkages between activities. Using this process may, for example, highlight that an organization has a logistics system which enables it to deliver its goods faster than its competitors, and therefore that this can be used in its marketing as a *unique selling proposition. Alternatively, it may highlight that cost savings could be made in its manufacturing process which may then enable an organization to sell its goods at a lower price, thereby improving value for its customers.

variable costs Costs that are directly related to an enterprise and that vary according to the scale of the enterprise. The most common form of farm accounting distinguishes between variable costs and *fixed costs which are the overheads of the business as a whole. Thus for an arable enterprise, the variable costs include seed, fertilizers, and chemical pesticides, whilst for a livestock enterprise they would include feedstuffs, medicines, seed, and fertilizers. The fixed costs would include machinery, labour, and other overheads such as rent and finance. These variable costs are deducted from the gross output to produce the *gross margin. Net margin or profit is then determined by deducting fixed costs.

VAT *See* VALUE ADDED TAX.

veal The meat of *calves as opposed to beef from mature *cattle, slaughtered at around six months of age. It usually comes from the male offspring from dairy herds which are otherwise unwanted. In the past the calves were reared in crates, but this practice was banned in the European Union in 2007. Prior to this, there was consumer resistance to veal meat, at least in the UK, but it is now becoming more common.

vendor Person or organization selling something, particularly property, in which case the vendor sells to a *purchaser.

venison Normal description of meat from deer, but can be used more widely to describe the meat from any furred *game. Venison has a dense structure and is a deeper colour than *beef or *lamb. It is nutritionally valued and has a much lower fat content than these other red meats.

vermicide A substance that is poisonous to worms. A drug that is used to kill *endoparasitic worms in animals, especially intestinal worms. The most commonly used are *anthelmintic drugs, usually administered as a drench.

Vesting Order (Northern Ireland) *See* COMPULSORY PURCHASE.

veteran tree These are not precisely defined, but are of a great age or a great age relative to others of the same species, and often have gone through cycles of *dieback and regrowth. The age is dependent on the species; for example, an 80-year-old aspen can be defined as a veteran, but a 200-year-old oak is only middle-aged. *Coppicing and *pollarding can significantly extend the age of such a tree. Britain has a relatively high proportion of veteran trees in relation to the rest of Europe, and they are commonly freestanding, especially in parkland. Frequently, high *natural capital values are attributed to veteran trees.

veterinary surgeon A university-trained person who is registered to treat animals. More commonly referred to as a 'vet', their work can involve treating farm livestock, and they can be a trusted source of information and guidance for farmers. This can be particularly important in the identification of notifiable diseases such as *bovine tuberculosis and *foot and mouth disease. Vets also work within the *Animal and Plant Health Agency, and have an important role in the control of these and other animal diseases.

(((⊕))) SEE WEB LINKS
• Website of the Royal College of Veterinary Surgeons (RCVS), the governing body of the profession in the UK.

vicarious liability Situation where one person is held legally liable for the actions of another. An employer can, in some circumstances, be held legally liable for the actions of their employee (or potentially of a contractor, or agent) if that employee carried out those actions as part of their employment. This means that, in some circumstances, an employer can be sued, or prosecuted under the criminal law, for something that their employee did (though it does need to be

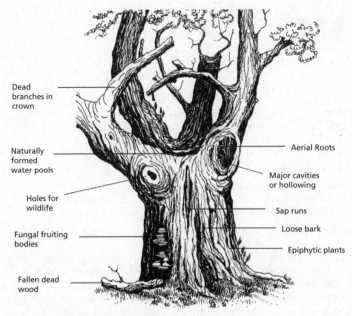

Dead
branches in
crown

Naturally
formed
water pools

Holes for
wildlife

Fungal fruiting
bodies

Fallen dead
wood

Aerial Roots

Major cavities
or hollowing

Sap runs

Loose bark

Epiphytic plants

Veteran tree. (North York Moors National Park.)

something that the employee did that was clearly related to his or her
employment, rather than for personal reasons of his or her own).

volunteer An unwanted plant that is self-sown rather than deliberately planted.
It is often used to describe plants growing from seeds shed by a previous crop;
for example, oilseed rape seedlings growing in a subsequent winter wheat crop.
See also ROGUE.

V

waiver The giving up of a legal right to do something by not asserting it, or abandoning it; for example, a landlord can **waive** their right to use *forfeiture (or the Scottish equivalent *irritancy) as a remedy for a tenant's *breach of covenant in certain circumstances (for further discussion of the role of waiver in forfeiture, *see* BREACH OF COVENANT).

walking up *See* ROUGH SHOOTING.

walnut (black, common) *See* TREE.

warble fly A large parasitic insect of the genus Hypoderma that attacks *cattle and *deer in particular. The fly lays its eggs on the hair of the animal and the resultant larvae bury into the skin, causing large painful lesions. Whilst infected cattle suffer, the main impact is to the *hide, reducing its value as leather. In 1978 an eradication programme was set up in the UK, making it a *notifiable disease in 1982 with compulsory treatment of infected cattle with a systemic insecticide. This was successful, and the warble fly was declared eradicated in 1990. It remains a notifiable disease in Scotland, but the regulations in England and Wales were revoked as from 1 April 2015.

warfarin A poison to kill rodents, particularly rats, mice, and squirrels. It was the first anticoagulant compound used as a *rodenticide, coming into commercial use around 1950. It acts by thinning the blood, thus preventing it from clotting. It is also used as a human medicine to thin the blood to prevent clots forming in the bloodstream. Some resistance has built up in rats, and it has now been largely replaced as a rat poison by newer chemicals known as second-generation anticoagulant rodenticides. *See also* PESTICIDE.

warmblood Not a breed of *horse, but a grouping of horse types that share characteristics. Essentially based on European horse breeds crossed with the hot-blooded Arabian horses. The combination produces excellent riding horses particularly suited to sporting events such as *eventing.

warrandice *See* CONVEYANCING.

waste 1. Objects or substances which the holder discards, plans to discard, or is required to discard (perhaps because something is out of date). In the *United Kingdom, the law defines what is considered to be waste, and sets out legal requirements in relation to how waste must be managed (for further information, *see* AGRICULTURAL WASTE REGULATIONS). At the time of writing, the legal definitions and principles relating to waste management in the United Kingdom are grounded in those set out in the **Waste Framework Directive** (which is a *European Union Directive). *See also* CONTROLLED WASTE; HAZARDOUS WASTE. **2.** *See* COVENANT.

waste carrier registration A form of registration with the relevant *United Kingdom government environmental body (the *Environment Agency in England, *Natural Resources Wales, the *Scottish Environment Protection Agency, or the *Northern Ireland Environment Agency), allowing someone to transport *waste as part of their business activities. In England, Wales, and Northern Ireland there are two tiers of waste carrier registration: **upper tier waste carrier** and **lower tier waste carrier**, with upper tier waste carriers entitled to transport a wider range of types of waste. Farmers in England, Wales, and Northern Ireland are required to register as waste carriers (normally lower tier) if they transport their *agricultural waste to licensed waste processers (as opposed to having it collected). In Scotland, people transporting agricultural waste must register as a **professional collector and transporter of waste**.

waste exemption Form of permission for businesses and organizations to carry out certain forms of lower-risk waste management activities in the *United Kingdom. Businesses and organizations that are required to comply with the *waste regulations (for further information, *see* AGRICULTURAL WASTE REGULATIONS) have to register waste exemptions with the relevant government environmental body (the *Environment Agency in England, *Natural Resources Wales, the *Scottish Environment Protection Agency, or the *Northern Ireland Environment Agency) if they wish to undertake specified waste management activities; for example, if farmers wish to burn their waste plant materials in the open (say from tree cutting). Each waste exemption has limits and conditions attached to it (for example, how much waste can be dealt with under the particular exemption). If the activity does not fit in under the limits and conditions (for example, the volume of waste exceeds that allowed under the exemption), then in order to continue to carry out the activity, an *environmental permit must be applied for from the respective government environmental body in England and Wales (or a *waste management licence and/or *pollution prevention and control permit in Scotland and Northern Ireland).

waste hierarchy *See* WASTE MANAGEMENT HIERARCHY.

waste management hierarchy A principle, incorporated into *European Union and *United Kingdom law, which sets out the order in which it is considered environmentally preferable to manage *waste. See the figure on p. 426.

waste management licence A form of permission required from the *Scottish Environment Protection Agency or the *Northern Ireland Environment Agency for certain types of *waste management activities in Scotland and Northern Ireland. Generally, businesses and organizations that are required to comply with the *waste regulations (for further information, *see* AGRICULTURAL WASTE REGULATIONS) must either register *waste exemptions (for the lowest-risk activities) or apply for a waste management licence (for higher risk activities) and/or, in some circumstances, for a *pollution prevention and control permit. In England and Wales, *environmental permits are broadly equivalent to waste management licences/pollution prevention and control permits.

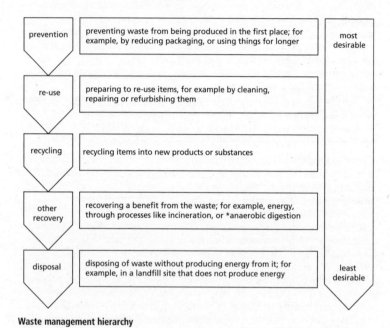

prevention	preventing waste from being produced in the first place; for example, by reducing packaging, or using things for longer	most desirable
re-use	preparing to re-use items, for example by cleaning, repairing or refurbishing them	
recycling	recycling items into new products or substances	
other recovery	recovering a benefit from the waste; for example, energy, through processes like incineration, or *anaerobic digestion	
disposal	disposing of waste without producing energy from it; for example, in a landfill site that does not produce energy	least desirable

Waste management hierarchy

waste regulations Regulations which set out legal requirements relating to how *waste is managed. *See also* AGRICULTURAL WASTE REGULATIONS; CONTROLLED WASTE.

waste transfer note Paper-based or electronic record which accompanies the transfer of certain types of waste in the United Kingdom, as defined in the *waste regulations, which include most types of non-hazardous agricultural and business waste. A waste transfer note is required to set out information including details of the type and volume of waste, when it was transferred, the waste producer, and to whom the waste was transferred. Copies are retained by both parties. The system is designed to reduce *fly-tipping by allowing waste to be tracked as it is transferred throughout the waste processing chain. A separate system, using *consignment notes, exists for movements of *hazardous waste.

wasting asset An asset that has a finite life at the end of which it has no (or very minimal) value. For example, at the end of a *lease, the lease will have no value as it will no longer give any permission to occupy a property. Similarly, some plant and machinery is considered to be a wasting asset; for example, a car will gradually reduce in value as it becomes older, until the end of its life when it will have almost no value. *See also* SINKING FUND.

water Transparent, colourless, odourless, liquid compound of hydrogen and oxygen (chemical formula H_2O) which is essential for living organisms to survive.

Water circulates around in what is known as the **water cycle**. In simple terms, water falls from clouds as **precipitation** (rain, sleet, or snow). Some falls directly into water bodies and *watercourses, whilst the rest falls onto land (either directly, or onto vegetation).

Water which falls onto land will either flow directly across its surface and into watercourses and water bodies, or will soak into the *soil. Most of the water that has soaked into the soil is either taken up by the roots of plants (with some of it then entering the air as a vapour via the process of **transpiration**), or it percolates slowly down through the ground (at which stage it is referred to as *groundwater) eventually finding its way into the oceans and seas.

Water on the Earth's surface (for example in oceans, seas, rivers, streams, lakes and ponds) is known as *surface water and evaporates, turning into water vapour which rises into the atmosphere. As it cools in the atmosphere it then starts to condense back into liquid droplets, thus forming frost, fog, dew, and clouds (which then form precipitation), thereby continuing the cycle. Surface water flows through watercourses (which can be either natural, such as rivers or streams, or man-made, such as a ditch). The area around a watercourse which collects the water that feeds into it is known as its *catchment area, with the dividing point between waters going into different river systems (usually a physical feature such as a ridge), being known as the **watershed**. Some watercourses have *dams placed in them to hold back the movement of water, thereby creating a **reservoir** (in effect, an artificial lake), so the water can easily be collected, and where necessary cleaned and treated, before use in, for example, the public water supply.

As groundwater moves down it often finds its way into rocks (known as **aquifers**). Water may emerge naturally at the surface from an aquifer as a spring (or a well may be drilled down to the aquifer to release the water). Where the aquifer contains water that is hot, then it can sometimes be used to provide *geothermal energy.

The ground level at which the soil and rocks underneath are permanently saturated with water is known as the **water table**, and it can vary over time with changes in precipitation levels, the amounts of water removed by vegetation, or the quantities of water taken via mechanisms such as wells.

Water flowing directly off land into water bodies or watercourses (without having soaked into the soil) is sometimes referred to as surface *runoff and presents a potential environmental hazard, as it can pick up soils and chemicals such as *nitrates, *ammonia, or *phosphates (from *fertilizers or *manure and *slurry) and carry them into watercourses where they can then cause issues such as *eutrophication and damage to *biodiversity. The pollution of a watercourse by cumulative low levels of nitrates or other pollutants running off land over a large area is often referred to as *diffuse pollution (i.e. low levels of pollution coming from many different sources). This contrasts with *point source pollution where a high level of pollution may come from an individual source (for example, a leak of slurry from one farm into a watercourse, causing *contamination).

The *European Union's *Water Framework Directive and *Nitrates Directive are focused around improving water quality in the European Union, and in the *United Kingdom there are several controls which aim to reduce the risks of water pollution from agriculture, particularly in relation to how fertilizers, manures, and slurries, sprays, *silage, and agricultural fuel oils are stored and used. These include *Nitrate Vulnerable Zone regulations, *cross-compliance rules, and the *SSAFO Regulations. New **Farming Rules for Water** were also introduced in England in 2018 which require farmers to take further actions to protect water quality through reducing soil *erosion and the improved planning and monitoring of the application of manure and fertilizers. Improving water quality is also one of the core objectives of the UK's *agri-environment schemes, as well as other mechanisms such as the *Catchment Sensitive Farming scheme.

Water quality is also significant for recreational users of water; for example, for people using water for activities such as *angling, sailing, windsurfing, waterskiing, canoeing, boating, rowing, and swimming.

As the movement of water through soil is one of the core elements of the water cycle, good soil management is key to the effective management of water. Soil which is suffering from *compaction, caused by the actions of animals or machinery compressing the soil particles together, does not allow water to easily soak through it (or roots to grow well within it). As a consequence, water often sits on the surface of compacted soil, resulting in weak plant growth and also causing potentially polluting runoff and localized surface-water *flooding. Surface water management issues can also be found in more urbanized areas where, for example, the paving of areas that were previously grass (in order to provide off-road parking) can result in significant increases in the volume of water flowing into surface water drainage systems (or **storm drains**). Where the storm drains were not designed to deal with such volumes they can become overwhelmed and overflow. Consequently, the planning system in the UK is increasingly requiring the incorporation of SUDS (*sustainable drainage systems) into modern designs (this may include mechanisms such as *rainwater harvesting, or alternatively permeable surface coverings such as gravel being used instead of, for example, *concrete).

The study of water and how it moves is known as **hydrology**. *Climate change modelling indicates that there may be more extreme weather events in future (for example, more spells of very heavy precipitation or high temperatures with little rain), and this has resulted in increased government focus in many parts of the world on whether approaches to future water management need to be altered, given that what models indicate are potentially greater risks of flooding as well as periods of water shortage.

In the UK this is translating into an increased focus on how water availability and flood risk are managed. This includes consideration of where new **hard-engineering flood defences** (physical structures such as walls) may be required, as well as a focus on increasing **soft-engineering flood defences** (for example, through encouraging the planting of more *trees to take up water, and supporting the use of land as flood meadows to accommodate flood water). In addition, potential flood risk forms part of the *planning decision-making process in relation to proposed *development. A number of parts of the country have also been identified as being under **water stress** (i.e. catchment areas where a high proportion of the water potentially available from precipitation is already being collected and used, and therefore where there is limited spare capacity for water supply and also a risk of potential environmental damage from over-exploitation). Taking water from surface or ground sources (i.e. abstraction), and the use of impoundment structures to store water, is controlled in the UK through a system of permissions (*see* ABSTRACTION LICENCE and IMPOUNDMENT LICENCE). Whilst most water in the UK is abstracted for the public water supply, electricity generation, and industrial uses, farmers can also hold abstraction licences to take water, for example, for use in the *irrigation of *crops. Improved water efficiency is also a focus of *building regulations in the UK, and the use of technologies such as rainwater harvesting and greywater harvesting (*see* *foul water) within building construction and refurbishment projects is increasingly common.

The water system is a user of energy as it is treated, pumped, and heated. Consequently, improving water efficiency is also beneficial in terms of reducing energy use. There are also several land-based water-related *renewable energy technologies: namely *hydro-electricity, water-source *heat pumps, geothermal energy, and also *anaerobic digestion where *sewage sludge can be used as a feedstock. Additionally, there is also a developing off-shore renewable energy sector including off-shore *wind turbines, and wave and tidal power technologies.

See also COASTAL CHANGE MANAGEMENT AREA; SUSTAINABILITY; WATER MEADOW.

watercourse A channel for a flowing body of water, which can be natural, such as a river or a stream, or man-made, such as a ditch or dyke. Its is most often open at the surface, but can be underground as a culvert. A watercourse may take seasonal flows of water, and may at times be dry. Owners of property next to a watercourse, or a property that has a watercourse that flows through it, can be described as *riparian landowners. Such a landowner has certain riparian legal rights and responsibilities.

w

waterfowl A group of birds of the order *Anseriformes* that include swans, ducks, and geese. They are characterized by having webbed feet, short legs, and mostly short, broad bills. Very similar-sounding **wildfowl** refers generally to a wild *game bird, but more especially relates to aquatic *quarry species such as duck or goose. *See also* WILDFOWLING.

Water Framework Directive (WFD) A European Council Directive 2000/60/EC (the 'WFD') establishing a framework for Community action in the field of water policy. The main purpose of the framework is for the protection of aquatic *ecosystems of inland waters (rivers and lakes), groundwater, estuaries, and coastal waters, and in addition, to promote sustainable water use and to reduce the impacts of floods and droughts. The WFD is implemented in each of the EU member states through domestic legislation; for example, in parts of the UK by means of the Water Environment (Water Framework Directive) (England and Wales) Regulations 2017. *See also* BUFFER STRIP; CATCHMENT SENSITIVE FARMING; NITRATES DIRECTIVE; NITRATE VULNERABLE ZONES.

waterlogging, soil A state when the *water table is so high that all of the air pockets between *soil particles are filled with *water, with any additional water consequently sitting on the surface of the soil rather than draining down into it. *Runoff is likely to increase when soil is waterlogged, potentially causing issues with *diffuse pollution of *watercourses and *erosion of soil. Most plants are also unlikely to survive lengthy periods of waterlogged soil, as the water displaces air from the pockets between the soil particles, and plant *roots are therefore unable to function properly (because they cannot obtain sufficient oxygen, and carbon dioxide does not diffuse away). The roots therefore rot, depriving the rest of the plant of water and nutrients. Soil that is heavily *compacted can be more prone to waterlogging (as can heavier soils such as *clays), and it often occurs after a period of heavy rainfall. *See also* PONDING.

water meadow A functioning water meadow is an area of *grassland subject to controlled irrigation which is achieved through a system of channels and sluice gates. An early form of agricultural intensification to promote growth of grass in early spring, they are used for making *hay or *grazing *livestock. The flow of water also carries *nutrients which fertilize the grassland. It was more common in Europe from the sixteenth century onwards, but is now mostly derelict, although some survive and are funded through *agri-environment schemes. Water meadows should not to be confused with flood meadows, which are grassland areas adjacent to a river and which may be subject to seasonal *flooding.

Water Resources (Control of Pollution) (Silage, Slurry, and Agricultural Fuel Oil) Regulations *See* SSAFO REGULATIONS.

watershed *See* WATER.

water table *See* WATER.

wayleave Agreement, normally between a utility company and a landowner in the United Kingdom (sometimes made voluntarily or sometimes compulsorily, under legal powers granted by government) which gives permission for the utility company to utilize the landowner's land in some way for a period of time; for example, to install, maintain, and inspect electricity poles and overhead cables. Depending on the type of wayleave it may last for a period of up to decades or may run from year to year.

Regular payments for the wayleave are typically made to the landowner (often on an annual basis). Whilst wayleaves can bind future owners, there are some circumstances in which a landowner can terminate a wayleave, though whether this is possible depends on the nature and circumstances of the particular wayleave.

Wayleaves are distinct from *easements (or in Scotland, *servitudes, which are also used by utility companies), because an easement is a normally permanent legal right which attaches to the land, and only a one-off payment is made at the point when it is created.

See also COMPULSORY PURCHASE.

weaner A young animal that no longer drinks its mother's milk. The young of farm animals are removed from their mothers to allow the mothers to breed again or to be milked. The term is usually applied to piglets up to 40 kg in weight that have been weaned.

weatherboarding *See* CLADDING.

weed Any unwanted plant, especially in a growing *crop. The same plant may not be considered a weed in other circumstances; for example, a wild flower may be welcome in a meadow but unwanted in an arable crop. Weeds are controlled by the application of *herbicides, most of which are selective, killing only the

Common broadleaved and grass weeds in the UK

Broadleaved Weeds		Grass Weeds
Black bindweed	Groundsel	Annual meadow grass
Campion	Hemp nettle	Barren or sterile brome
Chickweed	Knotgrass	Blackgrass
Cleavers (goose grass)	Mayweed	Common bent
Colt's foot	Nettle: annual, deadnettle	Common couch
Cornflower	Poppy	Creeping bent
Corn marigold	Ragwort	Meadow brome
Dock	Redshank	Onion couch or false oat grass
Fat hen	Shepherd's purse	Rough stalked meadow grass
Field pansy	Sow thistle	Soft brome
Fool's parsley	Speedwell	Wild oat
Forget-me-not	Thistle: creeping, spear	
Fumitory		

w

*target species. Common weeds of arable crops may be broad-leaved, such as cleavers, chickweed, or mayweed, or grass weeds, such as blackgrass, wild oats, or sterile brome. *See also* WEEDS ACT 1959.

Weeds Act 1959 This legislation makes it an offence to allow the spread of five injurious weeds: *ragwort, broad-leaved dock, curled dock, spear thistle, and *creeping thistle. Whilst it is not an offence as such to allow the weeds to grow, it is an offence to allow them to spread, i.e. to set seed. The Act gives powers to the Secretary of State at the *Department for Environment, Food and Rural Affairs (DEFRA) in England and appropriate authorities in other parts of the UK to take enforcement action by requiring the landowner or occupier to remove the plants or to have them removed at the expense of the landowner or occupier.

WEEE Acronym for waste electrical and electronic equipment. Under *European Union and *United Kingdom law, sellers of household electrical and electronic equipment have obligations in relation to the disposal of customers' old equipment. These can include taking back old items, providing information about how customers can recycle electrical or electronic equipment, or paying into an approved scheme which helps to fund recycling centres run by *local authorities.

weevil A small herbivorous beetle of which there are more than 60,000 species, mostly in the *Curculionoidea* family. They can be very destructive in many ways. There are those that eat stored produce such as grain, and others that eat plant leaves as adults, but grubs, such as that of the vine weevil, live in the soil and eat the roots. The cotton boll weevil caused havoc to cotton production before being eradicated in many states of the USA. It may be controlled by insecticides such as malathion or pyrethrins, by biological means such as pathogenic nematodes, or by genetic engineering of the host plant.

Weil's disease Also known as leptospirosis, a severe infection caused by *Leptospira* bacteria. The human symptoms of leptospirosis are similar to flu, with fever, headache, diarrhoea, and vomiting. Weil's disease is a particularly virulent form, usually caused by contact with rats or their faeces or urine, and may cause jaundice, swollen joints, and meningitis or encephalitis symptoms. If left untreated it may result in brain damage, kidney failure, internal bleeding, and even death.

weir A barrier constructed across a river to raise and regulate water levels. There are a range of different types of weir, but most of them commonly allow water to flow over the top edge and down to a lower level. Their main function is to help control flooding, to measure water flow rates, or to improve a river's navigability by boat. *See also* FLOOD.

well-being A description of a state of being well, contented, happy, healthy, and prosperous. It can apply to an individual or to a group, and is associated with a range of factors including health, diet, exercise, lifestyle, social networks, and physical environment. *See also* CULTURAL SERVICES.

Welsh Assembly *See* NATIONAL ASSEMBLY FOR WALES.

Welsh Government Devolved government for Wales, comprising the First Minister, Welsh Ministers, the Counsel General (the Welsh Government's chief legal adviser) and Deputy Ministers, all of whom are supported by civil servants. The First Minister is the leader of the Welsh Government and is appointed by Her Majesty the Queen following nomination by members of the **National Assembly for Wales**. The First Minister, in turn, appoints the Cabinet, Deputy Ministers, and Counsel General (with the approval of Her Majesty the Queen). The Government primarily comprises members of the National Assembly of Wales, which is the elected body that makes laws for Wales in devolved areas, and holds the Welsh Government to account. Elected members of the Assembly are identified by the letters AM (Assembly Member) after their name. The National Assembly of Wales has the right to pass laws in a range of areas, including *agriculture, *forestry, animals, plants, rural development, *ancient monuments and historic buildings, culture, economic development, education and training, environment, food, health and health services, highways and transport, housing, local government, social welfare, sport and recreation, tourism, town and country *planning, water and flood defence, and the Welsh language. For those areas where the United Kingdom Parliament has not given the National Assembly for Wales the right to make laws, the UK Parliament and Government is the main national law-making body.

Welsh Revenue Authority A *Welsh Government body responsible for collecting the taxes that the Welsh Government has the powers to collect. At the time of writing, these taxes are the **Land Transaction Tax** (*see* STAMP DUTY LAND TAX) and the **Landfill Disposals Tax** (*see* LANDFILL TAX). Other national taxes paid by individuals, businesses, and organizations in Wales continue to be collected by *Her Majesty's Revenue and Customs.

Western red cedar *See* TREE.

wet heating system A central heating system which is dependent on hot water being transferred around a system of radiators connected by pipes. This is distinct from *dry heating systems, which do not use water. *See also* COMBINATION BOILER; CONDENSING BOILER; SOLAR POWER.

wetland An area of land that is permanently or seasonally covered by water. **Fens** and **bogs** are permanently wet areas where the decomposition of plant material is slow, leading to accumulation of *peat. Fens develop where ground water is mineral rich and usually alkaline. Bogs are usually higher than the surrounding area, and develop where the groundwater is poor in nutrients and usually acidic. Where the peat rises above the water table so that wet *scrub and *woodland communities can develop, it is termed **carr**. **Marsh** describes an area of nutrient-rich wetland with similar types of vegetation found in bog communities, i.e. dominated by rushes and *sedges. It is a general term used to imply waterlogged *soil, and can commonly be used to refer to fens and bogs. **Swamps** are characterized by tall emergent vegetation including *trees, and are more commonly found associated with rivers and in *flood plains. Wetland areas have a limited value for productive *agriculture and *forestry, but are recognized as valuable *habitats and as *natural capital and for the associated benefits of *flood and *water quality management. *See also* LEAD SHOT; RAMSAR.

wet woodland Low-lying and poor draining areas of *woodland in which water-tolerant *tree species such as willow and alder thrive. Wet woodland is of more limited commercial value, but is an important *habitat for *species of woodland and *wetland.

WFD *See* WATER FRAMEWORK DIRECTIVE.

wheat A cereal, *Triticum aestivum* and other species, originally from the Middle East but now the third most commonly grown *cereal in the world after *maize and *rice. It is the most widely grown arable crop in the UK, usually sown in the autumn, though there are also spring-sown varieties. Milling varieties are grown for flour to make bread, biscuits, and other products, whilst other varieties are grown for animal feed. *Durum wheat, *Triticum durum*, is a species grown for flour to make pasta.

whip 1. A younger *sapling of about 0.6–0.9 metres in height with only a central stem and more limited side-branching. It is more commonly planted in high numbers to cover large areas. **2.** A spindly tree with only a small crown reaching into the *overstorey. **3.** Assistant to a huntsman. *See* HUNTING.

whitebeam *See* TREE.

White Paper A paper issued by government in the United Kingdom which sets out proposals for changes to, or the introduction of new, law. Once a White Paper has been issued it is normally followed by a *Bill, which is then considered and voted on by the elected representatives of either the *Houses of Parliament, the *National Assembly for Wales, the *Scottish Parliament, or the *Northern Ireland Assembly. Once the Bill has been approved, usually through a series of votes, and is given *Royal Assent (i.e. formal approval by the monarch), it becomes an *Act of Parliament or *Act of Assembly (i.e. law). A White Paper is sometimes preceded by a *Green Paper, which sets out the government's initial proposals on changes to the law and seeks comments from the general public or from interested parties such as professional bodies or interest groups.

Whitsunday *See* QUARTER DAYS.

wilderness An area of land that has had no, or limited, interference by human activity. They are important areas for *biodiversity conservation and for providing recreational benefits. Legislation may govern the type and scale of activity. Wilderness is also valued for spiritual, cultural, and aesthetic reasons. **Rewilding** is a parallel concept to describe a reduction in human intervention in some areas, in a planned way, and to promote natural environment processes towards a more resilient and sustainable form of land use with increased biodiversity. *See also* WILDWOOD; NATIONAL PARK; SITES OF SPECIAL SCIENTIFIC INTEREST.

Wildfowl and Wetland Trust A charity promoting the conservation of *wetlands, with a particular emphasis on aquatic birds, including ducks, geese, and swans. Set up in 1946 at Slimbridge, its first centre. There are now nine Wetland Centres across the UK involved in conservation, education, and research. *See also* WATERFOWL.

⟨⊕⟩ SEE WEB LINKS
• Website detailing the range of work carried out, and information for visitors.

wildfowling Shooting activity where the *quarry is most commonly *waterfowl such as wild duck or geese, but can include other aquatic bird species such as snipe. Shooting can take place in coastal areas such as below the high water mark on salt marshes and estuaries, or inland over rivers, ponds, or other wetland areas. A specific type of wildfowling is **flighting**, which usually involves regular feeding at a chosen location, very commonly a **flight** pond, in order to draw in ducks or other wildfowl. Models of ducks or **decoys** can also be placed in the selected area as a further attractant. Wildfowling requires the use of alternatives to the use of *lead shot, which is prohibited in the UK and many other countries where the shooting takes place over water and other wetland areas. *See also* GAME BIRD; OPEN SEASON.

wildlife A collective term for wild animals and plants, not under direct human control and usually in a natural or *semi-natural environment. More commonly used to refer to mammals and birds, including *game. The term *biodiversity is broader, and includes plants and animals and the variations between and within them.

Wildlife and Countryside Act 1981 This primary legislation (as amended) remains the main mechanism for legal protection of *wildlife in *Great Britain, although other legislation has since been passed. The Wildlife and Countryside Act also incorporates into British legislation a number of measures upgrading the protection of species to levels needed to comply with international conventions and European Union legislation. The various amendments to the text have included significant changes through, for example, the Countryside and Rights of Way Act (CROW) and from equivalent legislation in the devolved UK legislatures. There have also been changes made to the species listed in respective schedules of protected wild animals and plants.

⟨⊕⟩ SEE WEB LINKS
• Government webpage detailing the original Wildlife and Countryside Act 1981 and all subsequent amendments, including an outline summary of species protection.

wildlife corridor A link usually between similar *habitat areas which allows for easier movement of *species between the sites. Movement along wildlife corridors promotes an ability of species to accommodate environmental changes and a continuation of viable populations, so giving populations an increased resilience. They may be at the *landscape scale; for example, linking large areas of *woodland or *wetland, at the farm scale utilizing *field margins to connect areas within the farm and its immediate neighbours, or as part of an urban **green infrastructure** system. *See also* ECOLOGICAL NETWORKS.

wildlife management The management of wildlife to maintain a balance between its needs and the needs and objectives of people. Commonly an integral element of wildlife *conservation that can involve management of vegetation, but is often targeted on management of animal species. This can involve **culling** to reduce species populations or control of predators. Predator control is a key component of *game management undertaken by *gamekeepers. Wildlife management can be a controversial issue and can necessitate killing by whatever appropriate means, whether carried out by government, conservation organizations, or farmers.

Wildlife Trusts There are forty-seven individual Wildlife Trusts, operating as independent autonomous charities, that promote local *wildlife interests and practical land management and *conservation. They have a range of different histories and backgrounds to their formations and to their respective sizes and activities. Wildlife Trusts in England and Wales are now largely based on old county boundaries or small groupings of such counties. There are single Trusts covering Scotland and Northern Ireland respectively. The Royal Society of Wildlife Trusts is now its operating umbrella body, and the badger was adopted as its logo to establish a common identity. *See also* CONSERVATION VOLUNTEERS; ROYAL SOCIETY FOR THE PROTECTION OF BIRDS; WOODLAND TRUST.

(((∰))) SEE WEB LINKS

• Website detailing common objectives and areas of activity. It also identifies and links with all individual Wildlife Trusts.

wild oats *See* WEED.

wild service tree *See* TREE.

wildwood At its simplest it is *woodland unaffected by human activity. However, the very concept of wildwood is a challenging one, in that woodland unaffected to any degree by human influences risks creating a very rare category. Some island examples may be likely to provide some exceptions. *See also* WILDERNESS.

willow (white, crack, cricket bat, sallow (goat)) *See* TREE.

windbreak *See* WOODLAND.

windrow A row of cut plant material left in the field to dry. This may be *grass for *hay-making or an *arable crop such as *oilseed rape, which is cut a day or two before harvest to ensure even ripening.

windthrow The damage inflicted on *woodlands by excessive wind, resulting in trees being thrown over or uprooted. The risks of windthrow can be a constraint to the management of woodlands, and is a particular concern to upland commercial *forestry interests, where it is a major determinant of thinning and *felling programmes. The threats of wind damage can be assessed using the **windthrow hazard classification**.

wind turbine A piece of *renewable energy equipment used to harness energy from the wind and convert it into electricity. Wind turbines generally consist of a mast on which blades are mounted. These are driven around by the wind and in turn spin a shaft which connects to a generator which produces electricity. Wind turbines vary considerably in size (and therefore in the amount of electricity they generate), from very small turbines which may supply electricity to road signs, to those with blades of almost 90 metres in length. Wind turbines are commonly grouped together into **wind farms**, which may be located either **onshore** (i.e. on land) or **offshore** (at sea), and which feed electricity into national electricity grids meeting the electricity demands of thousands of homes and businesses. Technology in this area is advancing quickly, with turbines progressively becoming physically larger and able to generate more electricity.

It is critical that regular wind of a sufficient speed is available at a site where the installation of a wind turbine is being considered, because objects such as trees or buildings can interrupt the flow of wind to a particular location, thereby potentially rendering the wind turbine of little use. On-site **anemometry** is therefore often carried out prior to the installation of wind turbines at a site. This involves the installation of an **anemometer**, which measures the wind speeds at the location over a period of months in order to obtain a reliable understanding of the suitability of the site for wind turbines. *See also* CONTRACTS FOR DIFFERENCE; FEED-IN TARIFF; RENEWABLES OBLIGATION.

wood 1. Wood is more commonly used to refer to a defined area of trees, but can be interchangeable with *woodland. **2.** Wood as distinct from timber can describe the smaller *roundwood and other lower-value tree products that are used for supplying fencing, posts, *wood chip, *firewood, and *biomass.

wood chip Produced from low-value timber and *roundwood for production of *pulpwood and *medium-density fibreboard. A significant market is *woodfuel for the production of industrial energy. *See also* BIOMASS.

woodfuel Varying forms of *wood as a fuel to provide heat or generate electricity, and an important part of a *renewable energy supply. The commonest and simplest are logs and *firewood, though an increasingly used form of fuel is provided by *wood chip and *pellet.

woodland Can be more often used as a general description of a number of *woods in an area, e.g. farm or estate woodlands. The distinction between a wood, a woodland, and a *forest is essentially one of scale and is not strictly defined. Defined areas of woodland described by their function include

windbreak, which is woodland by virtue of its shape and topographical position, and which provides shelter, for example, for livestock. **Woodland blocks**, commonly rectangular or L-shaped, are found especially in upland grazing areas and allow livestock accessibility to find shelter from any wind direction. **Shelterbelts** also function as protection from the wind, but are normally a comparatively narrow strip of woodland. *See also* COPPICE; FOREST; ANCIENT WOODLAND; ANCIENT SEMI-NATURAL WOODLAND; WILDWOOD.

woodland blocks *See* WOODLAND.

Woodland Grant Scheme *See* FORESTRY GRANTS.

woodlands relief *See* INHERITANCE TAX.

Woodland Trust The UK's largest woodland conservation charity. It is a membership organization and owns or manages a range of sites, promoting woodland for the benefit of *wildlife and people. It has a particular focus on the protection and restoration of *Ancient Woodlands, the planting of native *trees and *woods, and encouraging *public access to its woods. *See also* CONSERVATION VOLUNTEERS; ROYAL SOCIETY FOR THE PROTECTION OF BIRDS; WILDLIFE TRUSTS.

((●)) SEE WEB LINKS
• Website with details of membership activities, and links to all its woodland sites.

woodlot A small area of woodland rented on licence that is managed usually by and for local people for small-scale productive forestry. They are generally managed fairly intensively and can bring local employment and other community benefits, as well as providing recreational opportunities. A Canadian-based initiative, but with more limited interest in Great Britain.

wood pasture A system where trees are widely spaced and *grassland or *heath grows between them. Primarily used for *grazing *livestock, with the *trees providing shelter and browsing, and for timber. The system usually reflects a long history of grazing and is now mainly restricted to *deer parks and *commons, with trees being managed by *pollarding. *See also* VETERAN TREE; AGROFORESTRY.

wood shake Fissures and splits in tree stems which can devalue the timber value, in some cases quite significantly. It is usually only detectable after the tree is felled, and increases as the timber dries or is sawn. It can be a particular concern with oak, and it is linked to genetic determinants and the *soil type in which it is grown.

wood shavings Used for livestock and poultry bedding but more especially for *horses and ponies. They are relatively dust-free and absorbent, and not palatable, which can be a problem with some other types of bedding. Their main disadvantage is that they does not compost readily, and disposal can be a difficulty.

Common woody shrubs in the UK

Alder buckthorn
Buckthorn
Box
Dogwood
Guelder rose
Hawthorn
Hazel
Juniper
Service tree
Spindle

woody shrubs Woody plants normally branching from the base, and in the height range of 1–5 metres. They can form or be part of the understorey. Native woody shrubs include, for example, *hawthorn and *blackthorn. *See also* SCRUB.

wool The fleece of *sheep or other animals such as goats or *camelids. The fibrous coat is removed from the animal by *shearing, and the wool is then cleaned and processed to produce a fibre that can be woven into cloth. The natural wool contains lanolin, a water repellent grease, that may be removed for finer wools or partially left in for more weatherproof clothing. Merino is a fine wool that comes from a specific breed of sheep, whilst lambswool comes from lambs. Mohair and cashmere come from goats, and other fibres come from alpacas and llamas. In the Middle Ages the UK was a major producer of wool, with exports to continental Europe, and great wealth was amassed. One of the results was investment in large churches, known as wool churches. With the advent of synthetic fibres, the value of wool has fallen and now may not cover the cost of shearing.

Work at Height Regulations Health and safety *regulations applying in the *United Kingdom which require a *risk assessment to be conducted and suitable safety equipment (for example, scaffolding) to be put in place to protect workers who are working at height (i.e. off the ground). *See also* HEALTH AND SAFETY AT WORK ACT 1974; HEALTH AND SAFETY AT WORK (NORTHERN IRELAND) ORDER 1978; HEALTH AND SAFETY EXECUTIVE; HEALTH AND SAFETY EXECUTIVE FOR NORTHERN IRELAND.

Working Time Regulations Law, applying in the *United Kingdom, setting out the maximum number of hours that an employee can be required to work in a week on average (although they can voluntarily opt to work more), together with rights to rest breaks and leave, and specific protections for night workers.

World Heritage Site Site designated by the United Nations Educational, Scientific and Cultural Organization (*UNESCO) for its international cultural and/or national significance as a site of 'outstanding universal value'. There are thirty-one sites in the United Kingdom, including Blenheim Palace, Stonehenge and Avebury, the Forth Bridge, Blaenavon Industrial Landscape, and Giant's Causeway and Causeway Coast. World Heritage Sites are covered by specific *planning policy, and *permitted development rights are often restricted. *See also* HERITAGE ASSET.

World Trade Organization (WTO) A global organization responsible for managing the rules of international trade between countries. These rules are based on the WTO agreements signed off by most of the world's trading countries. Its main and stated function is to facilitate multilateral trading systems so that trade flows as freely as possible. It is run by its member governments to enable importers and exporters to carry out their business. *See also* COMMON AGRICULTURAL POLICY.

((⊕)) SEE WEB LINKS
• Provides details of membership and documentation records.

World Wide Fund for Nature (WWF) The world's largest conservation organization that is focused on challenging threats to *biodiversity. It has an international membership and operates in more than one hundred countries, by supporting projects, often in partnerships. It was previously known as the World Wildlife Fund.

((⊕)) SEE WEB LINKS
• Website detailing targeted species, areas of work, and membership activities.

worming The process of treating farm animals to kill *endoparasitic worms, usually by drenching, squirting a liquid into the gullet of the animal, by injection, by tablet, or by pouring a liquid onto the skin. The most common class of drugs are *anthelmintics, such as Ivermectin and Panacur.

written representation *See* APPEAL, PLANNING; COMPULSORY PURCHASE.

WTO *See* WORLD TRADE ORGANIZATION.

WWF *See* WORLD WILDLIFE FUND.

WWT *See* WILDFOWL AND WETLAND TRUST.

years' purchase *See* DISCOUNTING.

yew *See* TREE.

YFC *See* YOUNG FARMERS' CLUB.

yield The annual income return from an investment relative to its *capital value, expressed as a percentage. For example, for an investment property it can be determined using the following equation:

$$\frac{Annual\ Market\ Rent}{Market\ Value\ of\ the\ Property} \times 100 = Yield$$

A yield can be thought of as broadly equivalent to the *interest received on a bank account, and is used for other types of investment, such as property. Like interest rates, the higher the yield the greater the income that will be received from the investment. Generally, the more risk attached to an investment, the greater the yield that will be expected, so an investor is likely to accept a lower yield (or rate of interest) for money invested in a bank than they would if investing in buying the *freehold of a shop. The bank is a much lower risk (in terms of receiving regular interest) than the shop (which, whilst potentially generating a higher income, could also be difficult to let if demand for shops in the area falls). Therefore, the investor will accept a lower rate of income (and hence yield) from the bank than the shop. There is a variety of different types of yield used in the valuation of property; however, the most common is the **all risk yield (ARY)**, which reflects all of the positive features of an investment plus the risks attached to it (for example, that rent levels may fall in future).

yield class An index that estimates timber yield for a particular site based on *tree species, height, and age. The volume production is assessed for a given stand of trees (e.g. cubic metres/hectare), and thus reflects the potential productivity of the site for the tree species growing on it. Tables of yield classes are published for different species.

Yorkshire boarding Wooden *cladding on a building, usually fixed vertically, on a building to house livestock. There may be gaps between the planks to allow ventilation.

Young Farmers' Club A rural youth organization aimed at, and for, people in *agriculture and those who have an interest and enjoyment in the *countryside and rural life. More commonly referred to as simply 'Young Farmers', its full title is the National Federation of Young Farmers' Clubs. It is comprised of more than 600 separate clubs in England and Wales, but does not cover the YFCs in Scotland and Northern Ireland. YFCs organize social events, competitions, training, and opportunities for working with local communities.

( ) SEE WEB LINKS

• Website of the YFC, with information on membership and regional and local offices.

zoning An approach used in the *valuation of property where high-street frontage is considered significant to its value (most commonly shops), the key principle being that the portion of the shop at the front is likely to be more valuable than that further back, as the larger the area of shop frontage the greater the number of customers (**footfall**) it is likely to attract. Hence, if comparing two shops with the same floor area, a deep, narrow shop with a small shop frontage is likely to have a lower value than a shop which is less deep but which has a longer frontage. This is addressed in a valuation by dividing the different areas of such a property into different **zones**, with the zone at the front of the property (say on the high street) being apportioned a greater value than those further back. Once each zone is apportioned a value, they are then all added up to give the overall value of the property. *See also* INVESTMENT METHOD.

Metric–Imperial Conversion

To convert metric to imperial:	Multiply by:	To convert imperial to metric:	Multiply by:
		Length	
centimetres to inches	0.39	inches to centimetres	2.54
metres to feet	3.2808	feet to metres	0.305
metres to yards	1.094	yards to metres	0.914
kilometres to miles	0.6214	miles to kilometres	1.609
		Area	
hectares to acres	2.471	acres to hectares	0.405
		Volume	
litres to UK gallons	0.22	UK gallons to litres	4.546
litres to UK pints	1.76	UK pints to litres	0.568
		Weight	
grams to ounces	0.0353	ounces to grams	28.35
kilograms to pounds	2.205	pounds to kilograms	0.454
kilograms to UK hundredweight (cwt)	0.01968	UK hundredweight (cwt) to kilograms	50.80
tonnes to UK tons	0.984	UK tons to tonnes	1.016

Appendix 2

Illustrated Building Terms

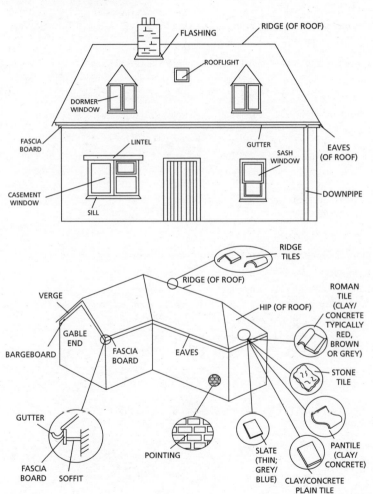

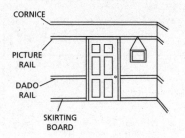

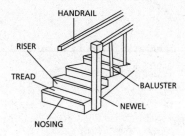

Oxford Quick Reference

A Dictionary of Chemistry

Over 5,000 entries covering all aspects of chemistry, including physical chemistry and biochemistry.

'It should be in every classroom and library ... the reader is drawn inevitably from one entry to the next merely to satisfy curiosity.'

School Science Review

A Dictionary of Physics

Ranging from crystal defects to the solar system, 4,000 clear and concise entries cover all commonly encountered terms and concepts of physics.

A Dictionary of Biology

The perfect guide for those studying biology — with over 5,800 entries on key terms from biology, biochemistry, medicine, and palaeontology.

'lives up to its expectations; the entries are concise, but explanatory'

Biologist

'ideally suited to students of biology, at either secondary or university level, or as a general reference source for anyone with an interest in the life sciences'

Journal of Anatomy

Oxford Quick Reference

A Dictionary of Psychology
Andrew M. Colman

Over 9,500 authoritative entries make up the most wide-ranging dictionary of psychology available.

'impressive ... certainly to be recommended'
Times Higher Education Supplement

'probably the best single-volume dictionary of its kind.'
Library Journal

A Dictionary of Economics
John Black, Nigar Hashimzade, and Gareth Myles

Fully up-to-date and jargon-free coverage of economics. Over 3,500 terms on all aspects of economic theory and practice.

'strongly recommended as a handy work of reference.'
Times Higher Education Supplement

A Dictionary of Law

An ideal source of legal terminology for systems based on English law. Over 4,800 clear and concise entries.

'The entries are clearly drafted and succinctly written ... Precision for the professional is combined with a layman's enlightenment.'
Times Literary Supplement

A Dictionary of Education
Susan Wallace

In over 1,000 clear and concise entries, this authoritative dictionary covers all aspects of education, including organizations, qualifications, key figures, major legislation, theory, and curriculum and assessment terminology.

Oxford Quick Reference

A Dictionary of Sociology
John Scott

The most wide-ranging and authoritative dictionary of its kind.

'Readers and especially beginning readers of sociology can scarcely do better ... there is no better single volume compilation for an up-to-date, readable, and authoritative source of definitions, summaries and references in contemporary Sociology.'

A. H. Halsey, Emeritus Professor, Nuffield College,
University of Oxford

The Concise Oxford Dictionary of Politics and International Relations
Garrett Brown, Iain McLean, and Alistair McMillan

The bestselling A–Z of politics with over 1,700 detailed entries.

'A first class work of reference ... probably the most complete as well as the best work of its type available ... Every politics student should have one'

Political Studies Association

A Dictionary of Environment and Conservation
Chris Park and Michael Allaby

An essential guide to all aspects of the environment and conservation containing over 9,000 entries.

'from *aa* to *zygote*, choices are sound and definitions are unspun'
New Scientist

Oxford Quick Reference

The Concise Oxford Companion to English Literature
Dinah Birch and Katy Hooper

Based on the bestselling *Oxford Companion to English Literature*, this is
an indispensable guide to all aspects of English literature.

Review of the parent volume:
'the foremost work of reference in its field'

Literary Review

A Dictionary of Shakespeare
Stanley Wells

Compiled by one of the best-known international authorities on the
playwright's works, this dictionary offers up-to-date information on all
aspects of Shakespeare, both in his own time and in later ages.

The Oxford Dictionary of Literary Terms
Chris Baldick

A bestselling dictionary, covering all aspects of literature, this is an
essential reference work for students of literature in any language.

A Dictionary of Critical Theory
Ian Buchanan

The invaluable multidisciplinary guide to theory, covering movements,
theories, and events.

'an excellent gateway into critical theory'

Literature and Theology

Oxford Quick Reference

The Oxford Dictionary of Art & Artists
Ian Chilvers

Based on the highly praised *Oxford Dictionary of Art*, over 2,500 up-to-date entries on painting, sculpture, and the graphic arts.

'the best and most inclusive single volume available, immensely useful and very well written'

Marina Vaizey, *Sunday Times*

The Concise Oxford Dictionary of Art Terms
Michael Clarke

Written by the Director of the National Gallery of Scotland, over 1,800 entries cover periods, styles, materials, techniques, and foreign terms.

The Oxford Dictionary of Architecture
James Stevens Curl and Susan Wilson

Over 6,000 entries and 250 illustrations cover all periods of Western architectural history.

'splendid ... you can't have a more concise, entertaining, and informative guide to the words of architecture.'

Architectural Review

'... definitions are not only elegantly concise, they often sparkle with sententious wit. Give me this pleasingly well written dictionary any day.'

Christopher Catling, *SALON: Society of Antiquaries of London Online Newsletter*

More History titles from OUP

The Oxford Companion to Black British History
David Dabydeen, John Gilmore, and Cecily Jones

The first reference book to explore the full history of black people in the British Isles from Roman times to the present day.

'From Haiti to Kingston, to Harlem, to Tottenham, the story of the African Diaspora is seldom told. This Companion will ensure that the history of Black Britain begins to take its rightful place in mainstream British consciousness.'

David Lammy, MP, former Minister for Culture

A Dictionary of World History

Contains a wealth of information on all aspects of history, from prehistory right up to the present day. Over 4,000 clear, concise entries include biographies of key figures in world history, separate entries for every country in the world, and subject entries on religious and political movements, international organizations, and key battles and places.

The Concise Oxford Dictionary of Archaeology
Timothy Darvill

The most wide-ranging, up-to-date, and authoritative dictionary of its kind.

'Comprehensive, proportionate, and limpid'

Antiquity

Oxford Quick Reference

A Dictionary of Marketing
Charles Doyle

Covers traditional marketing techniques and theories alongside the latest concepts in over 2,500 clear and authoritative entries.

'Flick to any page [for] a lecture's worth of well thought through information'
Dan Germain, Head of Creative, innocent ltd

A Dictionary of Media and Communication
Daniel Chandler and Rod Munday

Provides over 2,200 authoritative entries on terms used in media and communication, from concepts and theories to technical terms, across subject areas that include advertising, digital culture, journalism, new media, radio studies, and telecommunications.

'a wonderful volume that is much more than a simple dictionary'
Professor Joshua Meyrowitz, University of New Hampshire

A Dictionary of Film Studies
Annette Kuhn and Guy Westwell

Features terms covering all aspects of film studies in 500 detailed entries, from theory and history to technical terms and practices.

A Dictionary of Journalism
Tony Harcup

Covers terminology relating to the practice, business, and technology of journalism, as well as its concepts and theories, organizations and institutions, publications, and key events.

Oxford Quick Reference

Concise Medical Dictionary

Over 12,000 clear entries covering all the major medical and surgical specialities make this one of our best-selling dictionaries.

'"No home should be without one" certainly applies to this splendid medical dictionary'

Journal of the Institute of Health Education

'An extraordinary bargain'   *New Scientist*

A Dictionary of Nursing

Comprehensive coverage of the ever-expanding vocabulary of the nursing professions. Features over 10,000 entries written by medical and nursing specialists.

A Dictionary of Dentistry
Robert Ireland

Over 4,000 succinct and authoritative entries define all the important terms used in dentistry today. This is the ideal reference for all members of the dental team.

A Dictionary of Forensic Science
Suzanne Bell

In over 1,300 entries, this new dictionary covers the key concepts within Forensic Science and is a must-have for students and practitioners of forensic science.

Oxford Quick Reference

The Oxford Dictionary of Dance
Debra Craine and Judith Mackrell

Over 2,600 entries on everything from hip-hop to classical ballet, covering dancers, dance styles, choreographers and composers, techniques, companies, and productions.

'A must-have volume ... impressively thorough'
Margaret Reynolds, *The Times*

The Oxford Guide to Plays
Michael Patterson

Covers 1,000 of the most important, best-known, and most popular plays of world theatre.

'Superb synopses ... Superbly formatted ... Fascinating and accessible style'

THES

The Oxford Dictionary of Music
Michael & Joyce Kennedy and Tim Rutherford-Johnson

The most comprehensive, authoritative, and up-to-date dictionary of music available in paperback.

'clearly the best around ... the dictionary that everyone should have'
Literary Review

Oxford Quick Reference

The Concise Oxford Dictionary of Quotations
SIXTH EDITION
Edited by Susan Ratcliffe

Based on the highly acclaimed seventh edition of *The Oxford Dictionary of Quotations*, this dictionary provides extensive coverage of literary and historical quotations, and contains completely up-to-date material. A fascinating read and an essential reference tool.

Oxford Dictionary of Quotations by Subject
Edited by Susan Ratcliffe

The ideal place to discover what's been said about what, the dictionary presents quotations on nearly 600 areas of special interest and concern in today's world.

The Oxford Dictionary of Humorous Quotations
Edited by Gyles Brandreth

From the sharply witty to the downright hilarious, this sparkling collection will appeal to all senses of humour.

The Oxford Dictionary of Political Quotations
Edited by Antony Jay

This lively and illuminating dictionary from the writer of 'Yes Minister' presents a vintage crop of over 4,900 political quotations. Ranging from the pivotal and momentous to the rhetorical, the sincere, the bemused, the tongue-in-cheek, and the downright rude, examples include memorable words from the old hands as well as from contemporary politicians.

'funny, striking, thought-provoking and incisive ... will appeal to those browsing through it at least as much as to those who wish to use it as a work of reference'
Observer